W9-DGE-031

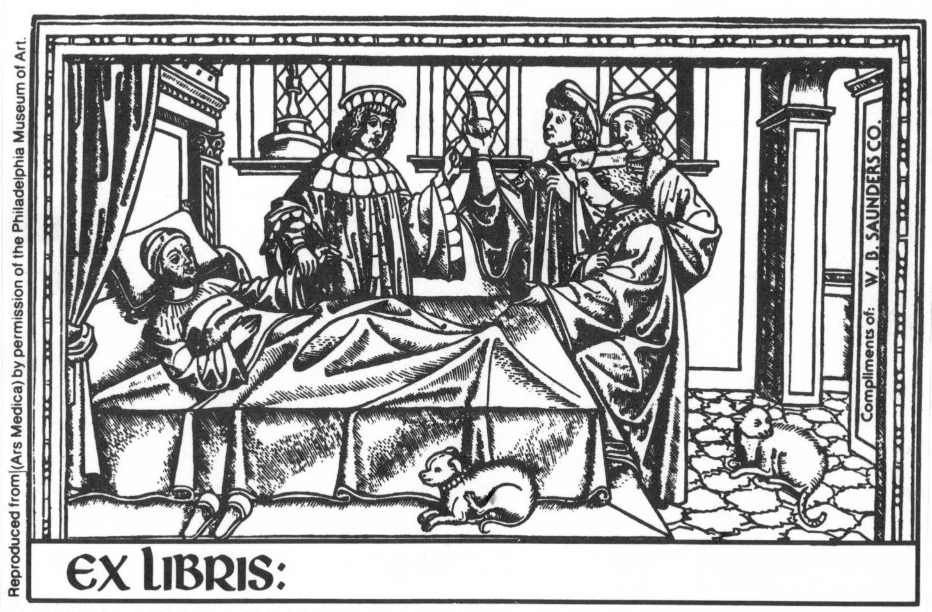
Reproduced from (Ars Medica) by permission of the Philadelphia Museum of Art.
Compliments of: W. B. SAUNDERS CO.
EX LIBRIS:

The Pain System
The Neural Basis of Nociceptive Transmission in the Mammalian Nervous System

Pain and Headache

Vol. 8

Series Editor
Philip L. Gildenberg, Houston, Tex.

S. Karger · Basel · München · Paris · London · New York · Tokyo · Sydney

The Pain System

The Neural Basis of Nociceptive Transmission in the Mammalian Nervous System

W.D. Willis, Jr.

Marine Biomedical Institute and Departments of Anatomy and of Physiology and Biophysics, University of Texas Medical Branch, Galveston, Tex., USA

152 figures and 4 tables, 1985

S. Karger · Basel · München · Paris · London · New York · Tokyo · Sydney

Pain and Headache

National Library of Medicine, Cataloging in Publication
Willis, William D., 1934 –
The pain system: the neural basis of nociceptive transmission in the mammalian nervous system
W.D. Willis, Jr. – Basel; New York: Karger, 1985.
(Pain and headache; v. 8)
Bibliography: p. Includes index.
1. Neural Transmission 2. Nociceptors – physiology 3. Pain – physiology
I. Title II. Series
Wl PA293 v. 8 [WL 704 W735p]
ISBN 3-8055-3930-4

Drug Dosage

The author and the publisher have exerted every effort to ensure that drug selection and dosage set forth in this text are in accord with current recommendations and practice at the time of publication. However, in view of ongoing research, changes in government regulations, and the constant flow of information relating to drug therapy and drug reactions, the reader is urged to check the package insert for each drug for any change in indications and dosage and for added warnings and precautions. This is particularly important when the recommended agent is a new and/or infrequently employed drug.

Printed in Switzerland by Thür AG Offsetdruck, Pratteln
ISBN 3-8055-3930-4

Contents

Preface IX

Chapter 1

Introduction 1

Conclusions 5

Chapter 2

Evidence for Nociceptive Transmission Systems 7

Nociceptors 7
Nociceptive Ascending Tracts 14
Participation of Higher Centers in Nociception 17
Experimental Approaches to the Investigation of Pain Transmission Systems . 17
Criteria for Identification of Nociceptive Neurons 18
Conclusions 19

Chapter 3

Nociceptors 22

Historical Overview 22
Nociceptors 23
Cutaneous Nociceptors 25
Aδ Mechanical Nociceptors 25
C Polymodal Nociceptors 28
Other Types of Cutaneous Nociceptors 33
Role of Cutaneous Nociceptors in Pain Sensation 34
First and Second Pain 36
Electrical Stimulation of Peripheral Nerve Fibers 37
Correlation of Human Pain Sensation with Responses of Nociceptors in Humans 38

Correlation of Human Pain Sensation with Responses of Nociceptors in Animals 40
Hyperalgesia 44
Chemical Basis for Activation and Sensitization of Nociceptors 49
Anatomical Distribution of Hyperalgesia 53
Muscle Nociceptors 54
Group III Muscle Receptors 54
Group IV Muscle Receptors 56
Role of Muscle Nociceptors in Pain 60
Joint Nociceptors 60
Role of Joint Nociceptors in Pain 62
Visceral Nociceptors and Referred Pain 63
Nociceptors in the Cardiovascular System and Their Role in Cardiac Pain . . 66
Nociceptors in the Respiratory System 66
Lung Irritant Receptors 67
Type J Receptors 67
Role of Respiratory Nociceptors in Pain 67
Nociceptors in the Gastrointestinal Tract 67
Role of Gastrointestinal Nociceptors in Pain 70
Nociceptors in the Genitourinary Tract 71
Role of Genitourinary Nociceptors in Pain 73
Conclusions 74

Chapter 4

Nociceptive Afferent Input to the Dorsal Horn 78

Historical Overview 78
Dorsal Root 82
Ventral Root 85
Segregation of Large and Small Afferent Fibers in the Dorsal Root 86
Lissauer's Tract 89
Terminations of Fine Afferent Fibers in the Dorsal Horn 92
Synaptic Endings of Primary Afferent Fibers in the Dorsal Horn 100
Transmitters Associated with Fine Afferent Fibers 105
Cell Types in the Dorsal Horn 112
Lamina I 112
Lamina II 119
Laminae III and IV 132
Lamina V 134
Lamina VI 134
Intermediate Region and Ventral Horn 134
Responses of Dorsal Horn Interneurons to Noxious Inputs Other than from Skin 135
Pharmacological Responses of Nociceptive Dorsal Horn Neurons 137
Conclusions 138

Chapter 5

Ascending Nociceptive Tracts

Ascending Nociceptive Tracts . . . 145
Historical Overview . . . 145
Spinothalamic Tract . . . 147
Cells of Origin . . . 147
Organization of the Spinothalamic Tract in the Spinal Cord and Brain Stem . . . 152
Thalamic Nuclei of Termination . . . 154
Identification of Spinothalamic Tract Cells in Physiological Experiments . . . 157
Responses of Spinothalamic Tract Cells to A and C Fiber Volleys . . . 159
Responses of Spinothalamic Tract Cells to Natural Forms of Stimulation . . 164
Responses of Spinothalamic Tract Cells Projecting to the Medial Thalamus . 172
Effects of Capsaicin . . . 175
Influence of Anesthesia . . . 175
Receptive Field Organization . . . 177
Prolonged Inhibition following Peripheral Nerve Stimulation . . . 181
Pharmacology of Spinothalamic Tract Cells . . . 184
Role of Spinothalamic Tract in Pain Transmission . . . 184
Spinoreticular Tract . . . 189
Cells of Origin . . . 189
Organization of the Spinoreticular Tract in the Spinal Cord and Brain Stem 191
Nuclei of Termination in the Reticular Formation . . . 191
Identification of Spinoreticular Neurons in Physiological Experiments . . . 193
Electrophysiological Response Properties of Spinoreticular Neurons and Role in Pain . . . 193
Spinomesencephalic Tract . . . 194
Cells of Origin . . . 194
Organization of the Spinomesencephalic Tract in the Spinal Cord and Brain Stem . . . 195
Nuclei of Termination in the Midbrain . . . 195
Identification of Spinomesencephalic Neurons in Physiological Experiments 195
Electrophysiological Response Properties of Spinomesencephalic Tract Cells and Role in Pain . . . 195
Dorsally Situated Ascending Pathways That May Be Nociceptive . . . 196
Spinocervical Tract . . . 197
Lateral Cervical Nucleus . . . 197
Cells of Origin . . . 197
Organization of Spinocervical Tract and Destination of Terminals . . . 199
Identification of Spinocervical Tract Neurons . . . 200
Response Properties of Spinocervical Tract Cells . . . 200
Postsynaptic Dorsal Column Pathway . . . 202
Cells of Origin . . . 202
Organization of the Postsynaptic Dorsal Column Pathway and Destination of Terminals . . . 202
Identification of Postsynaptic Dorsal Column Neurons . . . 205
Response Properties of Postsynaptic Dorsal Column Cells . . . 205

Role of the Spinocervical Tract and of the Postsynaptic Dorsal Column Pathway in Pain . . . 205
Conclusions . . . 206

Chapter 6

Nociceptive Transmission to Thalamus and Cerebral Cortex . . . 213

Historical Overview . . . 213
Thalamic Nuclei Receiving Nociceptive Input from the Spinothalamic Tract: Ventral Posterior Lateral Nucleus . . . 216
Evidence that Nociceptive Spinothalamic Tract Axons End in the VPL Nucleus . . . 224
Nociceptive Responses of Neurons in the VPL Nucleus . . . 226
Monkey . . . 226
Cat . . . 232
Rat . . . 235
Thalamic Nuclei Other than VPL Receiving Input from the Spinothalamic Tract . . . 239
Medial Part of the Posterior Complex (PO_m) . . . 239
Intralaminar Nuclei . . . 241
Nucleus Submedius . . . 244
Thalamic Reticular Nucleus . . . 245
Thalamic Nuclei in Which Alternative Nociceptive Tracts End . . . 245
Nociceptive Responses of Neurons in the Somatosensory Cerebral Cortex . . 246
Monkey SI Cortex . . . 247
Rat SI Cortex . . . 252
Monkey SII Cortex . . . 255
Cat SII Cortex . . . 256
Role of the Thalamus and Cortex in Pain . . . 257
Conclusions . . . 259

Chapter 7

Overview and Future Directions . . . 264

Nociceptors . . . 264
Input System . . . 267
Nociceptive Tracts . . . 273
Thalamocortical Mechanisms . . . 276
Plasticity . . . 279
Descending Control Systems . . . 281
Conclusions . . . 281

References . . . 282
Subject Index . . . 331

Preface

This review is intended to provide an overview of our current understanding of the nociceptive transmission system. However, because of limitations of time and space, the emphasis will be on nociceptive systems supplying the body. The trigeminal system will be discussed only when findings on the trigeminal system are helpful as a confirmation of principles pertinent to both the trigeminal and body nociceptive systems or where the evidence is most complete in the trigeminal system.

I would like to thank my colleagues who participed in the experiments described here that were done in my laboratory, including *A.E. Applebaum, J.E. Beall, R.N. Bryan, J.M. Chung, C.L. Clifton, R.E. Coggeshall, J.D. Coulter, K. Endo, Z.R. Fang, R.D. Foreman, K.D. Gerhart, G.J. Giesler, L.H. Haber, M.B. Hancock, Y. Hori, L.M. Jordan, D.R. Kenshalo, Jr., G.A. Kevetter, K.H. Lee, R.B. Leonard, R.F. Martin, R.J. Milne, B.D. Moore, R.F. Schmidt, B.J. Schrock, H.R. Spiel, D.L. Trevino, W.H. Vance, W.S. Willcockson,* and *R.P. Yezierski.* I would also like to express my appreciation to *Phyllis Waldrop* for typing the manuscript, to *Margie Watson* for proofreading, and to *Calvin Cargill* and *Helen Willcockson* for the illustrations.

The work that was done in my laboratory was supported by NIH grants NS 09743 and NS 11255 and by a grant from the Moody Foundation.

Galveston, 1984 *W.D. Willis, Jr.*

Chapter 1

Introduction

Like any other sensory modality, the sensation of pain generally depends upon the activation of a discrete set of neural pathways [*Kerr,* 1975a; *Perl,* 1971; *Price and Dubner,* 1977; *Zimmermann,* 1976]. These pathways include primary afferent fibers that terminate distally in nociceptors[1]. The activity evoked in nociceptors is transmitted to the central nervous system where it activates neural circuits in the spinal cord or in the trigeminal nuclei. Ascending tract cells then transmit information concerning the noxious stimuli to the brain stem, thalamus, and cerebral cortex for sensory processing. Sensory-discriminative processing of nociceptive signals is likely to be accomplished by brain mechanisms comparable to those utilized for processing of other sensory data, such as visual, auditory or tactile information [*Perl,* 1971; *Price and Dubner,* 1977]. However, noxious stimuli are the most effective stimuli for triggering the arousal response [*Magoun,* 1963], and they evoke complex reflex and motivational-affective behaviors that depend upon neural activity at several levels of the central nervous system, including the spinal cord, brain stem, and wide areas of the forebrain [*Hardy* et al., 1952a; *Melzack and Casey,* 1968; *White and Sweet,* 1955]. Pain in this wider context is a sensory experience associated not only with discriminative components like intensity, duration, location, and quality, but also such emotional reactions as suffering, anxiety, and depression, and reflex events including muscular contractions, circulatory responses, and hormonal changes. The processing of sensory-discriminative and motivation-affective information by the brain could be done either in a parallel fashion, or in series as illustrated diagrammatically in figure 1/1.

Another property of pain reactions is that they are highly labile [*Beecher,* 1959; *Melzack,* 1973; *Melzack and Wall,* 1965]. Under some conditions, a given noxious stimulus may evoke the full pain reaction, whereas in other circumstances the same individual may fail to experience pain at all.

[1] *Sherrington* [1906] defined nociceptors as sense organs that respond to noxious stimuli, and he defined noxious stimuli as those that either threaten or actually produce damage.

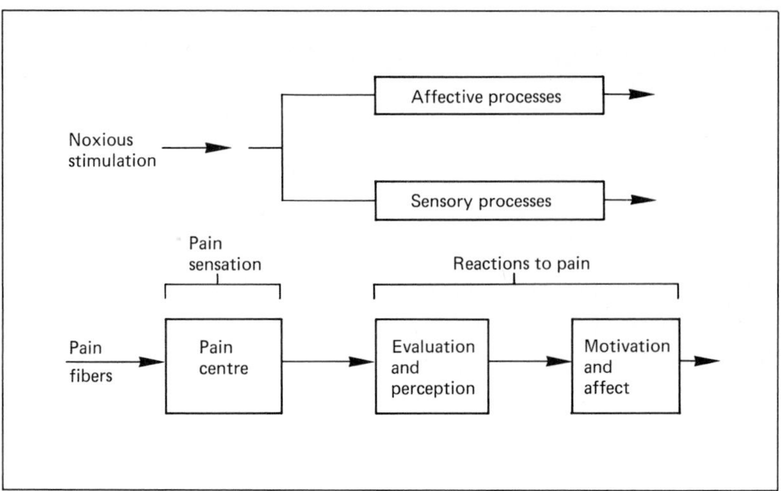

Fig. 1/1. Systems for processing sensory-discriminative and motivational-affective information. The mechanism above involves two parallel systems and is derived from Marshall's concept of pain, as described in *Melzack* [1973]. The mechanism below shows a serial concept, with motivational-affective changes occurring in response to pain sensation, as described by *Melzack and Casey* [1968].

Presumably this variability in the pain reaction is due in part to the operation of centrifugal control systems [see review by *Willis,* 1982]. Although it has been argued that the variable nature of the pain reaction casts doubt on the presence of discrete nociceptive pathways [*Melzack,* 1973; *Melzack and Wall,* 1965], another view is that there are specific nociceptive pathways whose activity is sometimes suppressed by the centrifugal control systems, depending upon as yet poorly defined behavioral states.

Figure 1/2 illustrates a concept of how interactions may occur between the sensory-discriminative and motivational-affective systems for pain and how these systems may be controlled at the input stage in the dorsal horn and also by pathways descending from the brain [*Melzack and Casey,* 1968]. In this diagram, the sensory-discriminative and motivational-affective mechanisms are shown as if they operate in parallel. However, each receives input from a 'gate control system' in the dorsal horn. Details about the 'gate control' theory have been discussed elsewhere [*Willis and Coggeshall,* 1978]. For present purposes, this system can be considered as a processor that allows inhibitory interactions in response to afferent inputs from

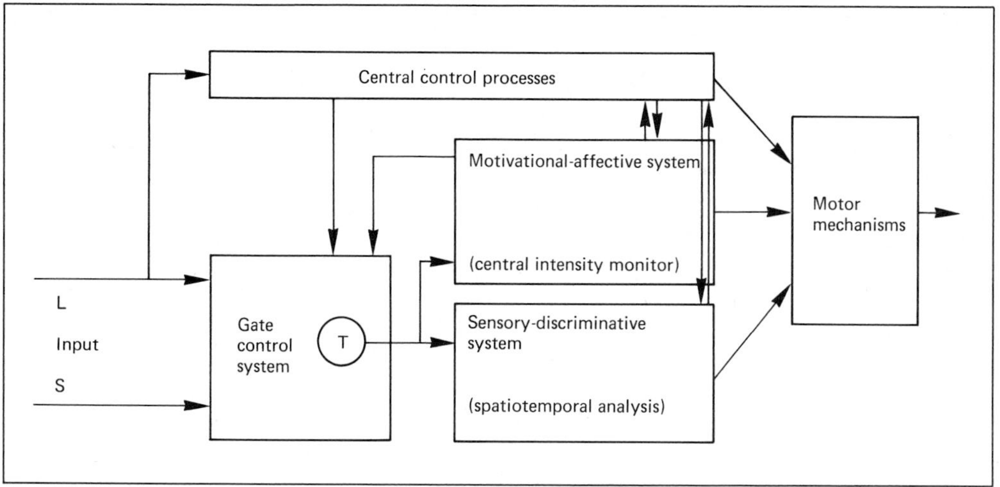

Fig. 1/2. Model of potential interactions between afferent input, a 'gate control' system in the dorsal horn of the spinal cord (and medulla), ascending pathways accessing sensory-discriminative and motivational-affective systems, central control systems, and motor output. Input is from large (L) and small (S) primary afferent fibers. Pain transmission to the brain is accomplished by transmission (T) cells. [From *Melzack and Casey,* 1968.]

various types of primary afferent fibers. The afferent input also has more direct access to the brain through the dorsal column-medial lemniscus pathway, and brain systems can operate back onto the dorsal horn or on the sensory-discriminative and motivational-affective systems within the brain. All of these systems have a behavioral read-out through motor pathways. The motor responses would include somatic and autonomic activity.

The description of pain so far applies to the ordinary kinds of pain experienced by almost everyone (except those who are congenitally insensitive to pain; see review by *Sweet* [1981a]). Pain due in a directly causal way to the activation of nociceptors is often called 'acute pain' or 'nociceptive pain' by physicians who treat pain states. However, pain can also result from injury to the nervous system (e.g., 'central pain' or 'deafferentation pain'; see *Cassinari and Pagni* [1969] and *Tasker* et al. [1983]. Generally, this type of pain does not seem to depend upon activity in nociceptors. It may result from changes either in the nociceptive pathways or in the centrifugal control systems affecting these. One theory is that the pain following nervous system injury reflects denervation hypersensitivity. Evidence that

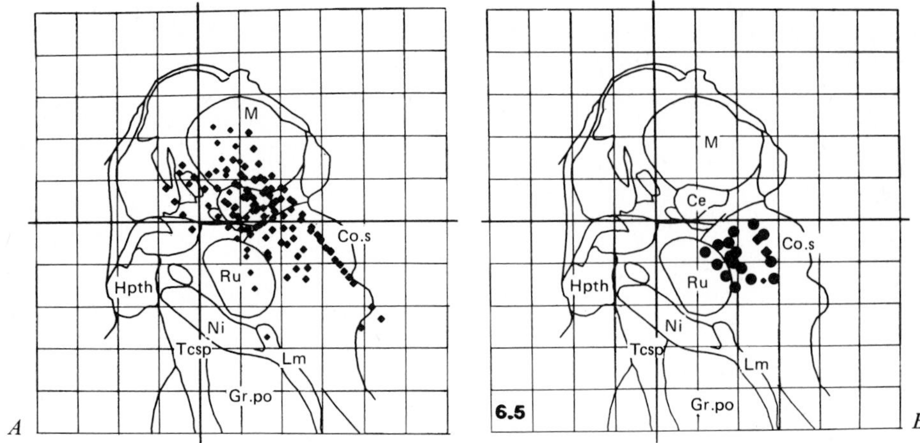

Fig. 1/3. Computer-generated plots of stimulus sites in the sagittal plane 6.5 mm lateral to the midline in human subjects undergoing stereotaxic surgery. In *A,* sites are shown that yielded no response in patients having pain other than 'deafferentation pain' or some other problem unrelated to pain. (It should be noted that other points in the midbrain in this plane elicit sensations of hot, warm or cold, responses that can be attributed to stimulation of the spinothalamic tract.) In *B,* the circles indicate points that when stimulated causes reproduction of the 'deafferentation pain' of these patients. [From *Tasker* et al., 1983.]

there is a change in the central nervous system circuits that process pain comes from observations of the results of stimulation in the medial part of the human midbrain and thalamus. In subjects who do not suffer pain, stimulation seldom evokes a sensation of pain; however, in patients that have long-standing 'deafferentation pain', it is common to evoke a pain sensation (fig. 1/3) [*Tasker* et al., 1983], suggesting that there is a plastic change in central nervous system circuits in cases of 'central pain'.

It may be that prolonged activation of nociceptors also produces plastic changes in the central nervous system. At any rate, chronic pain is a problem of a different order of magnitude than that posed by acute pain. Whereas acute pain can usually be managed well by standard medical or surgical therapies, chronic pain is poorly understood and often inadequately treated [*Casey* et al., 1979].

Fortunately, in the past decade, there has been a considerable expansion in research aimed at understanding pain. Part of this increased effect has been encouraged by such dramatic discoveries as the opiate receptors in the central nervous system [*Pert and Snyder,* 1973; *Simon* et al., 1973; *Terenius,* 1973], the endogenous opioid substances [*Cox* et al., 1976; *Goldstein,* 1976; *Goldstein* et al., 1979; *Hughes,* 1975; *Hughes* et al., 1975; *Li and Chung,* 1976; *Loh* et al., 1976; *Pasternak* et al., 1975; *Simantov and Snyder,* 1976; *Terenius and Wahlström,* 1975], and neural pathways that can produce analgesia [*Mayer and Liebeskind,* 1974; *Reynolds,* 1969; see reviews by *Basbaum and Fields,* 1978; *Fields and Basbaum,* 1978; *Mayer and Price,* 1976; *Willis,* 1982]. There have been many important advances in our understanding of the anatomy, physiology, and pharmacology of the neural systems involved in nociception. Hopefully, clues about the nature of chronic pain will also result from this expansion of research on pain mechanisms.

The purpose of this review is to provide an overview of current knowledge of the nociceptive pathways in the mammalian nervous system. Emphasis will be placed on what is known about neurons in the initial stages of the system, since little is known about higher processing of nociceptive information. Reviews of related topics have recently been published [*Besson* et al., 1982; *Cervero and Iggo,* 1980; *Chéry-Croze,* 1983; *Coggeshall,* 1980; *Fitzgerald,* 1983; *Kerr,* 1975a; *Perl,* 1984b; *Price and Dubner,* 1977; *Sinclair,* 1981; *Sweet,* 1981a, b; *Willis,* 1982, 1983; *Willis and Coggeshall,* 1978; *Yaksh and Hammond,* 1982].

Conclusions

1. The sensation of pain normally depends upon the activation of a discrete set of neural pathways made up of nociceptive afferent fibers, ascending tract cells, and neurons in the brain stem, thalamus and cerebral cortex.

2. Sensory-discriminative processing of nociceptive information is comparable to processing of data in other sensory systems.

3. However, painful stimuli are more effective than are many other kinds of stimuli in triggering arousal, a variety of somatic and autonomic reflexes, and motivational-affective reactions.

4. Reactions to painful stimuli are quite variable, due in part to the activity of centrifugal control systems.

5. Acute pain results from activation of nociceptors and the central pathways into which they feed and can usually be managed effectively by medical or surgical means.

6. Pain can also result from injury to the nervous system. This type of pain and that produced by prolonged input from nociceptors differs from acute pain. Chronic pain may depend upon alterations in the way the central nervous system processes information. Chronic pain is poorly understood and poorly managed.

7. Research on pain mechanisms is currently a very active field. Major discoveries in recent years include the opiate receptors, the endogenous opioid substances, and the endogenous analgesia systems. Hopefully, future efforts will lead to improvements in the management of pain, especially chronic pain.

Chapter 2

Evidence for Nociceptive Transmission Systems

In Chapter 1, the view was expressed that there are discrete neural pathways whose responsibility is to transmit information concerning noxious stimuli to the brain. Although transmission through these nociceptive pathways can be prevented or altered, for instance by centrifugal control systems, there are, nevertheless, neural pathways concerned with nociception that contribute to the experience of acute pain. This Chapter will be concerned with some of the evidence for this view. Reliance will be placed chiefly on studies in human subjects, since such subjects can provide verbal verification of their sensory experience. The main lines of evidence come from clinical and experimental observations in humans following either stimulation or ablative procedures. The latter include injections of local anesthetics.

Nociceptors

Historically, there has been a school of thought that stressed the role of a patterned input to the central nervous system as the chief determinant of pain, rather than the activation of specific primary afferent fibers of a nociceptor class [*Nafe,* 1929; *Sinclair,* 1955, 1967; *Weddell and Miller,* 1962]. However, it has been known since the 1930s that stimulation of fine afferent fibers, of Aδ or C caliber, produces pain in human subjects, whereas stimulation of large myelinated fibers results in sensations akin to touch (fig. 2/1) [*Heinbecker* et al., 1933, 1934; *Clark* et al., 1935; *Pattle and Weddell,* 1948; *Collins* et al., 1960; *Torebjörk and Hallin,* 1973; *Hallin and Torebjörk,* 1976]. The pain associated with volleys in Aδ fibers has a pricking quality, whereas that associated with volleys in C fibers is a prolonged burning sensation [*Torebjörk and Hallin,* 1973; *Hallin and Torebjörk,* 1976]. Furthermore, when local anesthetics block transmission in Aδ and C fibers, pain is abolished (fig. 2/2) [*Torebjörk and Hallin,* 1973; *Hallin and Torebjörk,* 1976]. This is the basis for the effectiveness of nerve blocks in

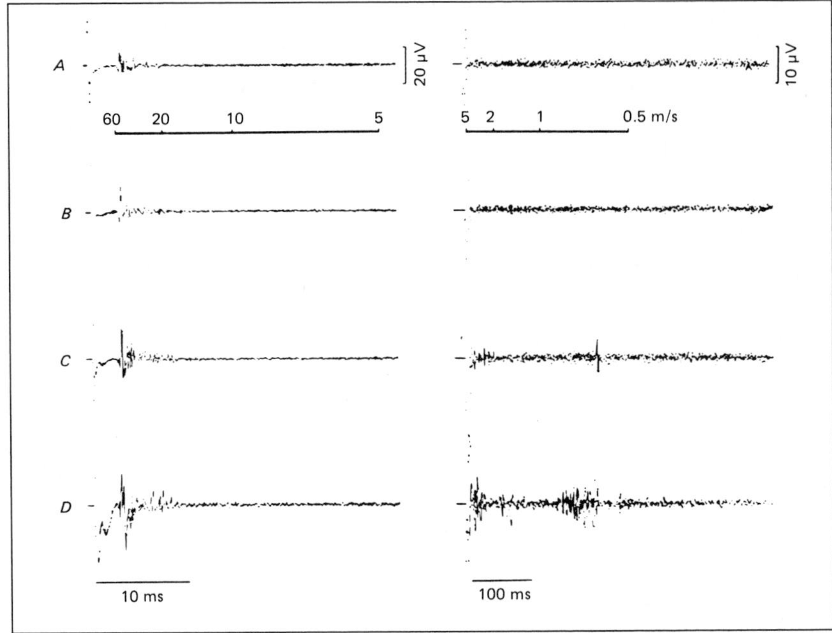

Fig. 2/1. Signal averaged recordings from the radial nerve in a human subject during electrical stimulation of the receptive field on the dorsal surface of the hand. The records in the left column were at a more rapid sweep speed than those in the right column. The horizontal bars in *A* show the conduction velocity spectra of the afferent fiber volleys. In *A*, the stimulus was above threshold for the A fiber volley, but below threshold for perception. In *B*, a higher stimulus strength activated more A fibers and there was a tactile sensation. In *C* and *D*, higher stimulus strengths activated Aδ and C fibers and caused pain. [From *Hallin and Torebjörk,* 1976.]

the clinical management of certain kinds of pain states [*Bonica,* 1953; *Wood,* 1978].

It has been suggested that since electrical stimulation strong enough to activate Aδ and C fibers will necessarily also activate the large myelinated fibers, it is inappropriate to conclude that the small fibers are specifically related to pain [*Melzack,* 1973]. Rather, '...pain results when the total afferent barrage in all fibres exceeds a critical level and the only way to exceed the level is by activation of small diameter fibres' [*Melzack,* 1973, pp. 105–106]. However, transmission in the myelinated fibers can be blocked, for instance by pressure, leaving conduction possible only in the unmyelinated fibers. In such experiments, pain of a prolonged, burning type can still be

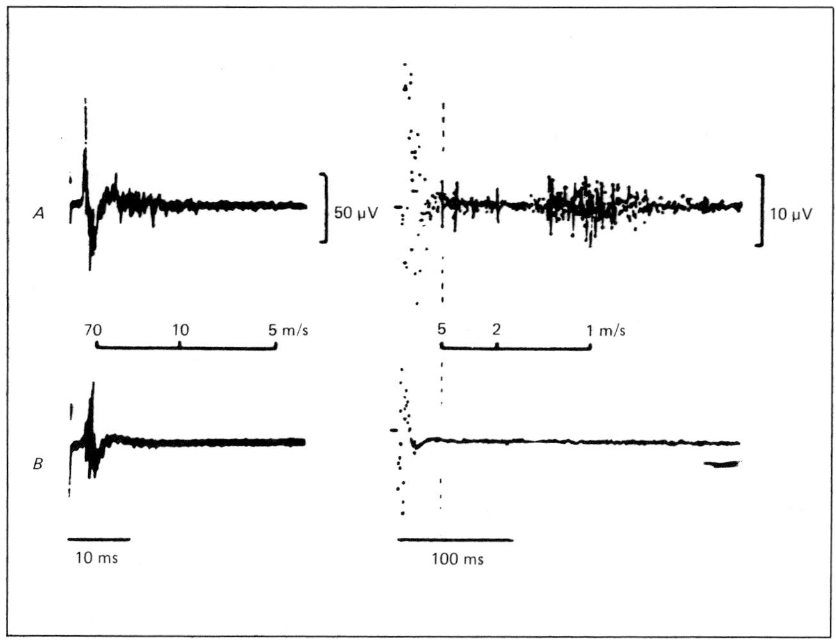

Fig. 2/2. Effect of local anesthetic (lidocaine, 0.25%) on conduction of nerve volleys in radial nerve of human subject and on sensation. The sweep speed was greater for the signal averaged records in the left column than in the right column. The horizontal bars in *A* show the conduction velocity spectra for the nerve volleys. The activity shown in *B* was evoked by strong electrical shocks applied to the dorsum of the hand. The stimuli produced a prolonged painful sensation. Lidocaine was then injected to block activity in small fibers between the stimulation and recording sites. Note that the Aδ and C fiber volleys disappeared following injection of the anesthetic. Sensory losses were first noted for warmth and also for the prolonged pain caused by stimulation. Cold sensation and pricking pain were soon also lost. Sometimes it was possible to demonstrate a block of C fiber activity while pricking pain was still present. After blockade of conduction in small fibers, touch was retained. [From *Hallin and Torebjörk,* 1976.]

provoked by cutaneous stimulation (fig. 2/3) [*Torebjörk and Hallin,* 1973]. Thus, unmyelinated fibers can produce pain in the absence of any input over large myelinated fibers. The other sensations that can be evoked in humans by stimulation of unmyelinated fibers are warmth and heat [*Clark* et al., 1935; *Hallin and Torebjörk,* 1976]. Further evidence that certain of the small afferent fibers are specifically involved in pain sensation comes from the results of single unit recordings and microstimulation experiments in humans.

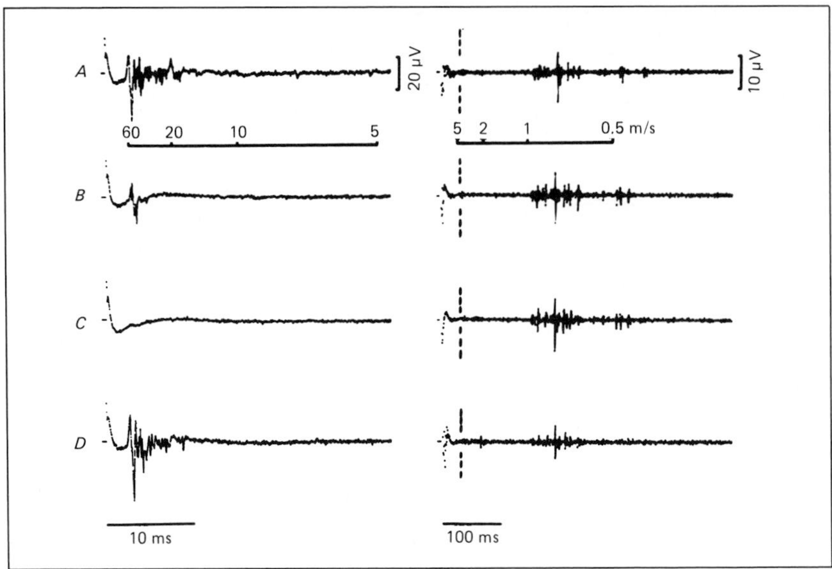

Fig. 2/3. Effect on sensation of blocking conduction in large afferent fibers by pressure. Recordings were made from the radial nerve during stimulation of the dorsum of the hand. Records were signal averaged. The left column is at a faster sweep speed than the right column. The horizontal bars in *A* show the conduction velocity spectra of the afferent nerve volleys. In *A,* the control records show activity in both large and small fibers. Pressure was then applied to the nerve, and as shown in *B* and *C* there was a progressive blockade of conduction in the A fibers. However, the C fiber volley was maintained. At this time, the receptive field felt numb, and no tactile sensations resulted from stimulation, using weak stimulus strengths that previously evoked tactile sensations. Strong shocks evoked a prolonged, burning sensation. In *D,* the changes are shown to be reversible. [From *Torebjörk and Hallin,* 1973.]

The concept that there are in fact specific nociceptive afferent fibers is supported by a great deal of evidence from animal experiments (see Chapter 3). At least two of the types of nociceptors that have been described in animal experiments have also been identified in recordings using the technique introduced by *Hagbarth and Vallbo* in 1967 [see also *Vallbo and Hagbarth,* 1968; cf. *Hensel and Boman,* 1960] from single units in human peripheral nerves [*Van Hees,* 1976; *Van Hees and Gybels,* 1972, 1981; *Gybels* et al., 1979; *Konietzny* et al., 1981; *Torebjörk,* 1974; *Torebjörk and Hallin,* 1972, 1974, 1976, 1979]. For example, the receptive fields of several C polymodal nociceptors are shown on a drawing of a human foot in

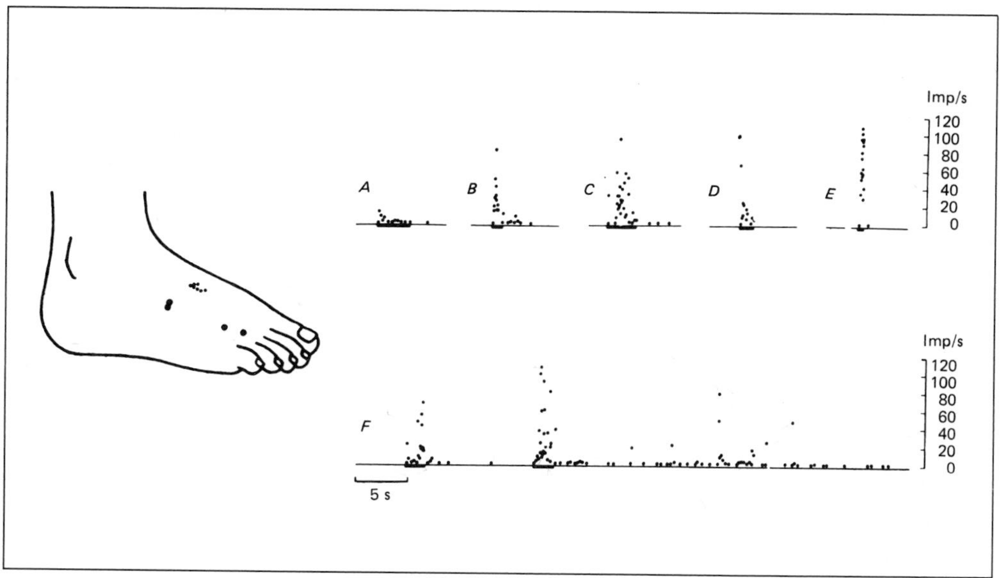

Fig. 2/4. Receptive fields of 5 C-polymodal nociceptors recorded from the human peroneal nerve are shown on the drawing of the foot at the left. The graphs show the instantaneous firing rate of one unit in response to the following stimuli: *(A)* pointed probe applied with a force of 2 g; *(B)* firm stroking; *(C)* squeezing the skin with forceps; *(D)* needle prick; *(E)* briefly touching the skin with a burning match; *(F)* puncturing the skin with a hypodermic needle at left and then injecting 0.02 ml of 5% KCl. [From *Torebjörk,* 1974.]

figure 2/4 [*Torebjörk,* 1974]. The activity of these units was recorded by microneurography from the peroneal nerve. The circles indicate the general locations of the receptive fields of 4 units, while the dots indicate 7 foci of sensitivity for another unit. The bursts of discharges of one of the units produced by a variety of stimuli are shown in the graphs. The C fiber was activated by noxious mechanical, thermal and chemical stimuli, and so it could be classified as a C polymodal nociceptor. Likewise, recordings have been made from human Aδ fibers that can be classified as mechanical nociceptors [*Konietzny* et al., 1981].

Recently, after recordings from a peripheral nerve demonstrated the receptive field of an individual C polymodal nociceptor, it has been possible to stimulate the same afferent fiber through the microelectrode (verified by comparison of the projected receptive field and that mapped while recording the activity of the afferent fiber) and to show that repetitive stim-

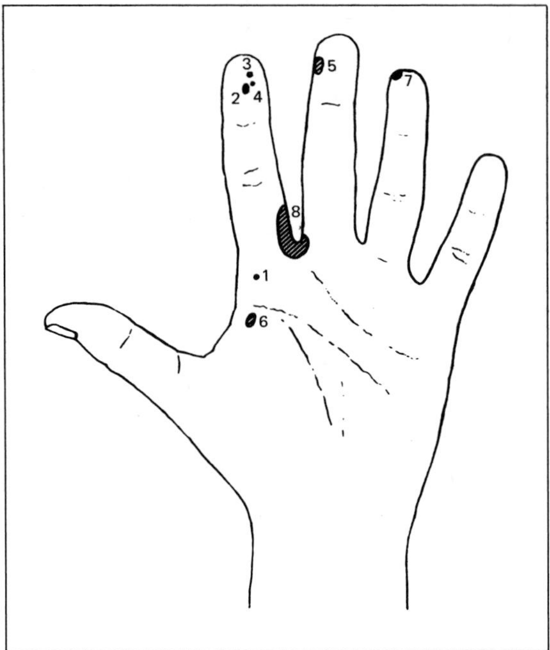

Fig. 2/5. Referred receptive fields recruited by progressively increasing stimulus intensities applied through a microelectrode inserted into a human median nerve. Submodalities recruited with increasing strengths were: (1) tap; (2) pressure; (3) tap; (4) sting; (5) pain; (6) pain; (7) pressure; (8) pain. [From *Torebjörk and Ochoa,* 1983.]

ulation of a single nociceptor can evoke a sensation of pain [*Torebjörk and Ochoa,* 1980, 1981, 1983; *Konietzny* et al., 1981]. Increasing the stimulus intensity recruits in all or nothing fashion sensation referred to new receptive fields, evidently by recruiting discharges in nearby afferent fibers (fig. 2/5) [*Torebjörk and Ochoa,* 1983].

Interestingly, some C polymodal nociceptors when stimulated evoke a sensation of itch, rather than of pain [*Torebjörk and Ochoa,* 1981]. Evidently, there are discrete sensory pathways for pain and for itch. Changing the stimulus frequency does not change the quality of the sensation from pain to itch or vice versa when stimulating the axon of a given C polymodal nociceptor [*Torebjörk and Ochoa,* 1981].

By contrast, stimulation of single cutaneous afferent fibers connected with other kinds of receptors does not evoke a sensation of pain, but rather sensations intuitively appropriate to their response properties. For example, a Pacinian corpuscle afferent stimulated at a high frequency evokes a sensation of vibration; stimulation of a Meissner corpuscle or of a hair follicle afferent at a moderate frequency causes a sensation of tapping or flutter; and stimulation of a type I (Merkle cell) afferent produces a sensation of sustained pressure [*Schady* et al., 1983; *Scharf* et al., 1973; *Torebjörk and Ochoa,* 1980, 1983; *Vallbo,* 1981]. There is no support from such studies for the argument that pain is produced by 'overactivation' of a tactile receptor.

Evidently, the stipulation of *Sinclair* [1967] that single fiber stimulation in humans would be required for there to be identification of stimulus specificity with modality specificity has now been met. *Sinclair* [1967, p. 12] states that:

> '...there is as yet no evidence that stimulation of a single fibre in isolation gives rise to a sensation of any kind, and until the experiment is done it is impossible to say what the results will be. To stimulate a single fibre in an intact human subject, to prove satisfactorily that only that fibre and no other has been stimulated, and to record a meaningful sensory judgement is an almost incredibly difficult technical feat, and it will be a long time before unequivocal evidence can be obtained.'

It appears that in fact only a single fiber has been stimulated in most instances where a direct correlation has been obtained between a single unit recording and the sensation evoked on stimulation through the recording electrode. It is admitted that it is difficult to be sure, in the case of C fibers, that stimulation is confined to single C fibers rather than to small numbers of C fibers [*Torebjörk and Ochoa,* 1980, 1983]. However, even with this reservation, it is clear that a sensation of pain can be produced by selective stimulation of at most a few C fibers.

Consistent with the proposal that there are specific nociceptive afferent fibers in peripheral nerves are reports of children who are congenitally insensitive to pain [*Baxter and Olszewski,* 1960; *Jewsbury,* 1951; *McMurray,* 1950]. Some of these cases appear to result from an absence of many of the fine nerve fibers that are normally found in peripheral nerves and Lissauer's tract [*Swanson,* 1963; *Swanson* et al., 1965; *Sweet,* 1981a]. Tactile sensibility appears to be normal, but pain is absent, as may be autonomic responses to stimuli that normally evoke pseudoaffective responses. It is not certain that there are no changes in the central nervous system as well as in the peripheral nervous system, but it is at least plausible to attribute the absence of pain in these cases to the absence of nociceptive afferents.

Nociceptive Ascending Tracts

It has been known since the observations of *Brown-Séquard* [1860], *Gowers* [1878], *Head and Thompson* [1906], *Petrén* [1902] and others, that pain in man depends upon the integrity of nerve fibers ascending the spinal cord on the side opposite to the application of a noxious stimulus. *Spiller* [1905] attributed the bilateral analgesia he found in a patient to interruption of both anterolateral spinal cord quadrants by tuberculomas. Based on this case, *Martin* performed the first deliberate cordotomy for the relief of pain in man, as reported by *Spiller and Martin* [1912]. Since that time, it has been repeatedly confirmed that a cordotomy will produce a complete loss of pain (and thermal) sensation on the contralateral side of the body beginning a few segments below the level of the lesion (fig. 2/6) [*Foerster and Gagel,* 1932; *Hyndman and Van Epps,* 1939; *Tasker* et al., 1976; *White and Sweet,* 1955, 1969]. It is true that in many cases pain returns after a variable period. However, the fact that pain disappears for months to years or even permanently suggests that fibers crucial to normal pain sensation have been interrupted[1].

It is not known why pain can return after an initially successful cordotomy. One possibility is that other pathways that are normally ineffective for transmitting pain become more effective with time after the cordotomy. Thus, although the most important pathway for pain sensation in the normal human subject ascends in the anterolateral quadrant of the cord on the side contralateral to the painful stimulus, there may be pathways in other sectors of the cord that are potentially of importance.

An unlikely alternative to the conclusions reached in the preceding paragraph that fibers ascending in the anterolateral quadrant mediate pain is that descending pathways that facilitate pain transmission are interrupted by cordotomy and that pain sensation actually depends on a path that ascends through another part of the spinal cord. However, this possibility is ruled out by observations made by *Noordenbos and Wall* [1976] that are complementary to those in cordotomy studies. A lady whose spinal cord was completely transected except for part of one anterolateral quadrant

[1] *Melzack and Wall* [1962] point out that the analgesia resulting from cordotomy need not necessarily be interpreted in the traditional way as the effect of interruption of a major pain transmission system. Alternative explanations include: (1) a decrease in the total number of neurons involved in signalling pain; (2) a change in temporal and spatial patterning of information sent to the brain; (3) a change in descending control of dorsal horn neurons; (4) an alteration in the relationships of several ascending pathways.

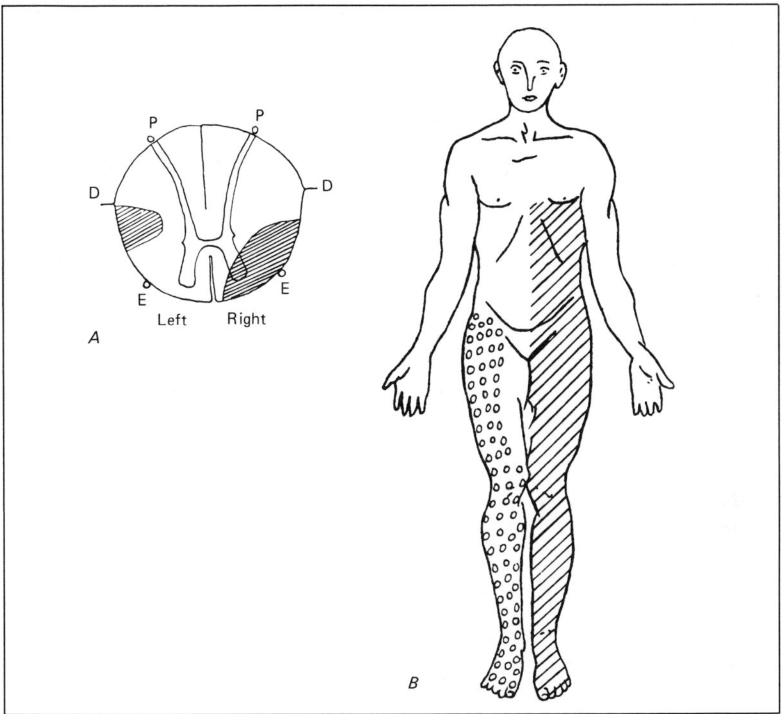

Fig. 2/6. Distribution of loss of pain sensation following cordotomy. The locations of bilateral partial cordotomies are shown in *A*. The distribution of sensory loss is shown in *B*. Hatching indicates a complete loss of pain and temperature sensations, whereas the circles indicate a partial loss of these sensations. [From *Hyndman and Van Epps,* 1939.]

could still feel pain following stimulation below the level of the lesion. Pricking pain was felt when a pin was applied on the side contralateral to the intact anterolateral quadrant, but an unpleasant sensation could also be produced by pin prick on the ipsilateral side. The point is that fibers of the anterolateral quadrant in the human are both necessary for pain, as shown by the effects of cordotomy, and sufficient, as shown in this case.

Not unexpectedly, the patient reported by *Noordenbos and Wall* [1976] was also able to sense temperature changes below her lesion on the contralateral side, a finding consistent with the effects of cordotomy, which causes thermanesthesia contralaterally. However, she had surprisingly good tactile capacity below the lesion, and she had proprioception in the ankle and knee joints ipsilateral to the lesion. Since cordotomies produce only slight

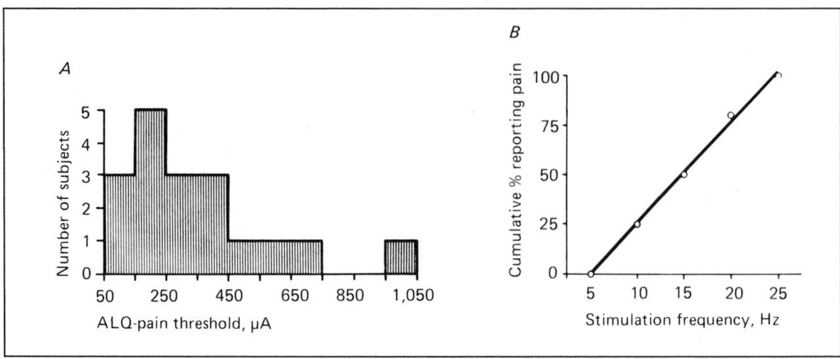

Fig. 2/7. Production of pain by stimulation of the anterolateral quadrant during high cervical percutaneous cordotomy in human patients. *A* shows the thresholds for pain in different subjects (1 s trains of 0.2 ms pulses at 50 Hz). *B* shows the proportions of patients reporting pain as stimulus frequency is varied (stimulus strength 15% above pain threshold at 50 Hz). [From *Mayer* et al., 1975.]

changes in touch and no detectable changes in proprioception [*White and Sweet,* 1955], these findings suggest that the anterolateral quadrant is necessary and sufficient for temperature sensibility, as well as for pain, and sufficient but not necessary for touch and for proprioception at some joints. An alternative explanation for the residual proprioception in the case described by *Noordenbos and Wall* [1976] is that fibers of the ipsilateral lateral funiculus that normally mediate proprioception from the lower extremity were left intact [see *Willis and Coggeshall,* 1978].

The concept that pain transmission in the human depends upon fibers ascending in the anterolateral quadrant of the spinal cord is consistent with observations of the effects of electrical stimulation of axons in the anterolateral quadrant. If sufficiently strong repetitive stimuli are applied in a conscious subject, that subject reports a sensation of pain (fig. 2/7) [*Mayer* et al., 1975]. This observation agrees with the reports of others [*Foerster and Gagel,* 1932; *Sweet* et al., 1950], although many neurosurgeons have noted that other sensations are more prominent when the anterolateral quadrant is stimulated. For example, a sensation of tingling warmth or cold or of burning is often noted during percutaneous cordotomy [*Mayer* et al., 1975; *Tasker* et al., 1976, 1983]. The particular sensation produced appears to depend upon the stimulus parameters [*Mayer* et al., 1975], with pain requiring higher intensities and frequencies of stimulation than thermal sensations.

Participation of Higher Centers in Nociception

There have been many clinical reports that lesions of the thalamus or of the cerebral cortex can interfere with pain, at least transiently (see Chapter 6). Of particular interest with respect to the multifaceted nature of the total pain reaction, as discussed in Chapter 1, are observations that lesions placed in the anterior and dorsal medial thalamus or in the frontal lobes may cause a dissociated loss of the motivational-affective aspect of pain without an obvious change in the sensory-discriminative aspect [*Freeman and Watts,* 1950; *Mark* et al., 1963]. Such observations not only support the idea that there are discrete neural pathways involved in mediating pain, but that there are subsystems concerned with different aspects of the total pain reaction.

Stimulation within specific sites in the brain can evoke pain sensations. For example, although uncommon, stimulation of the sensorimotor cortex in conscious subjects being treated surgically for epilepsy can sometimes elicit a sensation of pain [*Penfield and Boldrey,* 1937]. Furthermore, stimulation at some sites in the thalamus may cause pain [*Halliday and Logue,* 1972; *Hassler,* 1970].

Experimental Approaches to the Investigation of Pain Transmission Systems

It seems clear from the evidence derived from studies on humans that there are specific neural circuits that underlie pain sensation. There even appear to be at least partially separable circuits underlying the sensory-discriminative aspects of pain and the motivational-affective aspects. If these statements are true, then it should be possible to identify and to characterize the neurons in these circuits. The advantages of a full description of the neural circuits responsible for pain include the possibility of improved methods for intervention in cases of clinical pain in man and in animals.

Insofar as is feasible, experimental work should be done in humans to provide the most direct evidence linking the activity of particular neurons to pain sensation [*Sinclair,* 1981]. However, it is obvious that most experimental work on pain circuits will have to be done on animal subjects. Such experiments must be designed to minimize pain in the animal subjects by employing suitable preparations [*Zimmermann,* 1983]. For example, it is possible to examine the responses of neurons in some parts of the brain in animals that are fully awake and capable of normal behavior [e.g. *Hoffman* et al., 1981]. Such animals can be given control of the painful stimulus, so that they can elect to escape from stimuli of too high an intensity. Alternatively, stronger stimuli can be given to animals that are prepared in such a way as to avoid the experience of pain, using anesthesia or surgical procedures such as decerebration. In the case of anesthesia, it appears that while the responses of neurons at the highest levels of the nervous system to painful stimuli are dramatically affected by anesthetic drugs, there is much less change in the activity of neurons at the lower levels of the pain transmission system. Thus, it seems possible to do a reasonable analysis of the activity of nociceptors and perhaps of many of the spinal cord and brain stem neurons processing pain information in anesthetized animals. Decerebration allows the experimenter to avoid the effects of drugs, but introduces another complication in that decerebrate animals are characterized by a high level of tonic activity in descending inhibitory control systems originating in the brain stem and acting upon spinal cord neurons [see *Willis,* 1982]. Therefore, the activity of spinal cord circuits is likely to be quite different in decerebrate animals and in normally behaving animals, and conclusions about neural circuits in decerebrate animals should be made with caution. The problem of tonic descending inhibitory controls can be avoided in unanesthetized preparations by transecting the spinal cord, but the neural circuits that can be examined are then limited to the spinal cord.

Criteria for Identification of Nociceptive Neurons

Studies of the neural circuitry responsible for pain sensation begin with the assumption that it is possible to identify the responsible neurons. In humans, it may be possible to activate specific neurons and to evoke by the activity of such neurons a sensation of pain. In animals there can be no subjective report of pain, and so neural activity can only be related to the detection of noxious stimuli. Thus, animal experiments are concerned with nociception. There is no reason to think that experimental evidence concerning the neural basis of nociception in animals will be any less relevant to humans than, say, studies of photoreception in animals are to human vision. In either case, species differences are likely, but general principles may still be expected to emerge from the animal studies.

Price and Dubner [1977], in their review of evidence concerning neurons involved in the sensory-discriminative aspects of pain, list four lines of evidence that can be used to identify relevant neurons: (1) selective stimulation evokes an experience relatable to the sensory-discriminative aspects of pain; (2) the candidate neurons should respond either exclusively or differentially to noxious stimuli; (3) reduction in the responses of the neurons should reduce pain; (4) the neurons should belong to a neural circuit whose anatomical connections are appropriate for a role in the sensory-discriminative aspects of pain.

Price and Dubner [1977] then surveyed the evidence available concerning the role of primary afferent fibers, spinal cord and trigeminal neurons, reticular formation neurons, and neurons of the thalamus and cerebral cortex that might play a role in the sensory-discriminative aspects of pain sensation. Table 2/I is from their review and summarizes their conclusions about the neural circuitry of the pain transmission system. They did not address the question of neural circuits involved in the motivational-affective aspects of pain.

The Chapters that follow will review and update the evidence concerning the pain transmission system along lines similar to those followed by *Price and Dubner* [1977]. Chapter 3 will be concerned with nociceptors. Chapter 4 will review what is known about the terminations of primary afferent fibers, including nociceptors, in the spinal cord, and the activity of neurons of the dorsal horn. Chapter 5 will consider the role of neurons that project their axons into ascending tracts, including the spinothalamic, spinoreticular, spinomesencephalic, and other candidate nociceptive tracts. Chapter 6 will review the involvement of the thalamus and cerebral cortex in pain. Chapter 7 will provide an overview and suggest future lines of investigation.

Conclusions

1. Pain due to noxious stimuli depends upon conduction of afferent activity in fine fibers (Aδ and C fibers) in peripheral nerves. Such pain can be evoked by electrical stimulation and prevented by local anesthesia of fine afferent fibers.

2. Noxious stimuli in human subjects activate nociceptors of several of the same types that have been studied in experimental animals. Stimulation of individual or small numbers of nociceptive afferents repetitively can evoke a sensation of pain. Afferents from other receptor types evoke other sensations. Hence, there is no support for the notion that pain results from 'overactivation' of receptors other than nociceptors.

Table 2/I. Lines of evidence supporting the role of different neuron types in pain discrimination

Neuron type		Lines of evidence[1]			
		1	2	3	4
Mechanosensitive primary afferents	high-threshold Aδ		++		++
	low sensitivity/moderate pressure Aδ		++		++
	high-threshold C		++		++
	low-threshold Aβ/Aδ				+
	low-threshold C				
Thermosensitive primary afferents	cold Aδ		+		+
	warm Aδ/C		+		+
	high-threshold Aδ/C		+		
Mechanothermo-sensitive primary afferents	heat nociceptive Aδ	++	++	++	++
	polymodal nociceptive C	++	++	++	++
	mechano-cold nociceptive Aδ/C		+		
Spinal cord and brain stem	dorsal horn and trigeminal caudalis – WDR	++	++	++	++
	dorsal horn and trigeminal caudalis – NS		++	++	++
	n. gigantocellularis – NS and WDR	+	+	+	+
	other – NS and WDR		+		
Thalamic nuclei	ventroposterior – WDR		+		+
	ventroposterior – NS		+		+
	posterior group – WDR and NS		+		+
	medial group – WDR		+		+
	medial group – NS		+		+
Cerebral cortex	WDR		+		+
	NS		+		+

[1] Lines of evidence: 1 = selective stimulation produces pain; 2 = maximum response to noxious stimuli; 3 = selective manipulations reduce neural responses and pain; 4 = appropriate anatomical connections. WDR = Wide dynamic range; NS = nociceptive-specific; ++ = strong evidence; + = weak evidence. [From *Price and Dubner,* 1977.]

3. Congenital intensity to pain in some individuals may be due to a lack of nociceptive afferents.

4. Conduction of pain from the body in humans depends upon axons ascending in the anterolateral quadrant on the side opposite to that stimulated. This is shown by the contralateral analgesia produced by cordotomy

and by the sensation of pain on the contralateral side that may be evoked by electrical stimulation in the anterolateral quadrant. Furthermore, conduction in axons of the anterolateral quadrant is sufficient for the mediation of pain.

5. Higher processing of pain signals appears to involve the thalamus and the cerebral cortex, since pain can be at least transiently diminished by lesions in parts of these structures and pain can sometimes be evoked by stimulation in certain sites. There is in fact evidence for a difference in the brain systems involved in processing the motivational-affective and the sensory-discriminative components of the pain reaction, since lesions of the brain can interfere with one but not the other of these.

6. Pain mechanisms should be studied in human subjects insofar as possible because of the availability of a sensory report. However, most experimental work on pain will have to come from animal experiments. Suitable preparations are those that minimize the pain experienced by the animal but that interfere the least with the neural circuit operations.

7. Identification of the neural circuits for pain is more difficult in animals than in human subjects, since animal experiments can only demonstrate nociception. However, it can be anticipated that the nociceptive systems of animals will provide information useful for an understanding of the general principles of pain transmission.

8. Evidence that particular neurons are involved in a pain transmission system used for the sensory-discriminative aspects of pain include the following: stimulation causes pain or a related response; the neurons respond exclusively or differentially to noxious stimuli; a reduction in the activity of the neurons reduces pain; the neurons should belong to an appropriate anatomical system.

Chapter 3

Nociceptors

Historical Overview

Around the turn of the century, a number of investigators, including *Blix* [1884], *Donaldson* [1885] and *von Frey* [1894, 1895, 1906] observed that stimulation of discrete spots in the skin produced particular sensations. For example, it was possible to evoke a sensation of touch from some spots and cold, warmth (fig. 3/1A) or pain from other spots. The easiest explanation of these observations is that there are specialized sensory receptor organs in the skin (fig. 3/1B) and that different receptors are concerned with different sensory modalities. Unfortunately, *von Frey* made several errors in his proposals relating particular receptor organ types to specific sensory modalities. For example, he thought that cold was signalled by Krause's end-bulbs and warmth by Ruffini endings [*von Frey,* 1895]. These sensory receptors have since been shown to have a tactile function [e.g. *Chambers* et al., 1972; *Iggo and Ogawa,* 1977]. However, *von Frey* correctly attributed a tactile function to hair follicle afferents and Meissner's corpuscles and pain to free nerve endings.

Because of criticisms of *von Frey's* proposals concerning which sense organs gave rise to which sensations [*Lele and Weddell,* 1956; *Sinclair,* 1955, 1967; *Weddell and Miller,* 1962], support grew for the hypothesis that sensory modalities depend upon the pattern of peripheral input, rather than upon the activation of specialized endings [*Nafe,* 1929]. *Nafe's* theory was based on information derived from the advent of the single fiber recording technique [*Adrian,* 1926a, b; *Adrian and Zotterman,* 1926]. Interestingly, this same technique has permitted a reexamination of the issue of receptor specificity, and the extension of this technique to human subjects and the introduction of microstimulation of identified peripheral nerve fibers in human nerves has fully substantiated the basic position of *Blix and von Frey* that particular receptors are associated with specific sensations.

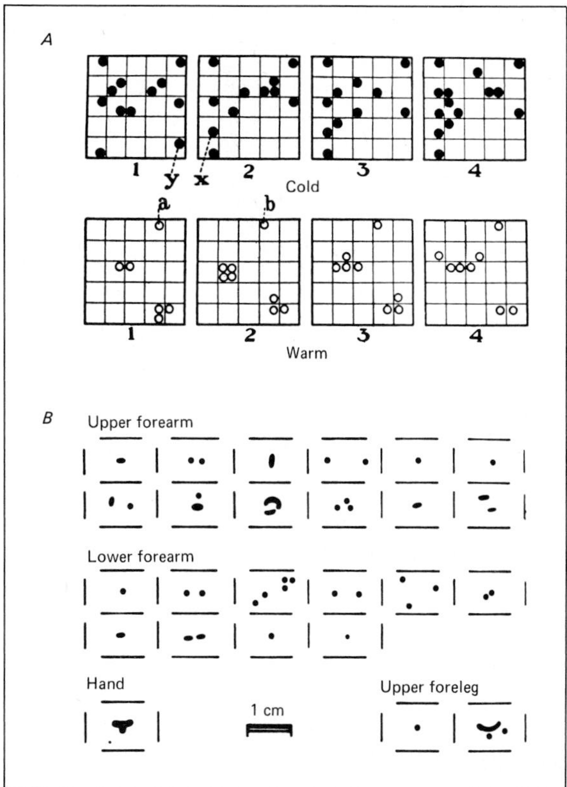

Fig. 3/1. Temperature spots and receptive fields of cold fibers. *A* shows maps of the distribution of cold and warm spots within 1 cm^2 on the skin of a human subject. The 4 maps were made on 4 different days. [From *Dallenbach,* 1927.] *B* shows the receptive fields of individual cold afferent fibers in the monkey. [From *Kenshalo and Duclaux,* 1977.]

Nociceptors

Although *Sherrington's* [1906] definitions of nociceptors as sensory receptors that signal noxious stimuli and noxious stimuli as those that threaten or produce damage are reasonable, the experimenter needs more specific guidelines for judging whether the activity recorded from a particular receptor is indeed produced by a nociceptor. *Burgess and Perl* [1973] provide a major criterion based on *Sherrington's* definitions: 'the ability of the sensory unit effectively and reliably to distinguish between noxious and

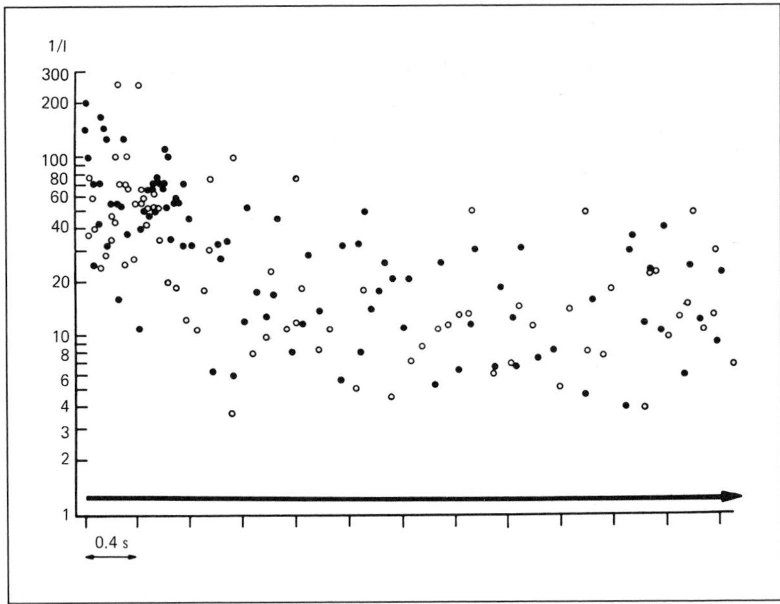

Fig. 3/2. Similar responses evoked in a slowly adapting mechanoreceptor by innocuous and noxious mechanical stimulation of the skin. The receptive field was on the glabrous skin in a monkey. A step stimulus was applied and the responses to moderate pressure (filled circles) and to noxious pressure (open circles) are shown. Instantaneous frequency is plotted on the ordinate and time on the abscissa. [From *Perl,* 1968.]

innocuous events in the signals it provides to the central nervous system'. Although some nociceptors might respond only to damaging stimuli, others would respond both to stimuli that threaten damage and to damaging stimuli, the difference being signalled by a higher discharge rate for the stronger stimuli. For cutaneous nociceptors, the thresholds for activation by mechanical stimuli may be orders of magnitude greater than for mechanoreceptors. More significantly, however, the responses of nociceptors increase as the stimulus strength is increased into a range that causes tissue damage. Mechanoreceptors, on the other hand, show no increased discharge but may instead show the same or even a decreased discharge as the stimulus becomes more intense (fig. 3/2) [*Burgess and Perl,* 1973; *Perl,* 1968].

Another criterion that can be added in those cases in which noxious stimulation or a disease process causes sensitization of a tissue is that the nociceptors may also demonstrate sensitization [*Bessou and Perl,* 1969;

Croze et al., 1976; *Fitzgerald and Lynn,* 1977]. Nociceptor sensitization is recognized by a lowered threshold and by an increased response to a particular suprathreshold stimulus.

Cutaneous Nociceptors

There appear to be two main classes of cutaneous nociceptors: Aδ mechanical nociceptors and C polymodal nociceptors.

Aδ Mechanical Nociceptors

These receptors derive their name from the conduction velocities of their axons, which are generally in the Aδ range (mean between 15 and 25 m/s; range 5–50 m/s), and from their specific sensitivity to mechanical but not thermal or chemical noxious stimuli [*Beck* et al., 1974; *Burgess* et al., 1968; *Burgess and Perl,* 1967; *Fitzgerald and Lynn,* 1977; *Georgopoulos,* 1976; *Hunt and McIntyre,* 1960; *Perl,* 1968]. These receptors have been described in the skin of cats [*Burgess and Perl,* 1967], rats [*Lynn and Carpenter,* 1982], rabbits [*Fitzgerald and Lynn,* 1977], monkeys [*Georgopoulos,* 1976; *Perl,* 1968], and humans [*Konietzny* et al., 1981], and they are present in both hairy and glabrous skin [*Perl,* 1968]. A proposed structural arrangement of an Aδ mechanical nociceptor is shown in figure 3/3 [*Perl,* 1984a].

The receptive field of an Aδ mechanical nociceptor consists of a group of 3–20 spots, each with a diameter of less than 1 mm^2, distributed over an area of 1–8 cm^2 [*Burgess and Perl,* 1973; *Perl,* 1968]. The threshold for activation of the receptor by applying a mechanical stimulus to one of the sensitive spots in the receptive field is higher than that of any of the sensitive mechanoreceptors by a factor of 5–1,000 times (fig. 3/4) [*Burgess and Perl,* 1973; *Georgopoulos,* 1976]. In the absence of stimulation, these receptors are silent [*Fitzgerald and Lynn,* 1977; *Perl,* 1968], although a slow background firing can develop after repeated stimulation [*Fitzgerald and Lynn,* 1977]. Many of the receptors discharge when stimulated with innocuous intensities, but the discharge rate increases as the stimulus intensity is raised to an overtly damaging level. Other receptors only respond to stimuli that damage the skin. According to *Burgess and Perl* [1973], Aδ mechanical nociceptors have the capability of detecting both stimulus velocity and position. There is an initial discharge that codes for stimulus movement, and then there may be a sustained discharge that signals the continuation of the stimulus.

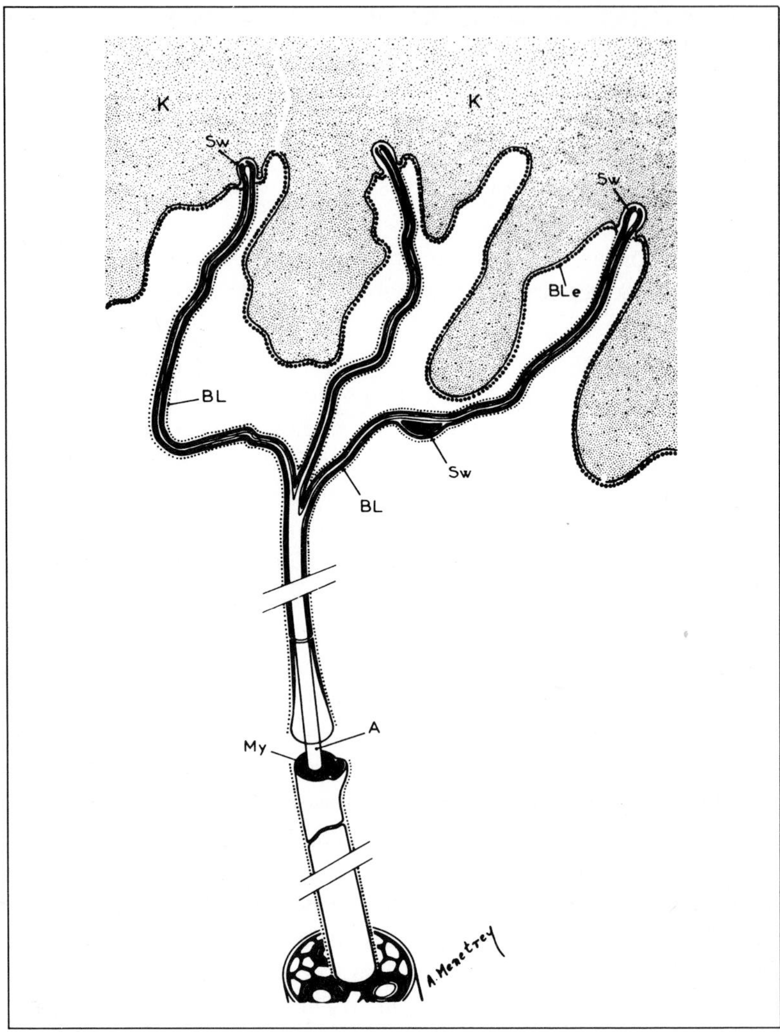

Fig. 3/3. Drawing of the arrangement of an Aδ mechanical nociceptor as it terminates in the skin. The axon loses its myelin sheath, and fine terminals associated with Schwann cell investments enter the epidermis. [From *Perl,* 1984.]

Figure 3/5 illustrates the receptive fields and the responses of Aδ mechanical nociceptors in the skin of the monkey [*Perl,* 1968]. Clusters of sensitive spots making up receptive fields for 3 different Aδ mechanical nociceptors are shown in figure 3/5D. The failure of a pressure stimulus to

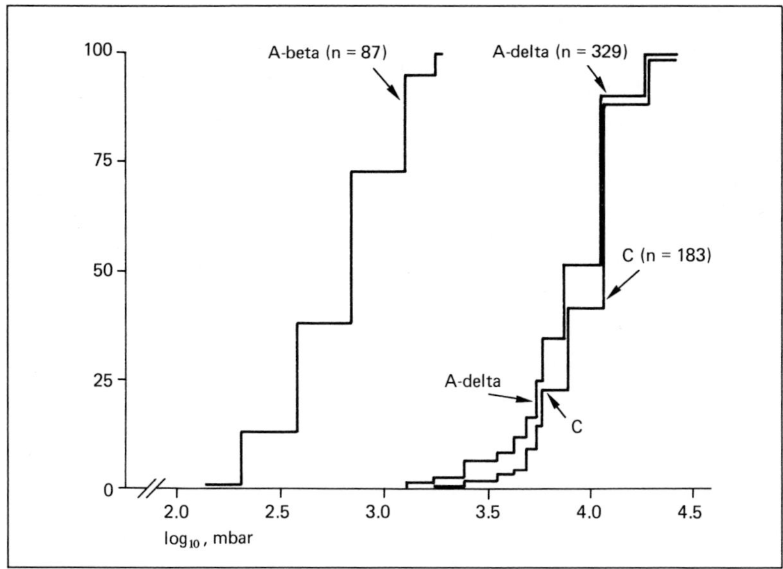

Fig. 3/4. Comparison of thresholds to mechanical stimulation of sensitive mechanoreceptors supplied by Aβ fibers and nociceptors supplied by Aδ or C fibers. [From *Georgopoulos,* 1976.]

activate an Aδ mechanical nociceptor is shown in figure 3/5A. The same receptor was excited by application of a needle to the receptive field (fig. 3/5B) and by pinching with serrated forceps (fig. 3/5C).

Repeated stimulation of a spot within the receptive field of an Aδ mechanical nociceptor may result in fatigue [*Burgess and Perl,* 1973; *Perl,* 1968]; that is, further stimulation of the same spot fails to evoke a discharge, unless a period of rest of 2–5 min is allowed [*Perl,* 1968]. However, stimulation of another spot within the same receptive field can still activate the receptor [*Burgess and Perl,* 1973].

Aδ mechanical nociceptors can be sensitized. The most dramatic evidence of this comes from experiments in which the receptive field is stimulated with noxious heat [*Fitzgerald and Lynn,* 1977]. The initial noxious heat stimulus may fail to evoke any discharge at all. However, repeated noxious heat stimuli may. An example of this is shown in figure 3/6. Three consecutive noxious heat stimuli were applied before the data shown in the figure were recorded; none of these stimuli had any overt effect. The fourth stimulus, shown in figure 3/6A, also failed to evoke a response. However, the next 3 stimuli resulted in discharges (fig. 3/6B–D). Excessive heating, for example to 55–60 °C, inactivates the receptors [*Fitzgerald and Lynn,* 1977].

Some Aδ nociceptors can be excited by noxious heat applied for the first time [*Georgopoulos,* 1976, 1977; *LaMotte* et al., 1982; cf. *Iggo and Ogawa,* 1971]. However, the thresholds tend to be higher than those of C polymodal nociceptors [*LaMotte* et al., 1982].

C Polymodal Nociceptors

These receptors are named for the facts that their afferent axons are unmyelinated (C fibers) and that they respond to noxious mechanical, thermal and chemical stimuli [*Bessou and Perl,* 1969; cf. *Iggo,* 1959; *Iriuchijima and Zotterman,* 1960]. The judgement that the axon is unmyelinated is based on measurement of the conduction velocity, assuming that velocities below 2.5 m/s are characteristic only of unmyelinated axons [*Gasser,* 1955]. C polymodal nociceptors have been demonstrated in the skin of cats [*Beck* et al., 1974; *Bessou and Perl,* 1969; *Iggo,* 1959], rats [*Fleischer* et al., 1983; *Lynn and Carpenter,* 1982], monkeys [*Beitel and Dubner,* 1976a,b; *Croze* et al., 1976; *Georgopoulos,* 1976; *Iggo and Ogawa,* 1971; *Kumazawa and Perl,* 1977; *LaMotte and Campbell,* 1978] and humans [*Torebjörk and Hallin,* 1974; *Van Hees and Gybels,* 1972] and they are present in both hairy and glabrous skin [*Bessou and Perl,* 1969].

The receptive fields of C polymodal nociceptors consist of 1 or 2 spots of 1–2 mm^2 [*Bessou and Perl,* 1969; *Burgess and Perl,* 1973] or areas of uniform sensitivity of up to 5 mm^2 [*Croze* et al., 1976; *Iggo and Ogawa,* 1971]. Thresholds for mechanical stimulation are similar to those of Aδ mechanical nociceptors (fig. 3/4). Some receptors discharge following non-damaging stimuli and others have thresholds so high that stimuli damage the skin. There is rarely any background discharge in the absence of skin damage [*Bessou and Perl,* 1969; *Kumazawa and Perl,* 1977; *LaMotte* et al., 1983]. In the case of C polymodal nociceptors that are excited by non-damaging stimuli, the discharge increases as stimulus intensity is elevated into the damaging range. Repeated mechanical stimulation causes fatigue [*Beitel and Dubner,* 1976a; *Kumazawa and Perl,* 1977].

Figure 3/7 shows an example of the responses of a C polymodal nociceptor to graded noxious mechanical stimuli [*Kumazawa and Perl,* 1977]. The stimuli were increased in intensity from A to D in figure 3/7. The stimuli in B–D were judged to be noxious, and in fact those in C and D were produced by penetration of the skin by the stimulator probe. It should be noted that there was a vigorous discharge at the time of stimu-

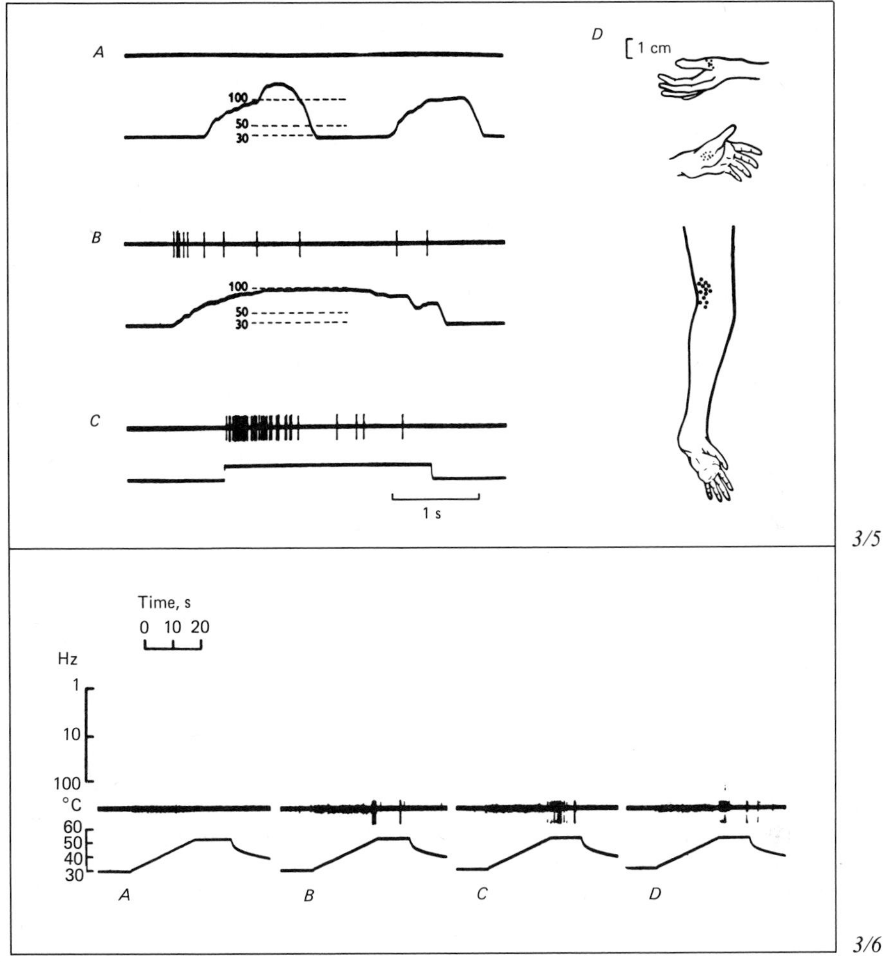

3/5

3/6

Fig. 3/5. Responses of an Aδ mechanical nociceptor and receptive fields of 3 different Aδ mechanical nociceptors in monkey skin. The responses in *A–C* were recorded from an Aδ mechanical nociceptor whose receptive field was in the glabrous skin of a monkey. In *A*, there was no response to pressure applied using a probe having a tip diameter of 2.2 mm (force indicated in lower trace). In *B*, the probe was a needle tip, and a response was produced. In *C*, the stimulus was pinching with serrated forceps. *D* shows the receptive fields of 3 different Aδ mechanical nociceptors; each receptive field consists of an array of sensitive spots. [From *Perl*, 1968.]

Fig. 3/6. Sensitization of an Aδ mechanical nociceptor by repeated noxious heating of the skin. Three heat stimuli (48–52 °C) were given before the recordings shown were made. The stimuli shown were 52 °C heat pulses. The stimulus in *A* failed to evoke a response, but the later ones in *B–D* produced discharges. [From *Fitzgerald and Lynn*, 1977.]

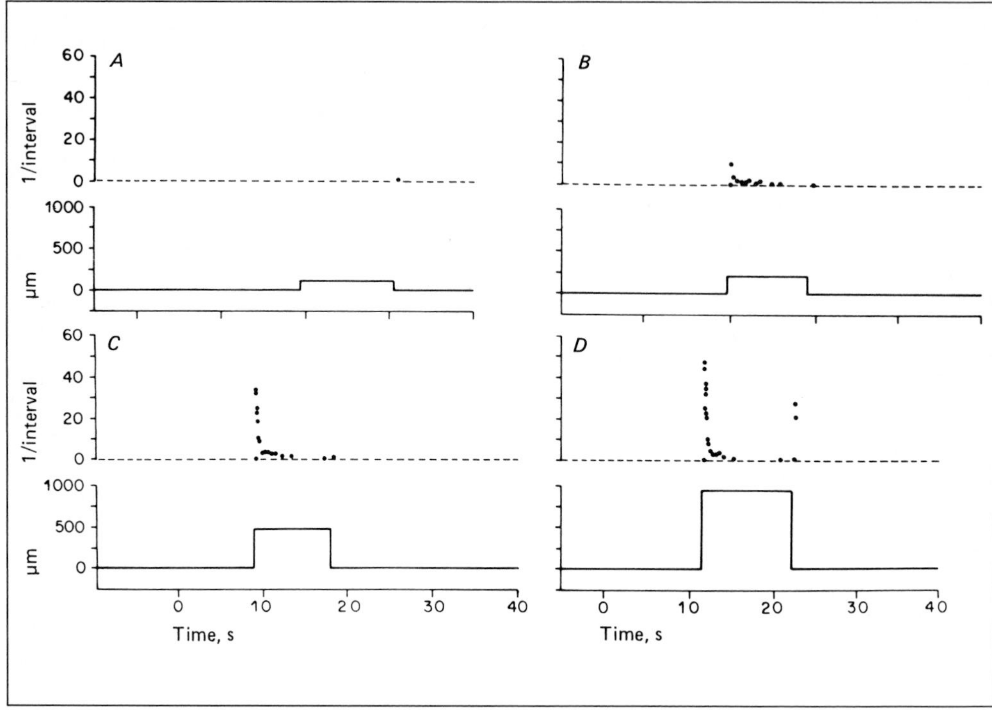

Fig. 3/7. A Responses of a C polymodal nociceptor to graded mechanical stimulation. The receptor was in monkey skin. The graphs plot instantaneous frequency against time; lower traces show probe excursion. The probe was pointed, and it contacted the center of the receptive field lightly. The stimuli used for *B–D* were noxious; those in *C* and *D* caused the probe to penetrate the skin. [From *Kumazawa and Perl,* 1977.]

lation but that the discharge adapted during maintenance of the stimulus. In figure 3/7D, the unit discharged also when the stimulus was removed. Thus, there are both dynamic and static components of the response. Repeated mechanical stimulation often causes a low frequency maintained discharge.

C polymodal nociceptors respond in a graded fashion to graded noxious heat stimuli [*Beck* et al., 1974; *Bessou and Perl,* 1969; *Croze* et al., 1976; *Iggo,* 1959; *LaMotte and Campbell,* 1978]. Thresholds for different units vary from 40 to 56 °C [*Bessou and Perl,* 1969; *Handwerker and Neher,* 1976; *Kumazawa and Perl,* 1977]. The discharge rate increases either linearly (fig. 3/8A, B) [*Beck* et al., 1974; *Croze* et al., 1976; *Torebjörk* et al.,

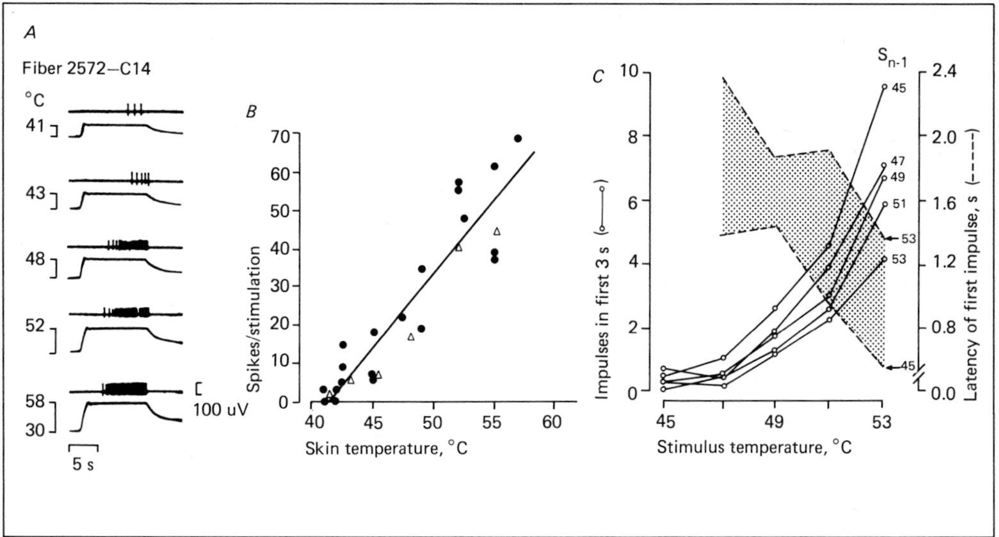

Fig. 3/8. Stimulus-response relationship for C polymodal nociceptors. The recordings in *A* were made from a C fiber supplying the skin of the cat. The progressively greater responses to graded noxious heat stimuli are evident. The graph in *B* shows that the number of discharges produced is a linear function of the temperature of the stimulus. [From *Beck* et al., 1974.] In *C,* the stimulus-response functions represent the mean responses of 10 C polymodal nociceptors in monkey skin; the curves show the responses obtained when the preceding stimulus was between 45 and 53 °C; the responses were systematically less when the preceding stimulus was high than when it was low. The curves, nevertheless, are all positively accelerating. The shaded area shows the decline in latency as stimulus intensity increases; again, the effects of previous stimulus strength are indicated. [From *LaMotte and Campbell,* 1978.]

1984] or with a positively accelerating stimulus-response curve (fig. 3/8C) [*Beitel and Dubner,* 1976b; *Croze* et al., 1976; *Georgopoulos,* 1976; *Handwerker and Neher,* 1976; *LaMotte and Campbell,* 1978; *LaMotte* et al., 1983]. Fatigue occurs with repeated noxious heat stimulation [*Beitel and Dubner,* 1976a; *LaMotte and Campbell,* 1978]. Deactivation results from excessive heating [*Croze* et al., 1976; *Kumazawa and Perl,* 1977].

In addition to noxious and thermal mechanical stimuli, noxious chemical stimuli are also effective. For example, locally applied bradykinin, histamine, acetylcholine, acids or potassium can all activate C polymodal nociceptors to varying degrees [*Bessou and Perl,* 1969; *Burgess and Perl,* 1973; *Fjällbrant and Iggo,* 1961; *Kumazawa and Perl,* 1977].

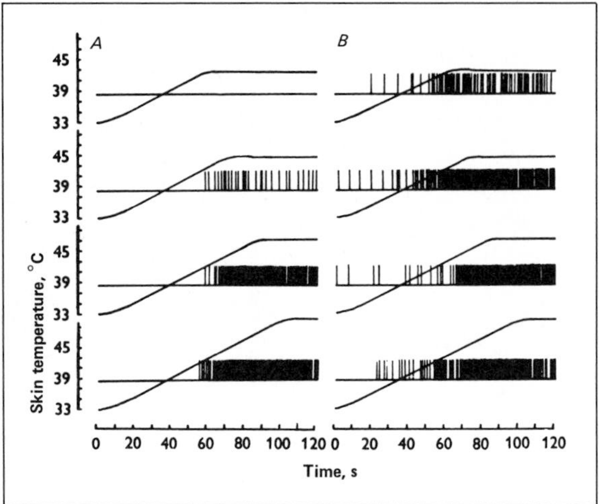

Fig. 3/9. Sensitization of a C polymodal nociceptor by repeated noxious heat stimuli. The discharges in column *A* were recorded from a C polymodal nociceptor in monkey skin. The graded series of noxious heat stimuli from an adapting temperature of 33 °C reached levels of 43, 45, 47 and 50 °C in the four recordings. These stimuli sensitized the receptor, since repetition of the same noxious heat stimuli disclosed a lowered threshold and an enhanced response, as seen in column *B.* [From *Croze* et al., 1976.]

Cold stimuli, even when in the noxious range, generally have little effect on C polymodal nociceptors [*Bessou and Perl,* 1969; *Burgess and Perl,* 1973; *Iggo and Ogawa,* 1971], unless the skin is first sensitized by heating [*Bessou and Perl,* 1969; *Kumazawa and Perl,* 1977]. C polymodal nociceptors are less sensitive to cold than are the cold nociceptors described by *LaMotte and Thalhammer* [1982].

Like Aδ mechanical nociceptors, C polymodal nociceptors can be sensitized [*Beck* et al., 1974; *Bessou and Perl,* 1969; *Croze* et al., 1976; *Kumazawa and Perl,* 1977]. For example, in figure 3/9 the responses of a C polymodal nociceptor to two series of noxious heat stimuli are shown [*Croze* et al., 1976]. A stimulus of 43 °C failed to excite the receptor on the first trial (fig. 3/9A, top record), and a stimulus of 45 °C produced a weak discharge (fig. 3/9A, second record). When the same series of noxious heat stimuli was repeated, there was a response to a 43 °C stimulus (fig. 3/9B, top record), and the response to the 45 °C stimulus was greater (fig. 3/9B, second record). According to *Lynn* [1979], the sensitization of C polymodal nociceptors by heat is not affected by interruption of the local blood supply.

Other Types of Cutaneous Nociceptors

In addition to Aδ mechanical nociceptors and C polymodal nociceptors, there are several other categories of cutaneous nociceptors. For example, some Aδ nociceptors appear to be sensitive to noxious heat [*Beck* et al., 1974; *Georgopoulos,* 1976, 1977; *Iggo and Ogawa,* 1971; *Kumazawa and Perl,* 1977; *Sumino* et al., 1973]. *Fitzgerald and Lynn* [1977] observed that many Aδ mechanical nociceptors fail to respond to noxious heating on the first trial of stimulation, but that after sensitization of the skin such receptors do respond consistently to noxious heat. It may be that many of the reports of responses of Aδ nociceptors to noxious heat were based on recordings from sensitized Aδ mechanical nociceptors. However, it appears that at least some Aδ nociceptors can respond to noxious heat even though there has been no previous heating of the skin [*Kumazawa and Perl,* 1977; *LaMotte* et al., 1982].

Conversely, there are nociceptors with unmyelinated afferent fibers that respond to mechanical noxious stimuli, but not to noxious heat or to irritant chemicals [*Bessou and Perl,* 1969; *Burgess and Perl,* 1973; *Georgopoulos,* 1976; *Kumazawa and Perl,* 1977; *LaMotte and Campbell,* 1978]. Thus, there are C mechanical nociceptors, in addition to C polymodal nociceptors. Interestingly, some of these C mechanical nociceptors supply subcutaneous tissue and so may be protected from thermal and chemical stimuli. When irritant chemicals are applied to broken skin, such receptors can be activated [*Bessou and Perl,* 1969], suggesting that they are at least potentially polymodal.

Some receptors appear to be specifically responsive to strong cold stimuli [*Bessou and Perl,* 1969; *Iggo,* 1959; *LaMotte and Campbell,* 1978; *LaMotte and Thalhammer,* 1982]. These high threshold cold receptors may have Aδ or C afferent fibers, and they are insensitive to mechanical stimulation [*LaMotte and Thalhammer,* 1982]. The receptive fields of high threshold cold receptors are areas ranging from 1 to 145 mm^2 [*LaMotte and Thalhammer,* 1982]. Figure 3/10 illustrates stimulus-response functions for several of these receptors. There are a number of differences between these cold receptors and the low threshold cold receptors described by several investigators [e.g. *Darian-Smith* et al., 1973; *Dodt and Zotterman,* 1952; *Dubner* et al., 1975; *Hensel and Iggo,* 1971; *Iggo,* 1969; *Kenshalo and Duclaux,* 1977; *Long,* 1977]. For example, the high threshold cold receptors do not have a background discharge at temperatures that typically activate low threshold cold receptors maximally. Furthermore, the high threshold cold receptors do not display the bursting discharge that is typical of low

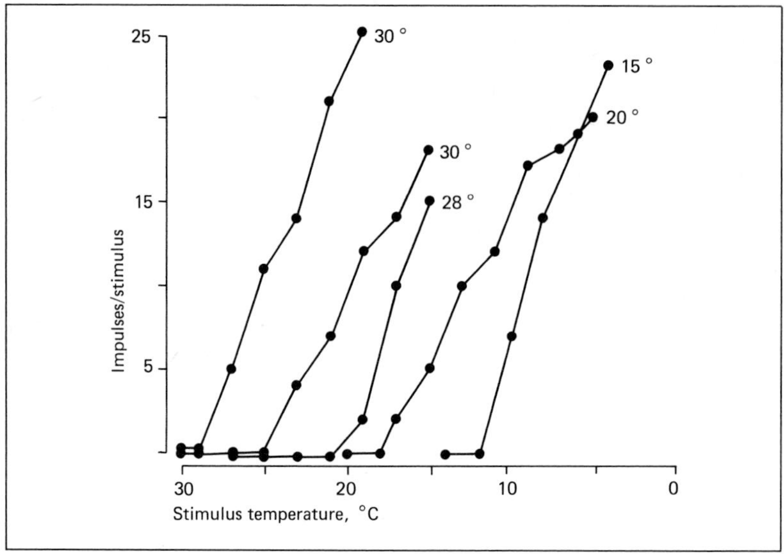

Fig. 3/10. Stimulus-response functions for high threshold cold receptors. The 4 curves to the left were from units supplying monkey skin and that at the right from the cat. Adapting temperatures are indicated at the top of each curve. [From *LaMotte and Thalhammer,* 1982.]

threshold cold fibers when exposed to a large cooling step, nor do high threshold cold receptors demonstrate a paradoxical response to noxious heat [*LaMotte and Thalhammer,* 1982].

Role of Cutaneous Nociceptors in Pain Sensation

There are at least two broad categories of cutaneous pain sensation: pricking pain and burning pain [*Lewis,* 1942]. Cold can also produce pain. Cold-induced pain has an aching quality [*Wolf and Hardy,* 1941], and so aching may be a third quality of cutaneous pain. However, *Wolf and Hardy* argued that the aching quality of cold pain depends upon the activation of subcutaneous receptors, perhaps associated with blood vessels. Localization, intensity, and duration of cutaneous pain can be well described. Thus, cutaneous nociceptors must be able to provide coded information sufficient to distinguish pain quality, spatial distribution, strength and temporal course. Interestingly, pricking pain can be evoked by a mechanical stimulus (such as a needle point contacting the skin), by noxious heat (for example,

Table 3/I. Peripheral receptors capable of being activated by a needle, a burning stimulator or a very cold stimulator [from *Chéry-Croze,* 1983]

Receptor		Response				
designation	afferent fiber	needle	hot stimulator		very cold stimulator	
			1 mm²	3–100 mm²	1 mm²	3–100 mm²
Mechanoreceptors						
Low threshold	Aδ	+	–	–	–	–
High threshold	Aδ	+	–	–	–	–
	C	+	–	–	if rate > 2 °C/s	
Mechanothermal nociceptors						
Hot	Aδ	+	+	+	–	–
Cold	Aδ + C	+	–	–	+	+
Thermal nociceptors						
Hot	Aδ + C	–	+	+	–	–
Hot + cold	Aδ + C	–	+	+	+	+
Polymodal nociceptors						
Hot + cold + chemical	C	+	+	+	+	+
Thermoreceptors						
Hot	Aδ + C	–	–	+	–	–
Cold		–	–	+	–	–
Cold	Aδ + C	–	+	+	–	+
		–	–	–	–	+

The + signs indicate a response on applying the stimulator and the – signs indicate absence of response.

by briefly touching the skin with a hot wire), or by electric shocks [*Lewis,* 1942; *Torebjörk and Hallin,* 1973]. Similary, burning pain can be elicited by noxious mechanical or thermal stimuli or by electric currents, provided that the stimulus duration is long enough [*Lewis,* 1942].

In normal skin, pricking pain appears to be mediated by Aδ nociceptors and burning pain by C polymodal nociceptors. The most direct evidence for this statement comes from experiments on human subjects, but important additional evidence is from experiments that correlate human sensory experience with observations of the activity of nociceptors in animals. However, a variety of receptors can respond to stimuli that evoke pricking or burning pain and so it may be premature to specify which particular receptors are actually causal (table 3/I) [*Chéry-Croze,* 1983].

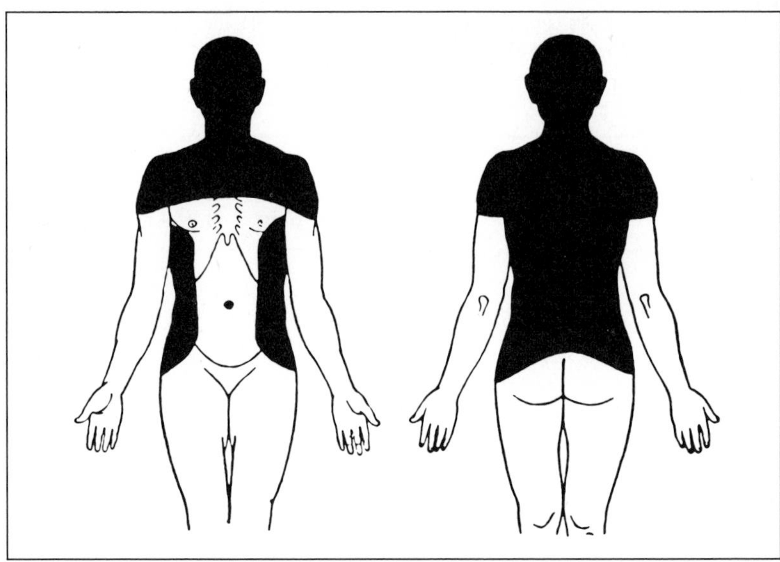

Fig. 3/11. A brief painful stimulus may give rise to two separate pain sensations when applied to the unblackened areas of skin, but to only a single sensation when applied to the black areas. The interval between the two sensations is greater for stimulus locations placed more distally. [From *Lewis and Pochin,* 1937.]

First and Second Pain

When certain areas of the body surface are stimulated, the subject may feel two distinct pains, called first and second pain based on their temporal sequence [*Goldscheider,* 1881; *Thunberg,* 1901; *Lewis,* 1942]. Not all subjects can make this distinction, and the stimulus conditions are critical [*Sinclair and Stokes,* 1964]. For instance, a double pain response is easier to obtain when heat is applied to the hairy skin than to the glabrous skin; the threshold for a double pain response in glabrous skin is above 51 °C, whereas it is lower in hairy skin [*Campbell and LaMotte,* 1983; *LaMotte* et al., 1982; cf. *Bigelow* et al., 1945; *Hardy* et al., 1952a]. Most investigators state that first pain has a pricking quality and second pain a burning quality [*Bigelow* et al., 1945; *Hardy* et al., 1952a; *LaMotte* et al., 1982; *Sinclair and Stokes,* 1964], although *Lewis* [1942] felt that both pains have the same quality. First and second pains are most readily distinguished following stimulation of the distal forelimbs, the lower extremities or the anterior wall of the lower trunk, and they are not distinguishable when stimuli are applied to the remainder of the trunk or to the head and shoulders (fig. 3/11)

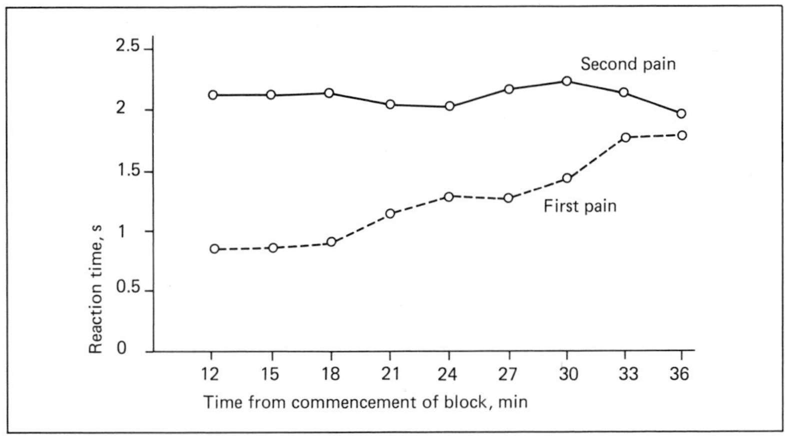

Fig. 3/12. Reaction times for first and second pain during progressive asphyxial block of peripheral nerve. [From *Sinclair and Stokes,* 1964.]

[*Lewis and Pochin,* 1937]. By timing the changes in latency of first and second pain as the stimulus was moved systematically over the body surface, *Lewis and Pochin* [1937] were able to estimate that the afferent fibers responsible for second pain conduct at 0.5–1 m/s. Those producing first pain conduct 20 times as fast (10–20 m/s) according to *Lewis* [1942]. *Campbell and LaMotte* [1983] have recently confirmed that the detection latency for first pain is brief enough to require the afferents responsible to conduct at 6 m/s or faster. *Lewis* argued that the ascending tracts involved in transmitting the information about first and second pain have the same conduction velocity and that both pains are dependent on the crossed anterolateral pathway since both are abolished by cordotomy [*Lewis,* 1942]. When conduction in peripheral nerves is blocked progressively using a blood-pressure cuff, the latency of first pain may either increase gradually or first pain may be blocked at times when the latency of second pain remains unchanged (fig. 3/12) [*Sinclair and Stokes,* 1964]. Evidently, different peripheral nerve fibers are responsible for first and second pain. These findings concerning first and second pain are consistent with the hypothesis that pricking pain is mediated by Aδ nociceptors and burning pain by C fibers.

Electrical Stimulation of Peripheral Nerve Fibers

As already discussed, electrical stimulation of the skin in human subjects can cause both pricking and burning pain [e.g. *Torebjörk and Hallin,*

1973]. These can be evoked individually by varying the stimulus strength or by blocking conduction in the large afferent fibers [*Hallin and Torebjörk,* 1976; *Torebjörk and Hallin,* 1973]. Furthermore, stimulation through a microelectrode placed close to the afferent fiber of a single Aδ nociceptor can evoke a sensation of either pricking pain in the case of an Aδ fiber or burning pain in the case of a C fiber [*Torebjörk and Ochoa,* 1980, 1981, 1983]. The pain is referred to an area of skin that is congruent with the receptive field of the nociceptor.

Correlation of Human Pain Sensation with Responses of Nociceptors in Humans

The pain spots of *von Frey* can be readily explained on the basis of the terminations of nociceptive afferent fibers in the skin. The thresholds for evoking pain by stimulating pain spots are around 0.5 g [*Foerster and Gagel,* 1932] when using von Frey filaments. The thresholds of Aδ mechanical nociceptors and of C polymodal nociceptors in human subjects extend down to a similar value (0.7–13.2 g) [*Torebjörk,* 1974; *Van Hees and Gybels,* 1972]. Both types of receptor have punctate receptive fields, and the large number of pain spots (40/cm^2, according to *Foerster and Gagel* [1932]) can easily be accounted for on the basis of the presumed large number of nociceptive afferent fibers in human nerves[1].

In several investigations, an attempt was made to relate the discharges of individual C polymodal nociceptors in human subjects to the occurrence

[1] The sural nerve in the human contains approximately 9,500 Aδ fibers and 35,000 C fibers [*Dyck and Lambert,* 1966; *Ochoa and Mair,* 1969]. Although sensitive mechanoreceptors with unmyelinated afferent fibers (C mechanoreceptors) are commonly found in the proximal cutaneous nerves of the cat [*Bessou* et al., 1971; *Douglas and Ritchie,* 1957; *Iggo,* 1960; *Iggo and Kornhuber,* 1977], such fibers have not yet been identified in human nerves [*Konietzny* et al., 1981; *Torebjörk,* 1974; *Van Hees and Gybels,* 1972], and only a few have been detected in monkey cutaneous nerves [*Kumazawa and Perl,* 1977]. Thus, most C fibers in distal cutaneous nerves of man are likely to be either C polymodal nociceptors, specific thermoreceptors (warm receptors), or sympathetic postganglionic fibers [*Torebjörk and Hallin,* 1976]. Assuming that 50% of the Aδ and C fibers in the human sural nerve are nociceptors, that each supplies only a simple punctate receptive field (actually, each Aδ nociceptor innervates up to 20 spots) and that the sural nerve innervates an area of 35 cm^2 [Privy Council, 1958], there would be more than 60 nociceptive endings/cm^2, a figure far in excess of the density of pain spots (40/cm^2) [*Foerster and Gagel,* 1932]. Presumably, the excess numbers of endings would be accounted for by multiple innervation of individual pain spots plus overlapping innervation of areas of skin other than that supplied exclusively by the sural nerve. Even if pain spots are supplied exclusively by Aδ mechanical nociceptors, there would appear to be enough fibers to account for all of them.

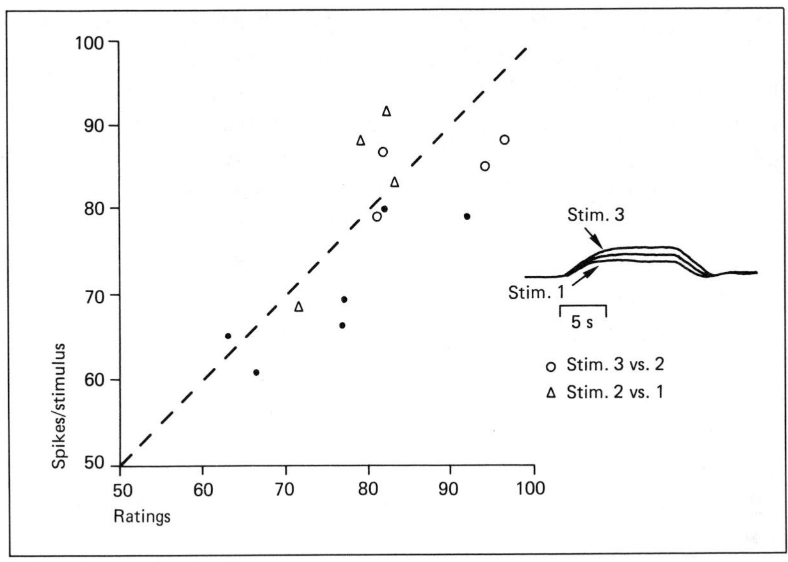

Fig. 3/13. Discrimination between noxious heat stimuli. The discharges of individual C polymodal nociceptors can discriminate as well as subjects can using pain ratings (equal discrimination shown by dashed line). [From *Gybels* et al., 1979.]

of pain sensation [*Gybels* et al., 1979; *Van Hees,* 1976; *Van Hees and Gybels,* 1972]. One or a few impulses in a particular afferent fiber did not cause any sensation. In fact, a considerable amount of summation appears to be required before a pain sensation is initiated [*Adriaensen* et al., 1980]. It should be noted that this situation is different from experiments in which single afferent fibers are stimulated electrically, since in the latter case the stimuli are repetitive at a controlled rate [cf. *Torebjörk and Ochoa,* 1980, 1981, 1983]. On the other hand, when graded thermal stimuli were applied to the skin, the responses of individual C polymodal nociceptors were able to discriminate stimulus intensity as well as the subjects could, using pain ratings (fig. 3/13) [*Gybels* et al., 1979]. However, the stimulus intensity was predicted better by the subjective ratings than by the discharges of individual C fibers, indicating that magnitude estimates would be improved if the central nervous system made use of information derived from a population of C fibers, rather than from a single fiber.

Recently, *Torebjörk* et al. [1984] have found a linear relationship between graded noxious heat stimuli and averaged C polymodal nociceptor discharges and a positively accelerating relationship between the same stimuli and magnitude estimates of pain in human subjects. Furthermore, pain threshold was often higher than the thresholds of individual C polymodal nociceptors. It was suggested that the magnitude estimates might reflect the input from progressively more receptors that would be recruited as threshold was exceeded either under the stimulus probe or in the adjacent skin.

Correlation of Human Pain Sensation with Responses in Nociceptors in Animals

Parallel psychophysical experiments in human subjects and single unit recording experiments in animals show a striking correspondence between the human sensory responses and the stimulus-response curves of C polymodal nociceptors when graded noxious heat stimuli are applied to normal skin (fig. 3/14) [*LaMotte and Campbell,* 1978; *LaMotte* et al., 1983; see review by *Chéry-Croze,* 1983]. By contrast, neither Aδ nociceptors nor specific thermoreceptors can account for the sensory data. For example, the thresholds of the population of Aδ nociceptors for heat stimuli are too high to account for the threshold for heat-evoked pain (fig. 3/15) [*LaMotte* et al., 1982]. Furthermore, warm receptors show U-shaped stimulus-response curves when heating is extended into the noxious range (fig. 3/16) [*Darian-Smith* et al., 1979; *Duclaux and Kenshalo,* 1980; *Handwerker and Neher,* 1976; *Hensel and Kenshalo,* 1969; *LaMotte and Campbell,* 1978; *Sumino* et al., 1973], and so the information they provide would be inappropriate to account for the psychophysical functions. Nor can the paradoxical responses of cold receptors [*Dodt and Zotterman,* 1952; *von Frey,* 1906] account for heat pain [*Dubner* et al., 1975][2].

Noxious chemical stimuli may evoke sensations of either pain or itch [*Becerra-Cabal* et al., 1983]. Such stimuli do not affect Aδ nociceptors, but they do activate C polymodal nociceptors. Actually, there now appear to be two different classes of C polymodal nociceptor. Both types respond to the same kinds of noxious mechanical and thermal stimuli, but only one is activated by itch-provoking stimuli; furthermore, when stimulated during microneurography, this class of C polymodal nociceptor evokes a sensation of itch [*Torebjörk and Ochoa,* 1981]. The other type of C polymodal nociceptor when stimulated causes pain.

[2] However, it has been suggested that a reduced firing, for example of warm receptors or even slowly adapting mechanoreceptors, could contribute to the signal for pain or to a reduction in noise against which the pain signal is detected [*Chéry-Croze,* 1983].

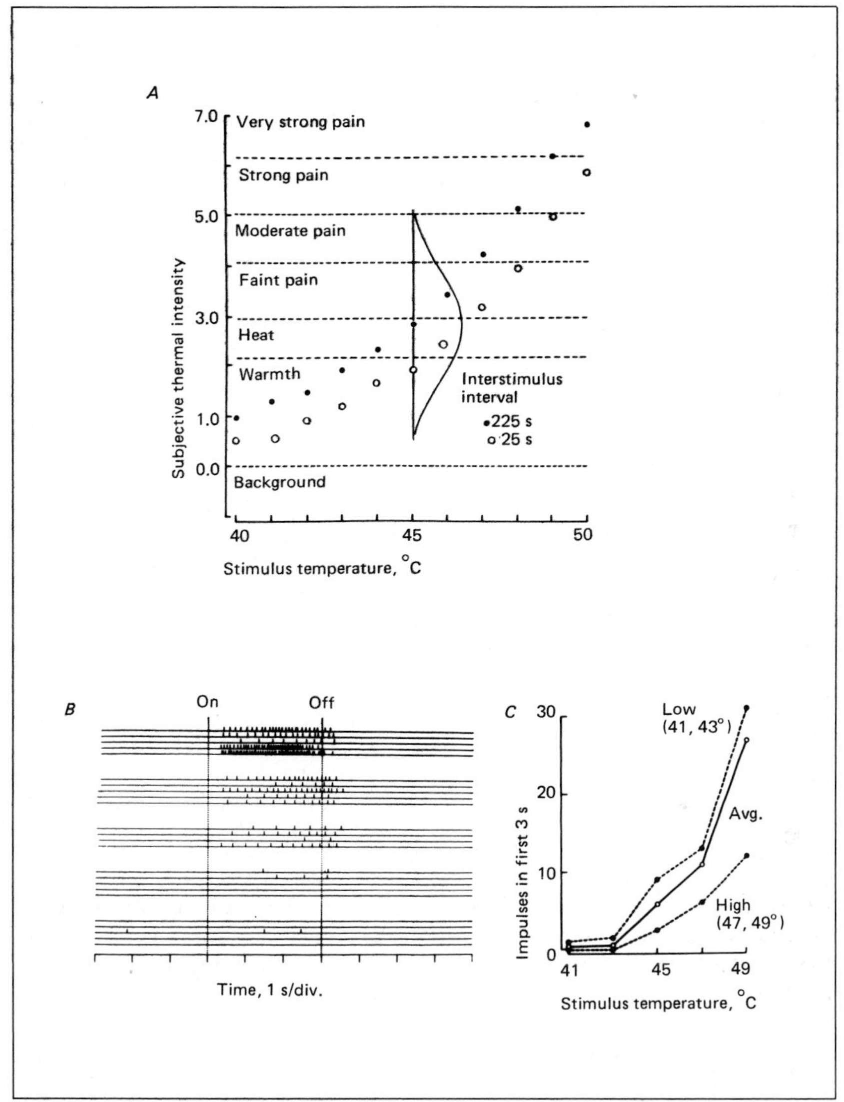

Fig. 3/14. The graph in *A* shows the intensity of thermal sensations produced by different thermal stimuli as judged by a human subject. Filled circles are from stimuli delivered at a long interval and open circles at a short interval. In *B* is shown the activity of a C polymodal nociceptor recorded in a monkey in response to graded noxious heat stimuli. In *C* are stimulus-response functions from the same data. The dotted curves show the effects of plotting the responses when the preceding stimulus was either low (41, 43 °C) or high (47, 49 °C). [From *LaMotte and Campbell,* 1978.]

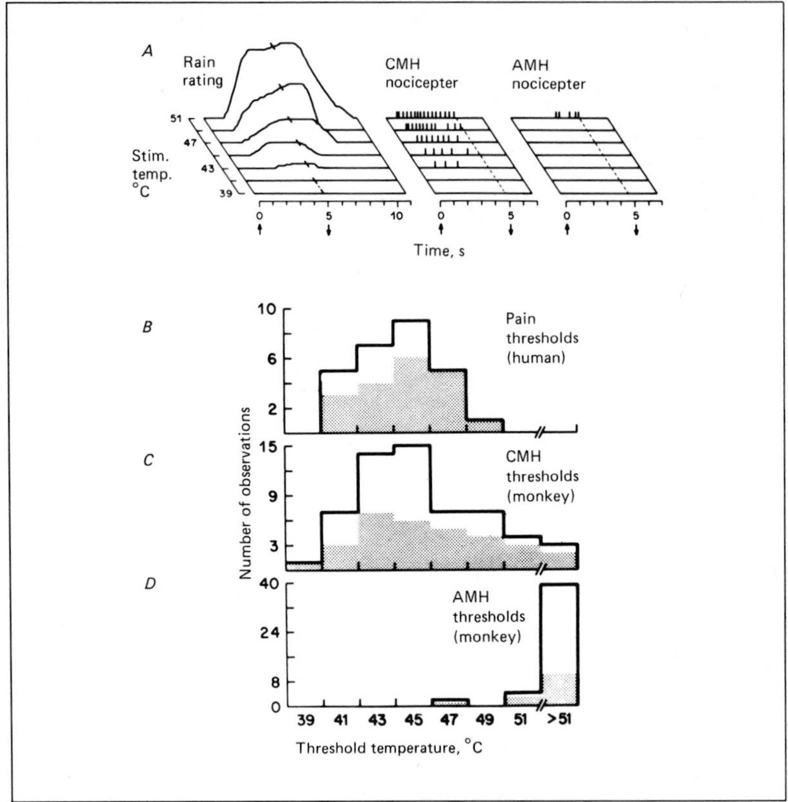

Fig. 3/15. Relationship between the responses of nociceptors and pain sensation. *A* shows the magnitude ratings of pain for a human subject and the responses of a C polymodal nociceptor and of an Aδ nociceptor to graded noxious heat stimulation of normal skin. *B* shows the distribution of pain thresholds for noxious heat in humans and of discharge thresholds for C polymodal and Aδ nociceptors in monkeys. [From *LaMotte* et al., 1982.]

Noxious cold produced by immersion of the hand in cold water causes an aching type of pain with a threshold of 18 °C [*Wolf and Hardy,* 1941]. *LaMotte and Campbell* [1982] argue that high threshold cold receptors are unlikely to account for this aching type of pain because these receptors are activated at a short latency (sometimes within a second) after application of cold stimulus, whereas a similar stimulus requires a minute to produce maximum pain [*Wolf and Hardy,* 1941]. Since blocking A fibers causes the

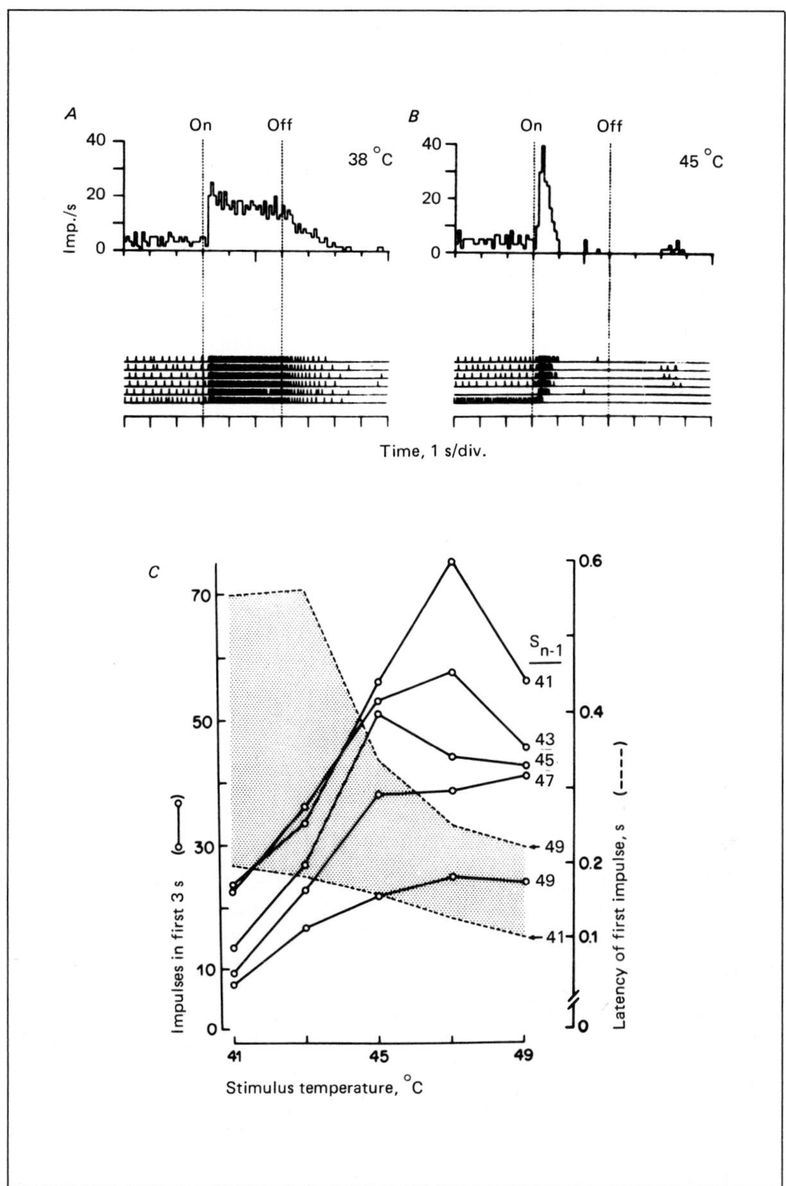

Fig. 3/16. Responses of warm thermoreceptors to noxious heat stimuli. *A* and *B* show the responses of a warm receptor to an innocuous heat pulse of 38 °C and to a noxious heat pulse of 45 °C. *C* shows the stimulus-response functions for 6 warm fibers over the range of 41–49 °C. [From *LaMotte and Campbell,* 1978.]

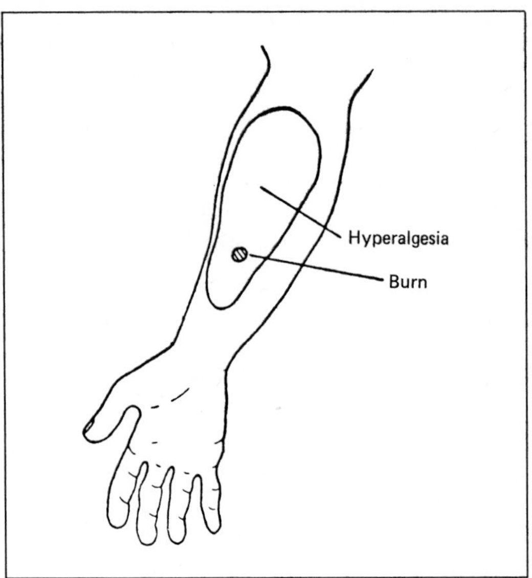

Fig. 3/17. Distribution of hyperalgesia. The damaged area, labeled 'burn', will show primary hyperalgesia if the nociceptors in the area are not inactivated but are rather sensitized. Surrounding the zone of hyperalgesia will be an area of 'secondary' hyperalgesia. [From *Hardy* et al., 1952a.]

pain associated with cold to disappear [*Wolf and Hardy,* 1941] or to change to a burning type of pain [*LaMotte and Thalhammer,* 1982], it is unlikely that cold-sensitive C polymodal nociceptors account for aching cold pain. Alternatively, as suggested by *Wolf and Hardy* [1941], the aching cold pain can be attributed to subcutaneous receptors. *Croze and Duclaux* [1978] find that more rapid cooling results in a pricking pain, with a threshold of 9.7 °C. Aδ fibers of the kind reported by *LaMotte and Campbell* [1982] could contribute to the pricking type of cold pain.

Hyperalgesia

Injury of the skin is frequently associated with hyperalgesia, which is characterized by spontaneous pain and by a reduced pain threshold. Hyperalgesia can be subdivided into primary hyperalgesia, which occurs in the damaged region, and secondary hyperalgesia which occurs in a surrounding zone (fig. 3/17) [*Hardy* et al., 1952a]. Recent evidence suggests that primary

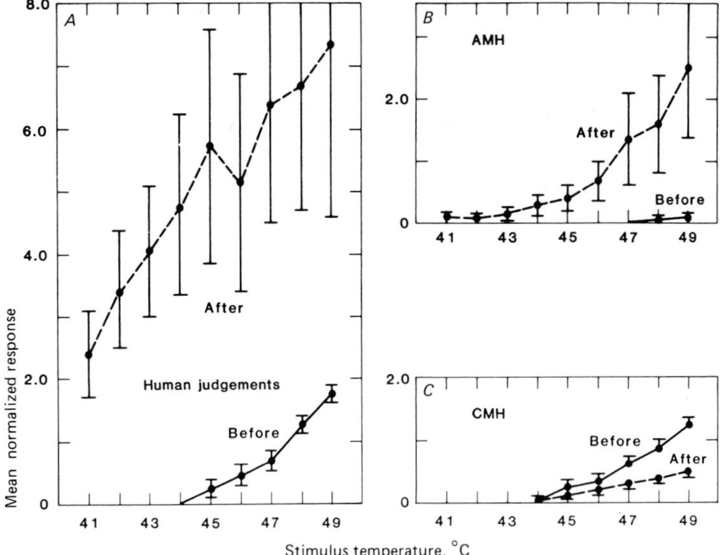

Fig. 3/18. Responses to noxious heat stimuli before and after thermal injury of the skin. In *A* are shown human judgements of pain intensity; in *B* are the mean responses of a population of Aδ nociceptors; and in *C* are the mean responses of a population of C polymodal nociceptors in monkey skin. [From *Meyer and Campbell,* 1981.]

hyperalgesia can be explained, at least in part, by sensitization of cutaneous nociceptors.

Meyer and Campbell [1981] report the results of experiments on the effects of a 53 °C, 30 s burn in producing hyperalgesia in human skin and sensitization of Aδ nociceptors (but not C polymodal nociceptors) in monkey skin. The stimuli were applied to glabrous skin. Figure 3/18 shows that before the burn, the human judgements of noxious heat intensity were paralleled by the responses of C polymodal nociceptors, but that Aδ mechanical nociceptors were practically unresponsive. After the burn, the human pain threshold was lowered, and the psychophysical function was shifted upward and to the left, indicating a greater amount of pain for each stimulus intensity. The responses of the C polymodal nociceptors were diminished after the burn, whereas the Aδ mechanical nociceptors were now responsive over a similar range as the human sensation. The conclusion that the Aδ nociceptors were responsible for the hyperalgesia was supported by the

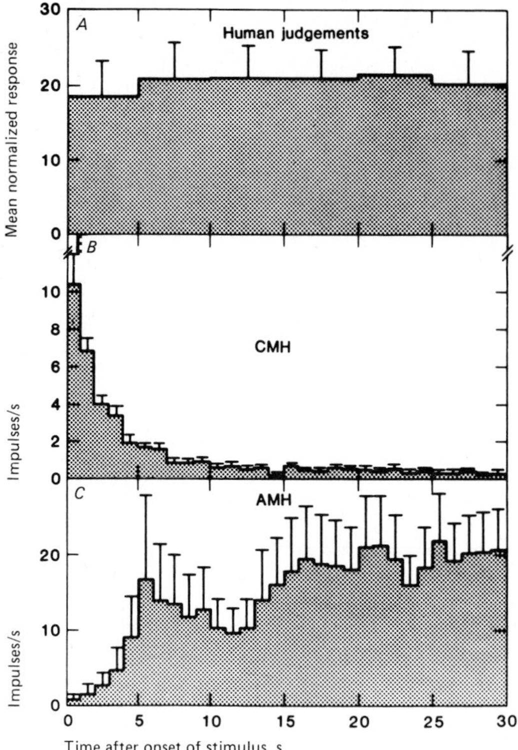

Fig. 3/19. Time course of responses to a noxious heat injury. The human pain judgements are shown in *A,* while the responses of C polymodal nociceptors and of Aδ nociceptors in the monkey are shown in *B* and *C.* [From *Meyer and Campbell,* 1981.]

results of pressure block of the A fibers in the peripheral nerve supplying the hyperalgesic area of skin. The pressure block eliminated the hyperalgesia, but not pain evoked by stimulation of adjacent undamaged skin. When activity of nociceptors was recorded during application of a prolonged noxious heat stimulus, the C polymodal nociceptors discharged vigorously initially, but then the discharge adapted. By contrast, the Aδ nociceptors discharged after a delay. This pattern is shown in figure 3/19. Evidently, the sustained pain during continued application of a noxious heat stimulus can be attributed to both C polymodal nociceptors and Aδ nociceptors.

LaMotte et al. [1982], on the other hand, provide evidence for a role of C polymodal nociceptors in hyperalgesia. In normal skin, it was noted that the

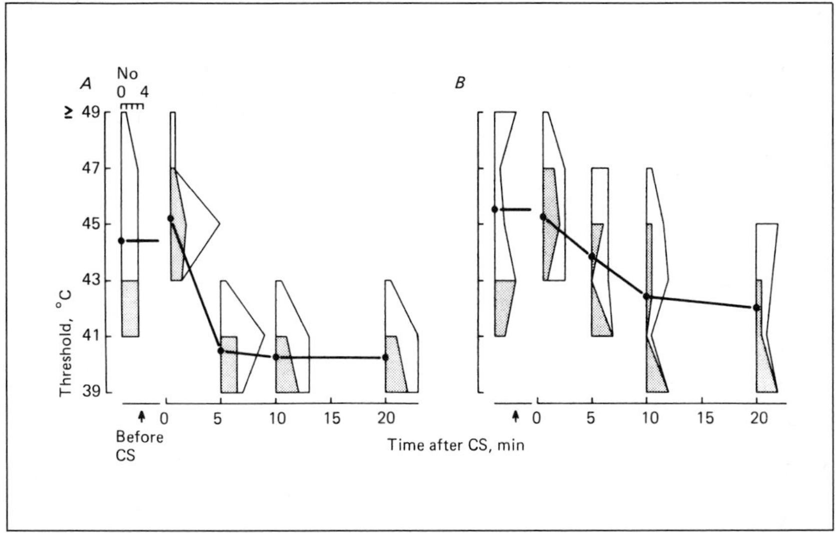

Fig. 3/20. Changes in the pain threshold of human subjects *(A)* and of the response thresholds of C polymodal nociceptors in monkeys *(B)* following heat injury (CS, arrow). The number of observations is shown by the width of each histogram. The shaded areas are for thresholds of 43 °C before injury. The circles and solid lines indicate means. [From *LaMotte* et al., 1982.]

human sensory responses correlated well with the responses of C polymodal nociceptors in monkey skin, but not with the responses of Aδ nociceptors. After heat injury, using a 50 °C, 100 s stimulus, there was an initial hypoalgesia followed by a hyperalgesia in the human subjects (fig. 3/20A). The same heat damage was then produced in monkey skin and the response changes in Aδ and C polymodal nociceptors were followed with time after the damage. Sensitization of the C polymodal nociceptors occurred, after an initial suppression of response (fig. 3/20B). Although sensitization was seen in some Aδ nociceptors, the time course of the changes did not match that of the alterations in human sensation. A compression block of the A fibers of the appropriate peripheral nerve in the human subjects did not alter the time course of hypo- and hyperalgesia produced by heat damage. The difference between this study and the one by *Meyer and Campbell* [1981] is attributed to the intensity of the stimulus. It is concluded that both C polymodal nociceptors and Aδ nociceptors are involved in hyperalgesia, the former in the case of mild injury and the later in the case of more severe injury.

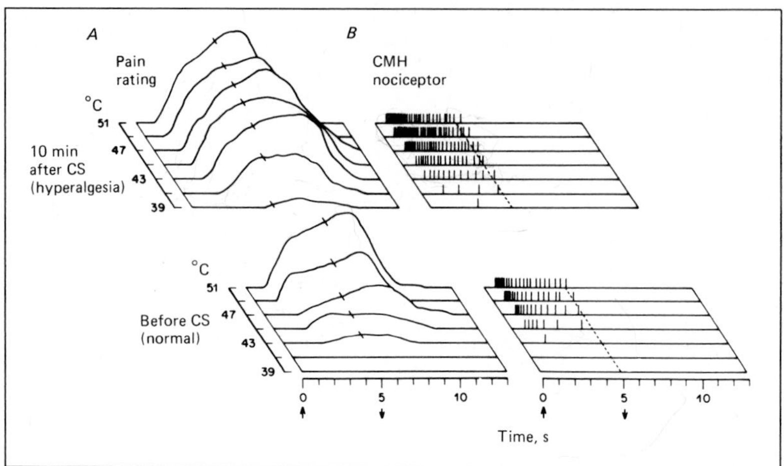

Fig. 3/21. Magnitude estimates of pain in human subjects *(A)* and responses of a C polymodal nociceptor in a monkey *(B)* to graded noxious heat stimuli before and after heat damage. The heat injury was produced by 50 °C heating for 100 s. The tests shown were made before and 10 min after the injury. The receptive field of the afferent fiber was on the hairy skin of a finger. The human subject was stimulated on the volar forearm. [From *LaMotte* et al., 1983.]

LaMotte et al. [1983] analyzed their data further to determine the effects of sensitizing heat stimuli on magnitude estimates of pain and on the stimulus-response curves of C polymodal nociceptors in hairy and glabrous skin. Whereas the magnitude estimates of pain were positively accelerating functions before sensitization, they become more linear or even negatively accelerating curves within 5–20 min after heat damage (fig. 3/21, 22A, C). On the other hand, the stimulus-response curves of C polymodal nociceptors in hairy skin did not show a shift that was parallel to the changes in the magnitude estimate curve (fig. 3/22B), and the C polymodal nociceptors in glabrous skin failed to become sensitized at suprathreshold temperatures (fig. 3/22D). *LaMotte* et al. [1983] argued against the possibility that the hyperalgesia following heat damage in their experiments could be explained by receptors other than C polymodal nociceptors or by cross species differences. Instead, they suggested that changes in central processing of the input from the population of nociceptors might account for the discrepancies. Findings consistent with these were obtained when recordings were made of C-polymodal nociceptor discharges and magnitude estimates of pain before and after heat damage in human subjects [*Torebjörk* et al., 1984].

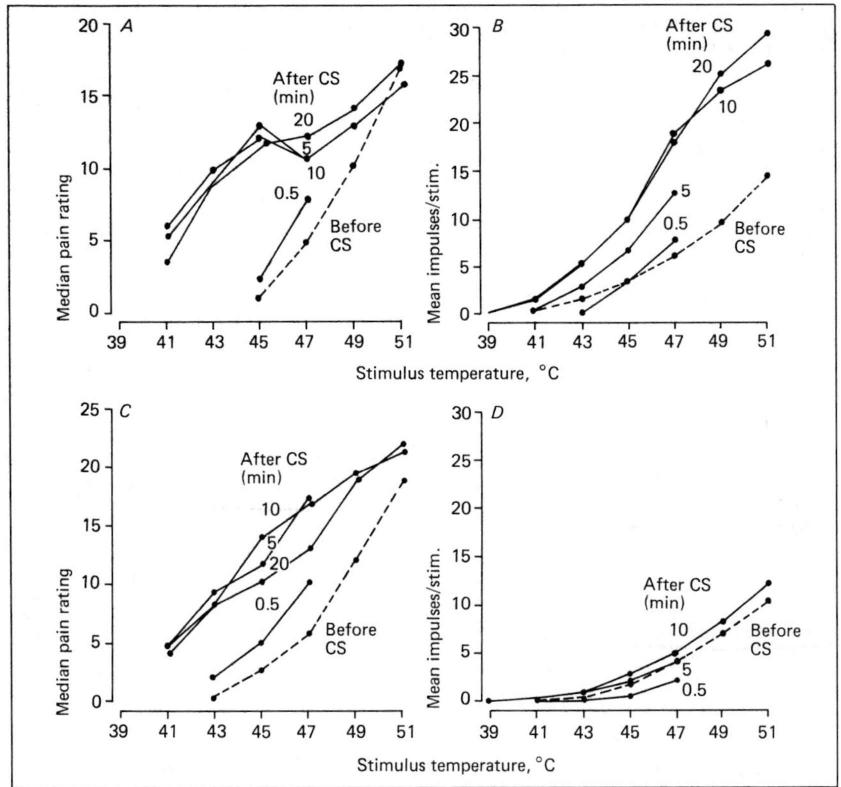

Fig. 3/22. Comparisons of averaged magnitude scaling functions with averaged stimulus-response curves of C polymodal nociceptors before and after heat injury of the skin. The curves in *A* were from 13 human subjects tested on the hairy skin of the forearm, whereas those in *B* were based on the responses of 12 C polymodal nociceptors supplying the hairy skin of the monkey. The heat injury was produced by a 50 °C, 100 s heat application. The curves in *C* were from 7 human subjects tested on glabrous skin of the thenar eminence, and those in *D* were from 11 C polymodal nociceptors supplying glabrous skin in monkeys. [From *LaMotte* et al., 1983.]

Chemical Basis for Activation and Sensitization of Nociceptors

Yaksh and Hammond [1982], in a recent review, distinguish between three types of action of chemical substances on pain transduction: (1) activation of nociceptive afferent fibers and production of pain; (2) facilitation of pain evoked by chemical or physical stimuli by sensitization of nociceptive terminals, and (3) production of plasma extravasation and thus increased access by blood-borne substances to nociceptive terminals.

It has been proposed that the responses of nociceptors depend upon the release of chemical substances following damage of the tissue innervated by the nociceptive terminals [*Lewis,* 1942; *Lim,* 1970]. Consistent with this hypothesis are observations that nociceptors can be excited by intra-arterial injections of such substances as acetylcholine, histamine, serotonin and bradykinin [*Beck and Handwerker,* 1974; *Fjällbrant and Iggo,* 1961]. However, the action of these substances is nonspecific, since they excite not only nociceptors but also slowly adapting mechanoreceptors. It may well be that some other chemical agent will prove to be a specific mediator for nociceptive responses, but for the present it can be stated that the hypothesis that nociceptors depend upon the local release of an algesic substance is unproven.

Another important substance that causes pain by activation of nociceptors, particularly of C polymodal nociceptors, is capsaicin. This substance is a vanillylamide derivative (8-methyl-N-vanillyl-6-nonenamide) and is the active ingredient of chili peppers [see review by *Fitzgerald,* 1983]. One effect of capsaicin administration is the production of pain. For example, local application of capsaicin or injection of small doses into the skin causes burning pain and hyperalgesia [*Becerra-Cabal* et al., 1983; *Carpenter and Lynn,* 1981]. The amount of pain depends upon the dose of capsaicin, and hyperalgesia may last up to 3 h. The mechanism of pain production may be a depolarization of the axons of C polymodal nociceptors, since these receptors are activated by capsaicin [*Foster and Ramage,* 1981; *Kenins,* 1982; *Szolcsányi,* 1977] and since capsaicin causes a calcium-dependent release of substance P (and somatostatin) from the spinal cord [*Gamse* et al., 1981; *Theriault* et al., 1979]. However, direct application of capsaicin to dorsal root ganglion cells in culture causes a hyperpolarization and an increased membrane resistance [*Godfraind* et al., 1981]. It may be that these ganglion cells give rise to axons other than those of C polymodal nociceptors. In another study, *Williams and Zieglgänsberger* [1982] found that capsaicin applied to dorsal root ganglion cells in vitro or in vivo caused a depolarization associated with a reduced membrane resistance. These changes were seen in dorsal root ganglion cells belonging to both 'A cells' and 'C cells', classified according to the conduction velocities of their axons. However, the capsaicin effect was reversible for the 'A cells' and irreversible for the 'C cells', suggesting a stronger and perhaps specific action on the 'C cells'.

The action of capsaicin on receptors seems relatively specific. Activation of warm receptors, as well as of C polymodal nociceptors has been

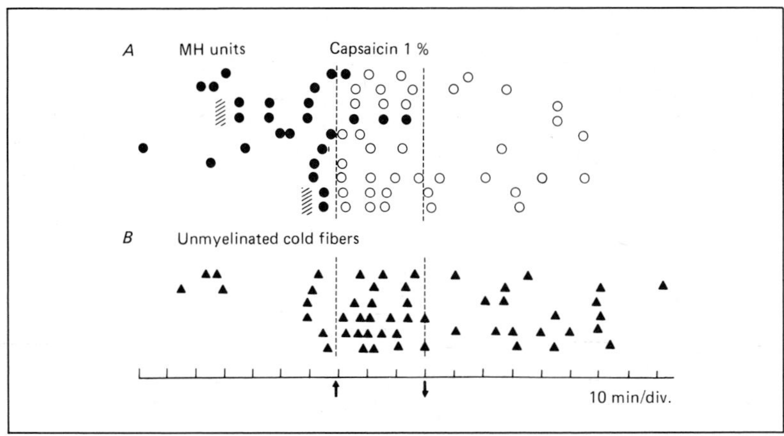

Fig. 3/23. Action of locally applied capsaicin in blocking discharges of C polymodal nociceptors but not cold receptors. *A* shows the discharges (filled circles) or failure of discharges (open circles) of C polymodal nociceptors (MH units) before and after capsaicin application to a peripheral nerve (hatched bars show pairs of units recorded in same filament). *B* shows the discharges of unmyelinated cold fibers (triangles); these were not blocked by capsaicin. The capsaicin was applied between the times indicated by the arrows. [From *Petsche* et al., 1983.]

reported, but cold receptors, several kinds of mechanoreceptors, and Aδ mechanical nociceptors are all unaffected [*Foster and Ramage,* 1981]. In another report, there was only a slight effect on warm receptors, and a single Aδ mechanical nociceptor was excited; the only consistent effects were a strong excitation of C polymodal nociceptors, with a latency of 0.5–3 min, lasting 20–30 min [*Kenins,* 1982].

Interestingly, after the acute effects of capsaicin administration, there is a partial analgesia. For example, when capsaicin is applied locally to a peripheral nerve, the axons of C polymodal nociceptors (but not cold receptors with unmyelinated axons) are blocked for hours (fig. 3/23) [*Petsche* et al., 1983; however, cf. *Wall and Fitzgerald,* 1981], and the terminals of the afferent fibers in the skin are unresponsive to mechanical stimuli and thresholds increase to noxious heat [*Kenins,* 1982; *Welk* et al., 1983]. Substance P (as well as cholecystokinin and fluoride-resistant acid phosphatase) is depleted from the afferent terminals in the spinal cord [*Ainsworth* et al., 1981], and axoplasmic transport of substance P (and somatostatin) is blocked [*Gamse* et al., 1982] after local application of capsaicin to a peripheral nerve. Systemic administration of capsaicin in either neonatal or adult rats also causes a depletion of substance P from nerve endings in the spinal cord and/or in the skin and other peripheral tissues [*Gamse* et al., 1980; *Helke* et al., 1981; *Jessell* et al., 1978; *Nagy* et al., 1981]. However, it should be emphasized that other peptides, like somatostatin, cholecystokinin, and vasoactive intestinal polypeptide, as well as fluoride-resistant acid phosphatase, are also depleted [*Gamse* et al., 1981, 1982; *Jancsó* et al., 1981; *Jancsó and Knyihár,* 1979; *Jessell* et al., 1978; *Nagy* et al., 1981; see also *Schultzberg* et al., 1982].

Behavioral results following capsaicin administration include a reduction in the responses to algesic chemicals and also under some experimental conditions to noxious heat [*Buck* et al., 1981; *Cervero and McRitchie,* 1981; *Gamse,* 1982; *Gamse* et al., 1982; *Holzer* et al., 1979; *Jancsó* et al., 1980; *Lembeck and Donnerer,* 1981; *Nagy* et al., 1980; *Pórszász and Jancsó,* 1959; *Szolcsányi,* 1977; *Szolcsányi* et al., 1975; *Yaksh* et al., 1979; however, cf. *Hayes* et al., 1981; *Hayes and Tyers,* 1980]. Visceral analgesia is especially profound [*Cervero and McRitchie,* 1982]. It is likely that depletion of substance P contributes to the analgesia produced by capsaicin.

A morphological correlate of the analgesic effect of neonatally administered capsaicin is a selective loss of unmyelinated dorsal root axons, although some small myelinated axons are probably also lost [*Jancsó* et al., 1977; *Jancsó and Király,* 1980, 1981; *Nagy* et al., 1980; *Lasson and Nickels,* 1980; *Scadding,* 1980]. Degeneration only occurs up to the 14th day of life in the rat [*Ainsworth* et al., 1981; *Jancsó and Király,* 1981]. However, this may be a reflection of reduced access to the spinal cord by the capsaicin, since degeneration of C fibers may occur in adult rats when capsaicin is given intrathecally [*Palermo* et al., 1981]. Autonomic C fibers do not appear to be affected [*Cervero and McRitchie,* 1982].

Another possible role for chemicals released by damaging stimuli is the sensitization of nociceptors. Evidence for such a role comes from the experiments of *Beck and Handwerker* [1974]. Intra-arterial injection of bradykinin might fail to discharge a nociceptor, yet the bradykinin could increase the responsiveness of the unit to a noxious heat stimulus. Clearly, stimuli strong enough to sensitize tissue could cause the release of substances that enhance the responsiveness of nociceptors [*Chahl and Iggo,* 1977]. However, the nature of the sensitizing agent(s) is unknown. Candidates besides bradykinin are prostaglandins and substance P [*Yaksh and Hammond,* 1982]. Neither prostaglandin nor substance P appear to activate nociceptive terminals directly, but prostaglandin sensitizes nociceptive terminals [*Chahl and Iggo,* 1977; *Handwerker,* 1976; *Moncada* et al., 1975; *Tyers and*

Haywood, 1979] and substance P causes plasma extravasation through an increase in the permeability of capillary walls [*Gamse* et al., 1980]. The action of substance P in promoting plasma extravasation would presumably increase the access of blood-borne substances to the terminals of nociceptors and might thus promote the activation of nociceptors. Following substance P depletion by capsaicin, vascular permeability is no longer altered by antidromic activation of C fibers [*Gamse* et al., 1980, 1982; *Jancsó* et al., 1977, 1980; *Szolcsányi* et al., 1975], and the flare reaction to cutaneous injury no longer occurs [*Carpenter and Lynn,* 1981].

Some of the processes just described can be antagonized by administration of suitable agents. For example, carboxypeptidase B (a bradykinin deactivator), indomethacin (a prostaglandin synthesis inhibitor), mepyramine (a histamine antagonist) and methysergide (a serotonin antagonist) reduce the responses of C polymodal nociceptors, especially when used in combination [*King* et al., 1976; *Perl,* 1976]. It may well be that the pain transduction mechanism involves multiple mediators.

Anatomical Distribution of Hyperalgesia

As mentioned earlier, in an area of damage, the skin may be sensitized and there is hyperalgesia in the damaged zone. This type of hyperalgesia is called primary hyperalgesia [*Hardy* et al., 1952a, b; *Lewis,* 1942]. In addition, the surrounding area of skin may also be hyperalgesic, which is referred to as secondary hyperalgesia. *Lewis* [1942] suggested that secondary hyperalgesia might be caused by a local system of nerve fibers called by him 'nocifensor' nerves. His evidence for this newly described set of nerves came from experiments involving injections of local anesthetics. *Hardy* et al. [1952a, b] repeated these experiments using local anesthetic injections and failed to support the notion of nocifensor nerves, nor is there any anatomical support for such a system [*Perl,* 1976].

In a reinvestigation of possible mechanisms of secondary hyperalgesia, *Fitzgerald* [1979] has found that C polymodal nociceptors can be sensitized not only if their terminals are in an area of damage but also if they innervate an adjacent region of skin. An increased sensitivity to noxious heat was demonstrated for C polymodal nociceptors with terminals 5 or 10 mm away from an area where a nick was placed in the skin with scissors. Injection of local anesthetics (but not of saline) between the nick and the receptive field of the nociceptor under investigation prevented the sensitization of the receptor. Sensitization could also be produced by antidromic activation of the sural nerve. The particular fiber being investigated was not antidromically activated, since it was included in a filament separated from the nerve for recording; thus, the antidromic activation of other fibers caused the sensitization. Possible mechanisms include axon reflexes or chemical release from nerve endings and diffusion into neighboring tissue.

In similar experiments, *Thalhammer and LaMotte* [1982] examined the sensitization of C polymodal nociceptor terminals following noxious heating of the skin. Half of the receptive field was in the heated area and half was outside it. Only the part of the receptive field that was heated showed sensitization. This result was consistent with the observation by *Fitzgerald* [1979] that a spread of sensitization was more difficult to demonstrate when noxious heating was used to produce damage, rather than mechanical injury. *Thalhammer and LaMotte* [1982] support the suggestion of *Hardy* et al. [1952a, b] that secondary hyperalgesia may depend upon central nervous system mechanisms.

Muscle Nociceptors

In addition to the well-studied mechanoreceptors [*Matthews,* 1972], muscle contains other types of receptors, including nociceptors. Candidate muscle nociceptors include group III (small myelinated) or group IV (unmyelinated) afferent fibers [*Mense and Schmidt,* 1977]. Presumably the muscle nociceptors are among the many free endings found in the connective tissue of the muscle, between muscle fibers, in blood vessel walls, or in tendons [*Stacey,* 1969].

Group III Muscle Receptors

Paintal [1960] first determined that group III fibers in muscles of the cat hindlimb could generally not be classified as stretch receptors, but instead were activated by locally applied pressure. The discharges tended to adapt within 0.5–2 s while a stimulus was maintained, although some had slower adaptation rates. Muscular contraction had a minimal effect or none at all, even during concomitant muscle stretch. Asphyxia had no effect. Injection of 6% NaCl was generally effective in exciting the group III afferents, whereas 0.9% NaCl usually had no effect. The possibility that these afferent fibers might have some role in pain originating in muscle led *Paintal* [1960] to suggest the designation of 'pressure-pain' endings for them.

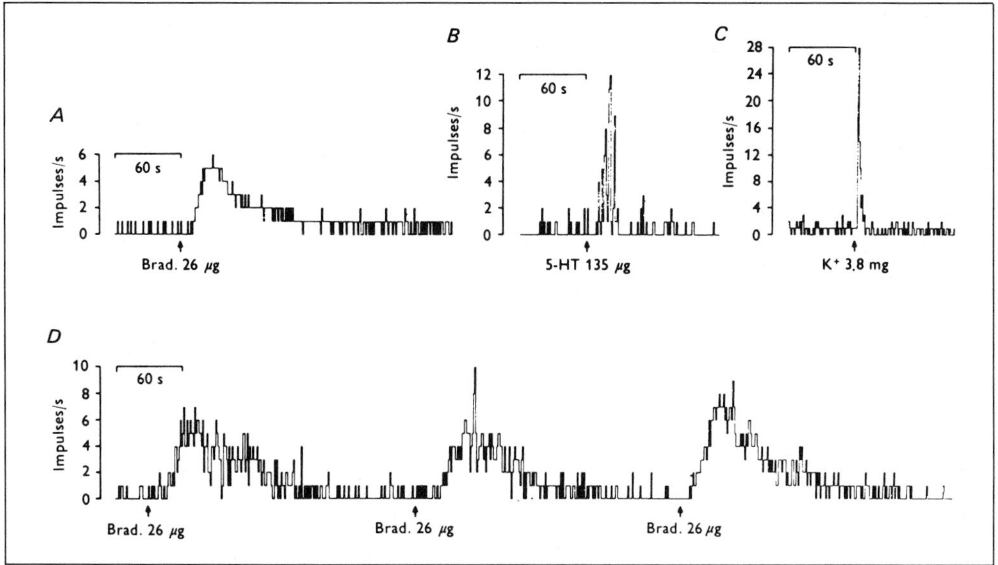

Fig. 3/24. Response of group III muscle afferent fibers to intra-arterial injections of algesic chemicals. *A–D* show the responses of different group III units innervating the medial gastrocnemius muscle to intra-arterial injections of bradykinin *(A)*, serotonin *(B)*, potassium ions *(C)* or repeated doses of bradykinin *(D)*. [From *Mense*, 1977.]

Group III muscle afferents were also investigated by *Bessou and Laporte* [1961]. They were able to distinguish between several types of pressure receptors. One kind was activated chiefly by pressure on the musculotendinous junction with a slowly adapting discharge. Other endings were activated by compression of the muscle belly. These afferents had very large receptive fields and were not affected by muscle contraction. Some fibers were excited by pressure applied to a limited region within a muscle. These had a rapidly adapting discharge. Others were activated by manipulation of the space between muscles or of the surface of a muscle. Finally, some fibers could be activated by muscle stretch. The possibility was discussed that some of the group III afferents participate in pain.

Mense [1977] examined the effects of intra-arterial injections of algesic chemicals upon the activity of group III muscle afferent fibers. Over 70% of the afferents could be excited by such injections. Positive results were obtained with bradykinin, serotonin and potassium ions (fig. 3/24A–C). Repeated injections of bradykinin had similar effects (fig. 3/24D).

Mense and Stahnke [1983] found that a high proportion of group III muscle afferent units responded to muscle stretch. Furthermore, many of these units were activated by muscle contractions as well. When ischemia was produced and contractions tried again, the group III fibers showed reduced responses. The amount of response to contraction was graded with the strength of the contraction. These properties suggested to *Mense and Stahnke* [1983] that group III muscle afferents are more likely to be activated during exercise than are group IV afferents (see below) and that their role may be ergoreceptive rather than nociceptive.

Group IV Muscle Receptors

The first studies of unmyelinated muscle afferent fibers were done by *Bessou and Laporte* [1958] and by *Iggo* [1961]. *Bessou and Laporte* [1958] showed that the C fiber compound action potential in a muscle nerve was reduced by collision with activity evoked by ischemic contractions of the muscle. In the study by *Iggo* [1961], individual C fiber recordings were made. The receptive fields were small areas of muscle or tendon. Adaptation was slow. Thresholds varied between 5 and 50 g. Muscle stretch or contraction had no effect. Most fibers were activated by warming or cooling the muscle. All of the fibers were excited by injection of a 5% NaCl solution. Ischemia had only a minimal effect, although some fibers showed a bursting activity during ischemia.

Mense and his colleagues have systematically investigated the response properties of group IV muscle afferent fibers. Several algesic chemicals were found to activate group IV muscle afferents when injected intra-arterially into the circulation of the muscle [*Franz and Mense,* 1975; *Fock and Mense,* 1976; *Hiss and Mense,* 1976; *Mense,* 1977, 1981; *Mense and Schmidt,* 1974]. The active agents included bradykinin, serotonin, histamine and potassium ions (fig. 3/25) [*Mense and Schmidt,* 1974]. Repeated doses of bradykinin produced similar actions, but repeated doses of serotonin resulted in tachyphylaxis to serotonin but not to bradykinin [*Hiss and Mense,* 1976; *Mense and Schmidt,* 1974], suggesting that these substances act at different sites on the membrane. Prostaglandin (PGE_2) and also serotonin were found to enhance the sensitivity of the group IV muscle afferents to intra-arterial doses of bradykinin (fig. 3/26) [*Mense,* 1981].

Franz and Mense [1975] have also described the responses of group IV muscle afferent fibers to other forms of stimuli than chemical. About half of their sample of units showed some background activity in the absence of any overt stimulation. Some fibers were found that showed such back-

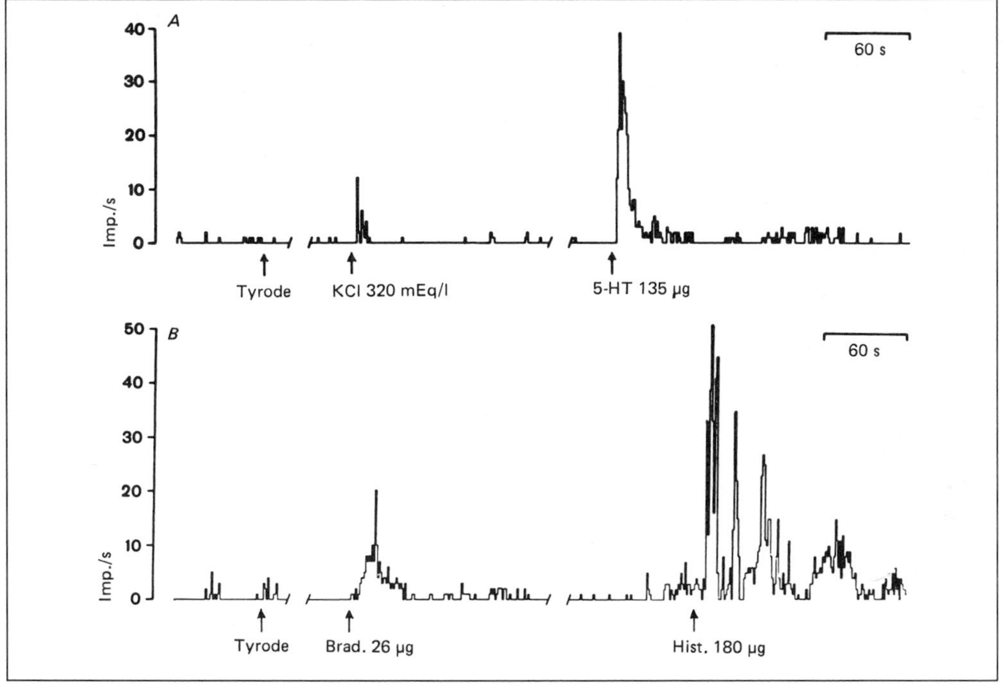

Fig. 3/25. Responses of a group IV muscle afferent to algesic chemicals. *A* shows the responses of potassium ions and serotonin and *B* shows the responses of bradykinin and histamine. The injections were made intra-arterially, and the unit was in the lateral gastrocnemius muscle. [From *Mense and Schmidt,* 1974.]

ground activity only after noxious stimulation of the muscle, but most were active even before the muscle was stimulated either mechanically or chemically. The fibers were activated with either gentle or strong mechanical stimuli, using forceps. Only a few units could be excited by weak mechanical stimuli. Many were activated by strong, presumably noxious, stimuli. Some could not be excited by any mechanical stimuli. The discharges of high threshold units adapted during maintained mechanical stimuli. Besides mechanical stimuli, many group IV muscle afferent fibers could be excited by thermal stimuli [*Hertel* et al., 1976]. Some units were responsive to warming, others to cooling. However, some units did not respond to temperature changes.

Kumazawa and Mizumura [1976, 1977a] examined the responses of group IV muscle afferent fibers and obtained evidence that these fibers

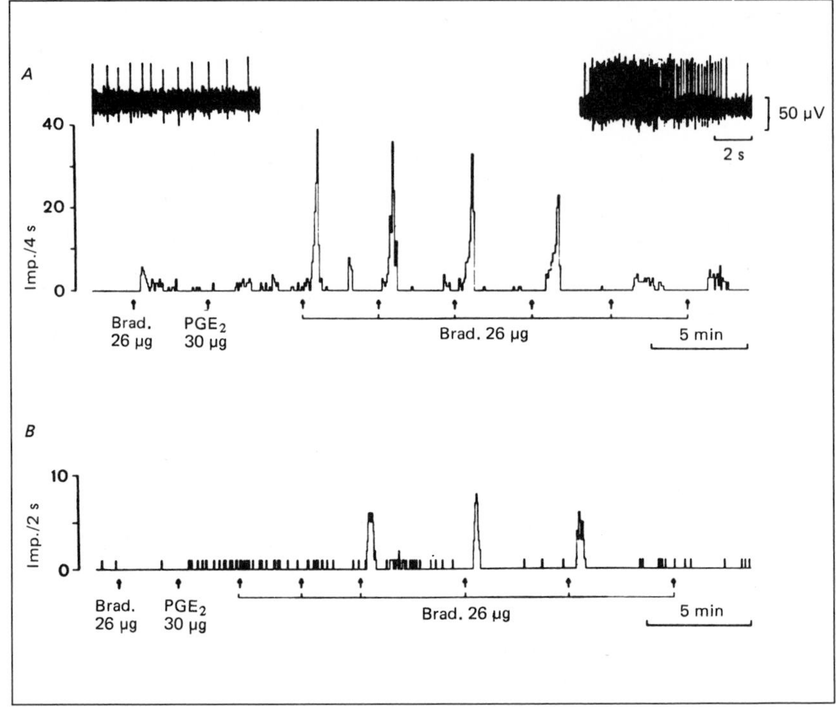

Fig. 3/26. A, B Sensitization of the response of a group IV muscle afferent unit to intra-arterial bradykinin injection by PGE_2. Note that the injection of prostaglandin has a weakly excitatory effect. [From *Mense,* 1981.]

often behave like C polymodal nociceptors innervating the skin. For example, all of the fibers that could be excited by noxious heating were also excited by mechanical stimuli and by chemicals, such as 4.5% NaCl, KCl, and bradykinin. Acetylcholine, histamine and sodium citrate were also effective in many cases.

Kniffki et al. [1978] investigated the effects of pressure, stretch, contraction and algesic chemicals on group IV muscle afferent fibers. One group of afferent fibers could be excited by algesic chemicals but not by muscular activity. These might be nociceptors. Other units responded to contractions of muscle but not to algesic chemicals. These might serve as 'ergoceptors' and function in triggering circulatory and respiratory adjustments to exercise. Still other group IV fibers showed properties of both types of receptor.

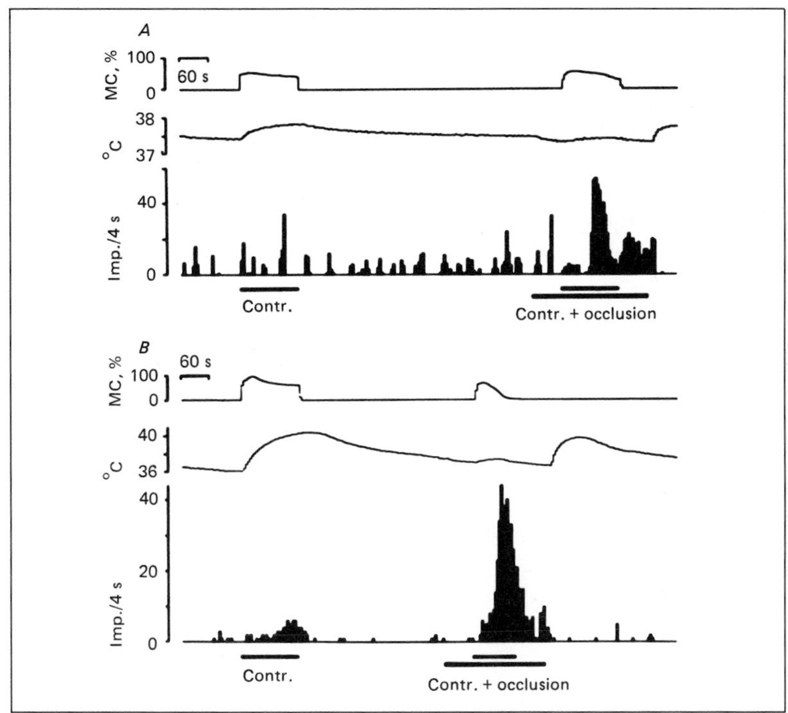

Fig. 3/27. Group IV muscle afferent units that respond most actively during ischemic contractions. The unit in *A* was classified as unresponsive to muscle contractions (MC) alone, but it discharged vigorously when contractions occurred during occlusion of the arterial blood supply of the muscle. The unit in *B* responded weakly to contractions alone but more to the combination of contractions and ischemia. [From *Mense and Stahnke,* 1983.]

Mense and Stahnke [1983] reexamined the responses of group IV muscle afferent units to mechanical stimuli, and found that only a few group IV units responded to muscle stretch or to muscle contraction. However, some group IV units discharged vigorously during contractions of muscle while the arterial circulation was occluded (fig. 3/27) [*Mense and Stahnke,* 1983]. Ischemia alone generally had no effect. It was suggested that the units that discharged during ischemia and muscle contraction might contribute to ischemic muscle pain. Algesic chemicals were effective in activating receptors that either did not or did respond during ischemic contractions, and so these substances did not help distinguish between different types of units.

Role of Muscle Nociceptors in Pain

Pain can be produced by noxious stimulation of muscle, fascia, and tendons [*Lewis,* 1942]. The quality of pain derived from muscle is unitary; according to *Lewis* [1942] it is 'disagreeable, it is rather diffuse and difficult to locate, it is continuous ... Its quality is indescribable. But the pain is distinctive in the sense that it is impossible to confuse pain from skin and pain from muscle when once you know the two'. For present purposes, the quality of muscle pain will be termed aching pain. Similar aching pains are produced by noxious stimulation of tendons and fascia. Stimuli that are particularly effective in evoking pain from muscle include injections of hypertonic sodium chloride or of isotonic potassium chloride solution, squeezing the muscle, or exercising muscle under ischemic conditions [*Lewis,* 1942].

It seems likely that muscle pain is due to the activation of group IV muscle afferent fibers, since (1) these often have a high threshold to mechanical stimulation; (2) they can be excited by algesic chemicals, including hypertonic NaCl solutions or KCl, and (3) are often excited by ischemic contractions. However, the possibility of an additional contribution by group III muscle afferent units, especially those with a high threshold, should not be ruled out at this time.

Although there have not been nearly so many studies of muscle nociceptors as of cutaneous nociceptors, it is clear that muscle and other forms of deep pain are more of a clinical problem than is cutaneous pain. For example, headaches and low back pain are among the most common human maladies. More research in this area is obviously needed.

Joint Nociceptors

There have been numerous studies of the activity of the mechanoreceptors supplying the knee joint [*Baxendale and Ferrell,* 1983; *Boyd,* 1954; *Boyd and Roberts,* 1953; *Burgess and Clark,* 1969; *Carli* et al., 1979; *Clark,* 1975; *Clark and Burgess,* 1975; *Eklund and Skoglund,* 1960; *Ferrell,* 1980; *Grigg,* 1975; *McCall* et al., 1974; *Millar,* 1973a; *Rossi and Grigg,* 1982; *Skoglund,* 1956]. Most of these have been concerned with proprioception and the controversy as to whether position sense is signalled by joint or muscle afferent fibers. Yet from nerve fiber counts, it is clear that the predominant population of joint afferents is formed not by the large afferents but rather by small myelinated (group III) and unmyelinated (group IV) fibers [*Freeman and Wyke,* 1967; *Gardner,* 1944; *Langford and Schmidt,* 1983; *Skoglund,* 1956]. Whereas the large joint afferents terminate in Ruffini- and Pacini-form endings, and in occasional Golgi tendon organs, the fine afferent fibers are connected with free endings [*Andrew,* 1954; *Boyd,* 1954; *Freeman and Wyke,* 1967; *Gardner,* 1944; *Samuel,* 1952; *Skoglund,* 1956]. Presumably, joint nociceptors are among the types of joint afferents that supply these free endings [*Gardner,* 1944].

Some recordings have been made from group III and IV joint afferents of the cat knee, both in normal animals [*Burgess and Clark,* 1969; *Clark,* 1975; *Ferrell,* 1980; *Schaible and Schmidt,* 1983a, b] and in animals treated chemically to produce an acute arthritis [*Coggeshall* et al., 1983]. Background discharges were observed while recording from some group III and IV units, even though care was taken to avoid traumatizing the joint [*Schaible and Schmidt,* 1983a]. Receptive fields were small, generally consisting of a single spot or groups of up to 4 spots. Thresholds for mechanical stimulation ranged from low to high, with the population of group III units having, on average, lower thresholds than group IV units. Graded strengths of mechanical stimulation evoked graded responses. Passive joint movements allowed a classification of the fine joint afferents into the following types: (1) units activated by innocuous joint movements; (2) units weakly activated by innocuous movements but strongly by noxious movements; (3) units that did not discharge in response to innocuous movements, but that were strongly activated by noxious movements (fig. 3/28), and (4) units that could not be activated by any joint movement, although they had distinct receptive fields [*Schaible and Schmidt,* 1983b]. It was suggested that the joint afferents that were excited chiefly by innocuous joint movements might contribute to deep pressure sensation, whereas those that were activated most prominently by noxious movements contribute to joint pain [*Schaible and Schmidt,* 1983b].

Joint inflammation was induced in the knee joint by injection of kaolin and carrageenan into the joint [*Coggeshall* et al., 1983]. Behavioral observations indicated that the joint became painful to the cat, and there was pathologic evidence that an acute arthritis had been produced. Recordings from fine joint afferents revealed the following changes from the pattern seen in normal joints: background discharges were both more common and at higher rates; furthermore, the background activity often consisted of bursts of discharges; a high proportion of afferents could be activated by innocuous joint movements; and low threshold units could be found in

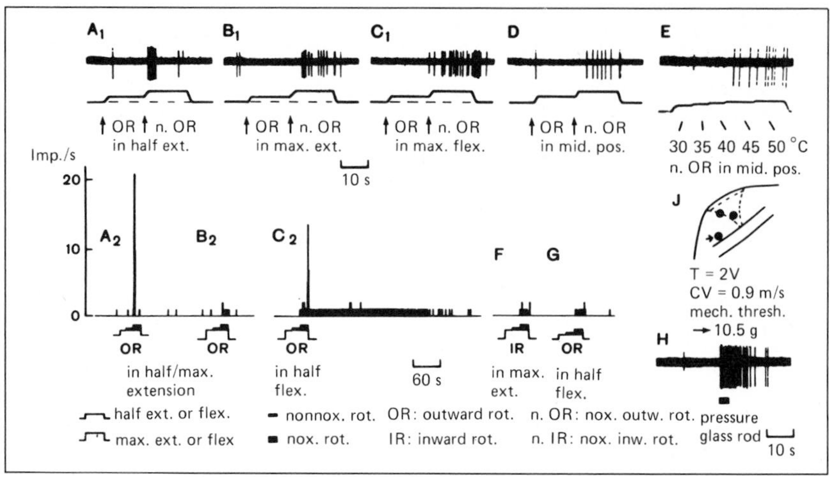

Fig. 3/28. Responses of a group IV joint afferent unit to movements of the joint. The unit was only activated by noxious movements. Its receptive field, J, consisted of 3 separate spots, and the lowest threshold to von Frey filaments was 10.5 g (at arrow in J). Intense stimulation produced afterdischarges, C_2 and H. [From *Schaible and Schmidt,* 1983b.]

areas of the knee joint that normally lacked these. By contrast, group II joint afferents showed the same response pattern with or without inflammation [*Coggeshall* et al., 1983].

Role of Joint Nociceptors in Pain

Joint pain is one type of deep pain, along with muscle and tendinous pain. In normal joints, pain can be produced by movements of the joint in excess of the normal range. However, in inflammed joints, even minimal movements can be painful. A role for some of the large joint afferents has been proposed for signalling the pressure felt as the extremes of the joint range are reached [*Clark,* 1975; *Clark and Burgess,* 1975]. A similar role has been suggested for some of the fine afferents, as well [*Schaible and Schmidt,* 1983b]. However, it seems plausible to regard many of the group III and IV joint afferents as nociceptors [*Clark,* 1975; *Schaible and Schmidt,* 1983b]. Their enhanced responsiveness during acute inflammation is consistent with the suggestion that they may play a role in the pain of arthritis [*Coggeshall* et al., 1983].

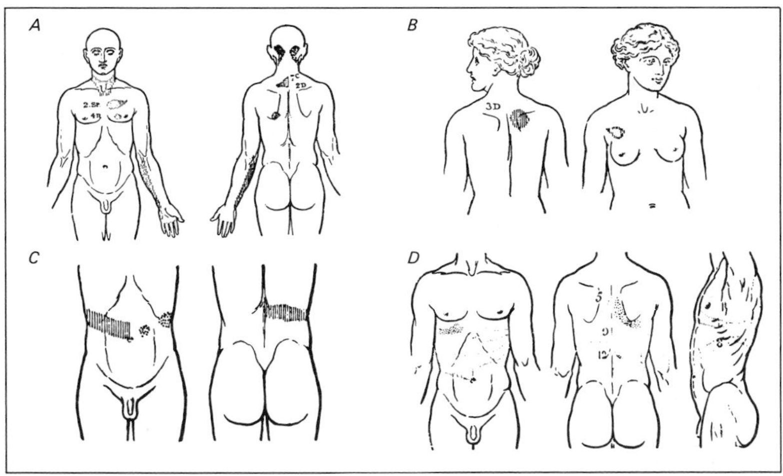

Fig. 3/29. Patterns of referred cutaneous tenderness. *A* shows the areas of cutaneous tenderness during an attack of angina in a man with aortic disease. *B* shows areas of tenderness in a person with acute bronchitis. *C* shows the distribution of tenderness in a patient with acute gastritis. *D* illustrates the distribution of tenderness in a person with colic from gallstones. [From *Head,* 1893.]

Visceral Nociceptors and Referred Pain

By contrast to our knowledge of cutaneous nociceptors and our expanding knowledge of nociceptors in muscles and joints, we still know very little about visceral nociceptors, despite the great clinical importance of visceral diseases that cause pain.

Before considering what is known of visceral nociceptors, however, several features of visceral pain should be noted. The most important feature of visceral pain is that often it cannot be localized by a human subject to a particular viscus [see review by *Cervero,* 1983b]. Instead, visceral pain is commonly referred to a somatic region [*Head,* 1893; *Lewis,* 1942; *MacKenzie,* 1893; *Ruch,* 1946]. Not only is pain referred to the body surface, but that area to which pain is referred may become tender [*Head,* 1893], as shown in figure 3/29. Another feature of visceral pain is the triggering of autonomic and somatic reflexes.

Why should visceral pain be so poorly localized? One reason may be that many of the afferents that supply the viscera are used for reflex purposes and not for sensation [*Cervero,* 1983b; *Leek,* 1972; *Ruch,* 1946]. Several theories have been proposed to account for pain referral. *Head* [1893] thought that there might be a convergence of cutaneous and visceral input onto the same structures in the spinal cord. *MacKenzie* [1893, 1909] developed this idea further into the notion that the activation of visceral afferents produced an irritable focus within the spinal cord. The irritable focus would then facilitate the transmission of information from cutaneous receptors. *MacKenzie* did not think that visceral nociceptors could by themselves activate central neurons, since he believed that there was no real visceral pain sense ('splanchnic pain'). *Ruch* [1946], on the other hand, accepted evidence for splanchnic pain, in addition to referred pain. Splanchnic pain indicates that visceral afferent fibers must be able to activate spinal cord nociceptive tract cells, such as those of the spinothalamic tract, not just facilitate their activity. *Ruch* proposed that many spinothalamic tract cells receive a convergent input from both the skin and viscera, the pattern of input following a dermatomal organization. Input from visceral nociceptors would activate the shared spinothalamic tract cells, but the brain would misinterpret the source of the input, based on learned experience from the more common episodes when cutaneous nociceptors activated the same pathway. *Ruch's* theory is called the 'convergence-projection theory'.

An alternative explanation of referred pain was offered by *Sinclair* et al. [1948], who suggested that some afferents might branch to supply both visceral and somatic structures. Visceral disease would cause antidromic activation of the branches to the skin, as well as an input to the spinal cord. The antidromic activity might cause the release of a substance in the skin that would secondarily excite cutaneous nociceptors to cause pain and also lower their threshold to cause tenderness. Referral of pain arising in viscera could also result from misinterpretation, since the previous learning process would attribute input over branched nociceptive afferents to the more commonly damaged cutaneous area supplied.

Evidence in support of the convergence-projection hypothesis will be given in Chapter 5. However, it should be pointed out here that there is now some evidence in support of the notion of *Sinclair* et al. [1948] that individual nociceptors might innervate both somatic and deep structures. For example, *Mense, Light* and *Perl* [cf. *Perl,* 1984] have observed collateral projections of some mechanical nociceptors to receptive fields in both skin and muscle (fig. 3/30), and *Bahr* et al. [1981] have been able to activate dorsal root ganglion cells antidromically from both the sympathetic chain and a cutaneous nerve. *Pierau* et al. [1982] have demonstrated that individual dorsal root ganglion cells can sometimes be activated by stimulation of

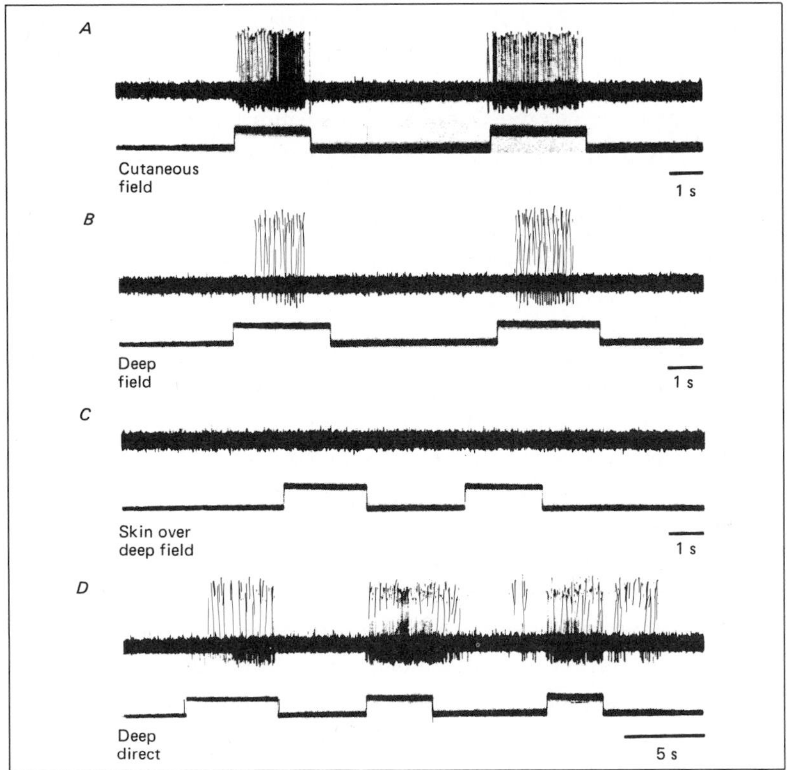

Fig. 3/30. Responses recorded from a mechanical nociceptor with a receptive field in the skin and another in fascia. In *A* the cutaneous field was stimulated by pinching with forceps. In *B* the stimulus was applied to deep tissue through the skin, but away from the cutaneous field. In *C*, no effect was produced by stimulating just the skin over the same location as in *B*. In *D* the fascia was stimulated directly after an incision was made. [From *Perl,* 1984.]

two different peripheral nerves (32/250 cases tried). This observation did not depend upon spinal cord mechanisms, since it could still be shown after dorsal rhizotomy. *Taylor and Pierau* [1982] were also able to label some dorsal root ganglion cells doubly by fluorescent dyes transported from two different peripheral nerves. It would be premature to speculate how extensive such collateral innervation may be. However, branching of dorsal root sensory fibers in the spinal nerves occurs [*Langford and Coggeshall,* 1981], and so it is possible for a given afferent neuron to gain access to quite separate peripheral nerves.

Nociceptors in the Cardiovascular System and Their Role in Cardiac Pain

A variety of receptor types in the cardiovascular system have been reviewed by *Paintal* [1972, 1973] and by *Howe and Neil* [1972]. These receptors include baroreceptors and other mechanoreceptors, as well as chemoreceptors. In addition, there are nociceptors innervating the heart. It is possible that these include both Aδ and C fibers, since afferent fibers of both size categories can become active during coronary occlusion [*Brown,* 1967; *Brown and Malliani,* 1971; *Uchida and Murao,* 1974c]. For example, *Brown* [1967] showed that the Aδ component of the antidromic compound action potential recorded from the left inferior cardiac nerve of the cat could be reduced by collision with activity originating from the heart during coronary occlusion or following injection of 5% NaCl solution into the myocardium. However, following coronary occlusion, the Aδ fibers are activated with a cardiac rhythm, whereas the C fibers are not [*Uchida and Murao,* 1974c], raising the possibility that the Aδ fibers may be mechanoreceptors rather than nociceptors.

Since the pain of angina pectoris [*Blumgart* et al., 1940] is presumably due to the activation of cardiac nociceptors during ischemia, the observation that ischemia causes the release of prostaglandin from the heart [*Block* et al., 1975; *Wennmalm* et al., 1974] suggests a mechanism by which cardiac nociceptors could become sensitized. Other candidate algesic substances that might contribute to angina are bradykinin, potassium and hydrogen ions [*Staszewska-Barczak* et al., 1976; *Uchida and Murao,* 1974a, b, 1975]. Bradykinin application causes an excitation of both mechanoreceptors and chemoreceptors [*Baker* et al., 1980].

The afferent fibers that cause angina enter the central nervous system over the sympathetic nerves, since interruption of the cardiac sympathetic nerves of the appropriate part of the sympathetic trunk abolishes angina [*Lindgren and Olivecrona,* 1947; *White and Bland,* 1948].

Nociceptors in the Respiratory System

Fillenz and Widdicombe [1972] have reviewed the sensory receptors of the respiratory system. Much of the afferent activity arising from receptors of the lungs and airways is involved in reflex actions. The conscious awareness of lung distension appears to derive from the chest wall. However, breath holding and lung disease can produce an unpleasant sensation that may be akin to pain, and this sensation may be mediated by the vagus nerves [*Guz* et al., 1966]. The most likely receptors include the lung irritant receptors and the J receptors.

Lung Irritant Receptors

Aδ sized nerve fibers innervate the epithelial lining of the respiratory passages down to the level of the respiratory bronchioles. The receptors formed by these fibers include the irritant and cough receptors. Irritant receptors are found in the lungs and cough receptors in the larynx and trachea. These receptors can be activated by mechanical stimuli, but they are named for their response to irritant aerosols and gases. Single fiber recordings have been reported by *Widdicombe's* group and others [*Mills* et al., 1969; *Sellick and Widdicombe,* 1969; *Widdicombe,* 1954].

Type J Receptors

These receptors are called J receptors [*Paintal,* 1969] because of their location, which is thought to be near the pulmonary capillaries ('juxtapulmonary capillary receptors'). These receptors were first detected by *Paintal* [1955], who observed a discharge when the lungs were deflated. They can also be activated by large inflations of the lungs and by capsaicin [*Coleridge* et al., 1965], as well as by pulmonary congestion, pulmonary edema, microembolism, and inhalation of irritants [*Paintal,* 1969]. Most of the afferent fibers from J receptors are unmyelinated axons.

Role of Respiratory Nociceptors in Pain

Dyspnea and pain originating from mechanical or chemical damage to the respiratory tract is presumably mediated by Aδ and C fibers from lung irritant and J receptors [*Fillenz and Widdicombe,* 1972; *Paintal,* 1969, 1973]. The pathway that the fibers take to reach the central nervous system is through the vagus nerve.

Nociceptors in the Gastrointestinal Tract

The gastrointestinal tract is innervated by afferent fibers that play a major role in reflex activity and normally a minor role in sensation. However, although visceral sensations are usually minimal, the occasional sensory results of abdominal disorders result in the powerful sensory experiences of nausea and abdominal pain. Helpful reviews of some of the work on gastrointestinal sensory functions may be found in works by *Leek* [1972], *Morrison* [1977], *Newman* [1974] and *Paintal* [1973].

The classes of receptors found in the gastrointestinal tract, according to *Leek* [1972], include slowly adapting mechanoreceptors, rapidly adapting mechanoreceptors, and chemoreceptors. An experimental problem has thus been to identify a class of nociceptors. The reason for this difficulty may be that visceral pain in humans generally accompanies disease, whereas experimental studies are usually done on healthy animal preparations. When normal human abdominal viscera are manipulated in conscious subjects, it is difficult to elicit a sensation of pain, even when the viscera are cut, crushed or burned [*Newman,* 1974]. Pain occurs in one (or more) of the following circumstances [*Leek,* 1972]: (1) irritation of the mucosa or serosa; (2) torsion or traction applied to the mesentery [cf. *Sheehan,* 1932]; (3) distension of a viscus; (4) powerful contractions of a viscus; (5) impaction.

Most of the single unit studies that have been done so far on gastrointestinal receptors have provided little evidence for the existence of a separate class of nociceptors [e.g. *Bessou and Perl,* 1966; *Floyd* et al., 1976a; *Gammon and Bronk,* 1935; *Iggo,* 1955, 1957a, b, 1966; *Mei,* 1970; *Paintal,* 1954a, b, 1957; *Ranieri* et al., 1973; *Schofield,* 1960]. However, nociceptors were observed in the multifiber recordings of *Gernandt and Zotterman* [1946], and it may well be that mesenteric receptors described by *Morrison* and his colleagues [*Floyd and Morrison,* 1974; *Floyd* et al., 1976b; *Morrison,* 1973, 1977] are nociceptors. These afferents respond best to overdistension or powerful contractions of the bowel and to traction on the mesentery, actions known to cause pain, and the afferents were contained in splanchnic nerves, which are the route followed by fibers responsible for gastrointestinal pain [*Newman,* 1974]. C polymodal nociceptors supply the mucosa of the anal canal [*Clifton* et al., 1976]. However, the anal canal is pain sensitive throughout its length [*Duthie and Gairns,* 1960; *Lewis,* 1942], and so its innervation may be more like skin than like most of the gastrointestinal tract.

Recently, *Cervero* [1982a] has developed a favorable preparation for the study of nociceptors supplying the biliary system. The biliary system is distended through a cannula inserted into either the gallbladder or the common bile duct. Controlled changes in pressure allow graded stimulation of receptors supplying the biliary system. Activity of the receptors can be

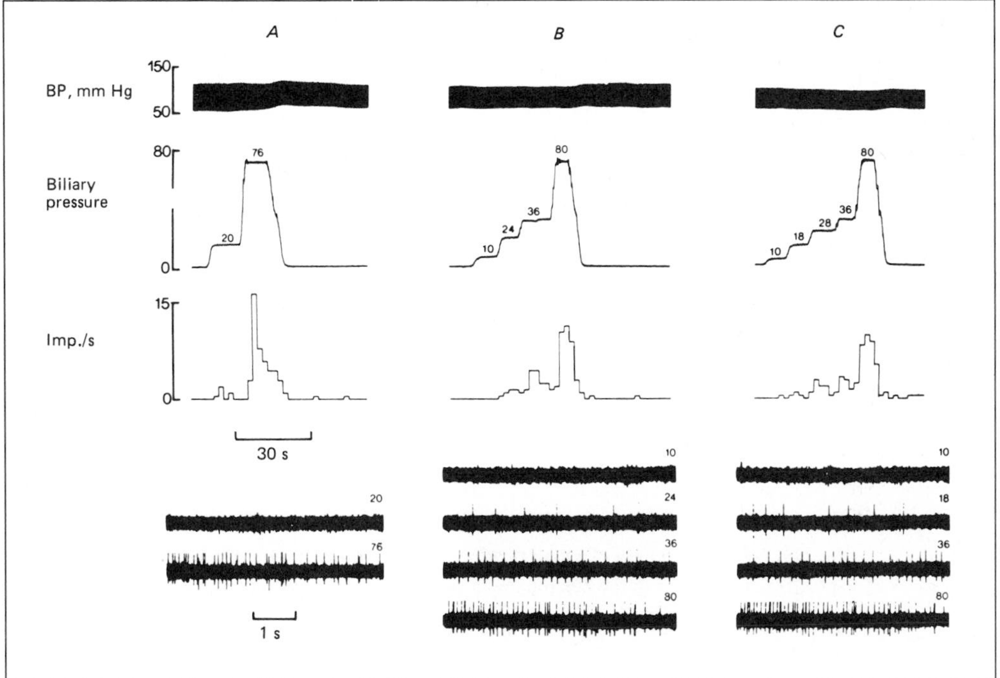

Fig. 3/31. Activity of three different high threshold biliary units is shown in columns *A–C.* Upper traces are systemic blood pressure. The second row of traces are biliary pressure. The third row are rates of discharge of the afferent units. The actual recordings of activity at different biliary pressures are shown at the bottom of each column. [From *Cervero,* 1982a.]

recorded from the splanchnic nerves or the biliary plexus. Systemic blood pressure recordings permit an assessment of whether or not stimulus intensity is sufficient to evoke a pseudoaffective response and hence to be judged noxious. Afferent fibers could be classified as low threshold units or high threshold units. Low threshold units were activated by pressures that did not produce blood pressure changes, whereas high threshold units were excited best by pressures that caused an increase in blood pressure (fig. 3/31) [*Cervero,* 1982a]. When receptive fields could be mapped, they usually consisted of a single spot-like area in the wall of the biliary system (fig. 3/32).

Some investigators have studied the action of bradykinin in exciting visceral mechanoreceptors [*Floyd* et al., 1977; *Guzman* et al., 1962; cf. *Lim*

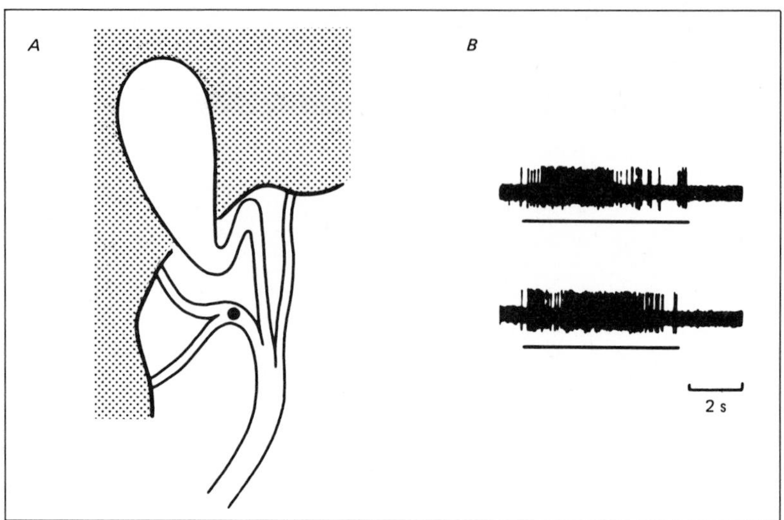

Fig. 3/32. Receptive field of a high threshold biliary afferent unit. The receptive field was a spot-like area on one of the hepatic ducts, as shown in *A*. The responses of the unit to probing are shown in *B*. [From *Cervero,* 1982a.]

et al., 1962]. Although bradykinin causes pain when injected intra-arterially, its activation of sensory receptors does not define the receptors as nociceptors. The action of bradykinin is not limited to the excitation of nociceptors; rather, bradykinin activates both nociceptors and slowly adapting mechanoreceptors [*Beck and Handwerker,* 1974; *Fjällbrant and Iggo,* 1961]. Perhaps the use of sensitizing chemicals, such as the prostaglandins, would reveal the presence of nociceptors by causing their activation by stimuli that had been ineffective before application of one of these agents. Another difficulty in such experiments is that bradykinin may produce indirect effects secondary to muscle contractions [*Floyd* et al., 1977].

Role of Gastrointestinal Nociceptors in Pain

A particularly instructive case history that provides information helpful for an understanding of visceral pain was that of the patient, Tom, studied by *Wolf* [1965]. Tom had a gastric fistula that was made surgically to allow him to be fed, since his esophagus had sealed following scalding by hot soup. The gastric mucosa was partially exposed at the rim of the fistula, and it was possible to introduce instruments directly into the stomach. The color, wetness and turgor of the mucosa could easily be observed as they changed with time and circumstance. It was noted, for instance, that the state of the gastric mucosa could be correlated with Tom's psychologic state, becoming red and turgid when Tom was excited or angry and pale, dry and sticky when he was frightened or depressed. The threshold for pain was lowered when the mucosa became hyperemic and engorged. Under normal conditions, pinching or pulling the mucosa was not painful, but the same stimuli did produce pain when the mucosa was red and swollen.

The possibility of provoking sensation was tested by applying suitable stimuli to Tom's gastric mucosa or stomach wall. Touch was not detected, but pressure was, when applied at an intensity of about 30 g/cm^2. The source of the pressure stimulus was poorly localized. Pressure sensation originated from both the stomach wall and from the mucosa. Temperature stimuli between 18 and 40 °C were not detected, but temperatures outside this range were felt as cool or warm sensations. Pain was not produced by pinching the mucosa, by electrical stimulation, or by applying solutions of alcohol, HCl, and NaOH or a suspension of mustard when the mucosa was not hyperemic or engorged. However, if the mucous coating were removed and the mucosa became reddened, pain was evoked by these stimuli. Pain could also be produced by strong pressure applied to the stomach wall, and the threshold for this pain was lowered if the stomach contracted. Pain resulted from strong contractions of the stomach, and the threshold diminished if the mucosa was red.

Undoubtedly, the pain described above was due to the activation of nociceptors, probably of several types (e.g., mucosal, gastric wall). Unfortunately, little is known about gastric nociceptors or about nociceptors supplying other parts of the gastrointestinal tract.

Nociceptors in the Genitourinary Tract

As for the gastrointestinal tract, studies of afferent activity arising from the urinary bladder have been unable to distinguish a distinct class of nociceptive afferents, although such afferents presumably exist [*Floyd* et al., 1976a, b; *Iggo,* 1955; *Winter,* 1971; however, cf. *Talaat,* 1937]. One difficulty is the lack of information about the parameters of natural stimulation that are noxious in animals.

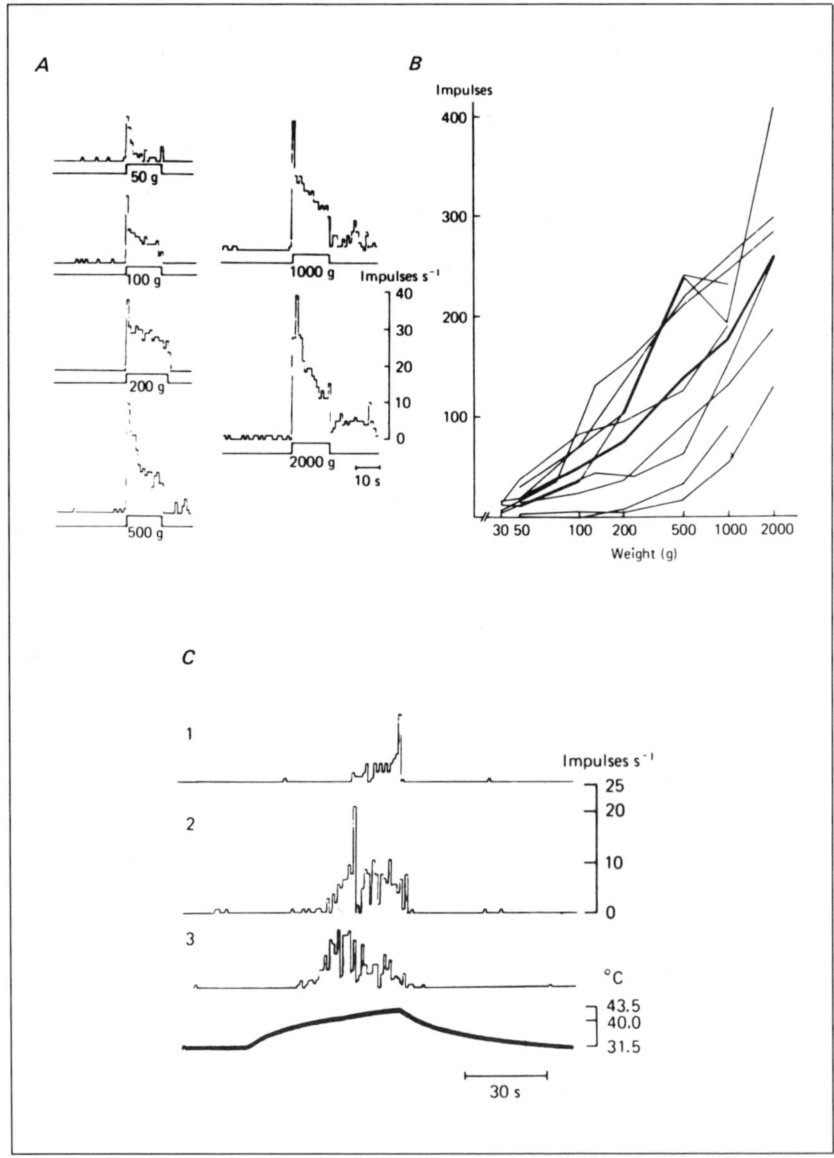

Fig. 3/33. A shows histograms of the responses of a polymodal nociceptor supplying the dog testicle. The weights indicated were applied during the times shown. *B* shows stimulus-response curves for 9 units. The thick line is the mean curve. *C* shows the responses of a polymodal unit to three trials of heat stimuli, repeated in 7-min intervals. [From *Kumazawa and Mizumura,* 1980b.]

Recordings have been made from afferents that were presumably nociceptive supplying other elements of the genitourinary system. For example, nociceptors have been described supplying the urethra [*Clifton* et al., 1976; see also *Floyd* et al., 1976b], although *Todd* [1964] observed mechanoreceptive activity from urethral receptors that he thought signalled flow of urine.

The best evidence concerning visceral nociceptors innervating the genitourinary system comes from investigations of the testis. Small myelinated afferent fibers have been described that respond to mechanical stimulation of the testis [*Peterson and Brown,* 1973]. Some of the these were identified as nociceptors on the basis of an afterdischarge. Polymodal nociceptors have been found to supply the testis [*Kumazawa and Mizumura,* 1977b, 1979, 1980a, b]. These afferent fibers have properties that are very similar to those of C polymodal nociceptors innervating the skin, but they include both Aδ and C fibers. Receptive fields consist of multiple spots (up to 9), either on the testis, the epididymis or both [*Kumazawa and Mizumura,* 1980b]. Presumably the afferents studied by *Peterson and Brown* [1973] would have included polymodal nociceptors if suitable stimuli had been tried. The testicular polymodal nociceptors respond to intense mechanical stimuli, noxious heat and also algesic chemicals (fig. 3/33) [*Kumazawa and Mizumura,* 1980b, 1983]. The units become sensitized with repeated noxious heat stimulation (fig. 3/33). A major advantage of the polymodal nociceptor supply of the testis is that the afferents and the testis can be removed together to be studied in vitro without noticeable changes in the properties of the afferent fibers [*Kumazawa and Mizumura,* 1983].

Role of Genitourinary Nociceptors in Pain

Distension and contraction of the bladder, ureter, or urethra can cause pain, as can inflammation of these structures [*Learmonth,* 1931; *McLellan and Goodell,* 1943]. These effects are comparable to those already described for the gastrointestinal tract. In both systems, it is still unclear if pain results from activation of a distinct class of nociceptors. A similar problem exists for pain arising from the female reproductive tract [*Floyd* et al., 1976a]. However, good evidence is available for the existence of specific nociceptors of a polymodal type innervating the testis. Testicular pain is well documented not only from trauma in normal individuals but also from cases of spermatic cord torsion and from experimental testicular compression in man [*Skoglund* et al., 1970; *Woollard and Carmichael,* 1933].

Conclusions

1. *Blix, von Frey* and others around the turn of the century discovered that there are discrete sensory spots in the skin that when stimulated gave rise to such sensations as touch, warmth, cold and pain.

2. Although the particular sensory receptors responsible for these different sensations were in part misidentified, experiments in recent years have shown that the basic hypothesis that each sensory spot is associated with a particular kind of nerve terminal that in turn evokes a specific sensation is correct.

3. The recognition of nociceptors in recordings from single sensory units depends upon operational criteria, the most important of which is the ability of the receptor to distinguish between innocuous and noxious intensities of stimulation. Sensitization of the receptor may be a useful ancillary criterion.

4. There are two main classes of cutaneous nociceptors: Aδ mechanical nociceptors and C polymodal nociceptors, supplied respectively by fine myelinated and unmyelinated afferent fibers.

5. Aδ mechanical nociceptors supplying the skin have a receptive field that is distributed over several spots located within an area of several square centimeters. The threshold to mechanical stimuli is often orders of magnitude higher than that of sensitive mechanoreceptors, and the discharges produced by graded stimuli are increased as stimulus intensity is raised to a damaging level. These receptors may show fatigue with repeated stimulation, but more striking is their ability to be sensitized by skin damage. For example, noxious heat usually fails to excite these receptors, but after sensitization noxious heat is a good stimulus for them.

6. C polymodal nociceptors have receptive fields that consist of only 1 or 2 spots or of areas of skin. They can be activated by noxious mechanical, thermal or chemical stimuli. Like Aδ mechanical nociceptors, their discharge is distinctly greater as stimulus strength increases into a damaging range, and they show fatigue and also sensitization.

7. Other kinds of cutaneous nociceptors include heat-sensitive Aδ nociceptors, C mechanical nociceptors, and perhaps high threshold cold receptors (although it is not clear that these are nociceptors).

8. Cutaneous pain sensation has at least two different qualities: pricking pain and burning pain. Cold pain has an aching quality, but this could be due to activation of subcutaneous receptors. Pricking pain appears to be due to excitation of Aδ nociceptors and burning pain to C polymodal nociceptors.

9. Stimuli applied to certain parts of the body surface with a sufficient intensity may evoke a sequence of first and then second pain. First pain has a pricking quality and is due to the activation of Aδ nociceptors; second pain has a burning quality and is caused by the excitation of C nociceptors.

10. Electrical stimulation of afferent nerve fibers, either as a population or as individual fibers, produces pricking or burning pain, according to whether Aδ or C nociceptors are activated.

11. The pain spots of *von Frey* are most readily explained on the basis of the terminations made in the skin by nociceptors.

12. Good correlations are found between human sensation and the responses of C polymodal nociceptors of monkeys when graded heat stimuli are applied to the skin, suggesting that there is a causal relationship, assuming a valid cross-species comparison.

13. Some C polymodal nociceptors produce itch, rather than pain.

14. It is unclear what nociceptors are responsible for cold pain. High threshold cold receptors could contribute to a pricking component of cold pain, and C polymodal nociceptors might be responsible for a burning quality. However, the aching quality may depend upon other, subcutaneous receptors.

15. Hyperalgesia often follows damage to the skin. Hyperalgesia consists of pain and a reduced pain threshold. The area directly damaged may develop primary hyperalgesia, and a surrounding area may develop secondary hyperalgesia.

16. Depending on the parameters of the heat injury used to damage the skin, primary hyperalgesia can be explained by sensitization of either Aδ nociceptors or C polymodal nociceptors. However, there may be an additional contribution to hyperalgesia from changes in the central processing of nociceptive signals.

17. There is so far insufficient evidence to support the idea that all nociceptors respond to the release of chemicals from adjacent cells damaged by noxious stimuli. However, a number of chemical substances have been identified that do excite nociceptors. These are generally nonspecific and activate other kinds of receptors as well (e.g., acetylcholine, histamine, serotonin, bradykinin, potassium ions).

18. An example of a substance that may have a more specific effect is capsaicin, which excites C polymodal nociceptors but not cold receptors with C fiber afferents. Capsaicin causes pain when injected into the skin. Later effects of capsaicin include a reduced sensitivity to pain, especially

that evoked by chemical or thermal stimuli. This analgesia (or rather hypalgesia) may be due to substance P depletion, although capsaicin depletes other substances as well (somatostatin, fluoride-resistant acid phosphatase).

19. Besides exciting nociceptors, chemicals may sensitize them. For example, bradykinin and prostaglandin cause a sensitization of C polymodal nociceptors.

20. A third effect of chemical agents that may alter the properties of nociceptors is a change in vascular permeability. For instance, substance P causes an increased vascular permeability when released from nerve terminals in the skin. An increase in vascular permeability may increase access of blood-borne chemicals to nerve endings.

21. There is no evidence for the system of 'nocifensor' nerves that *Lewis* suggested to explain the distribution of secondary hyperalgesia. The best current explanations include axon reflexes, diffusion of chemicals within the skin, or changes in central processing.

22. Muscle nociceptors may include afferents of both the group III and group IV sizes, although evidence is better for group IV muscle nociceptors. It may be that group III muscle afferents are ergoceptors. Group IV muscle afferents that respond best to ischemic contractions are the best candidates to explain the aching pain associated with exercising ischemic muscle.

23. Joint nociceptors appear to include fibers of both group III and IV caliber. These afferents can be sensitized by chemical substances that produce acute inflammation, and so they may well be involved in the pain of arthritis.

24. Visceral nociceptors have not been so well studied as other types of nociceptors. One feature of visceral pain that must be explained is its referral to somatic structures. The theory that there might be a peripheral mechanism of referred pain has received recent support from evidence that some nociceptors branch to innervate both skin and a deep structure. Axon reflexes and release of algesic substances in the skin could help account for referred tenderness.

25. Some evidence suggests that cardiac nociceptors might include Aδ fibers, as well as C fibers. A release of bradykinin or of prostaglandin from the heart during ischemia may contribute to the sensitization of nociceptors and the production of angina pectoris.

26. There appear to be at least two types of nociceptors supplying the respiratory passages: lung irritant receptors and type J receptors. The lung irritant receptors have Aδ fibers, and they innervate the epithelial lining of

the respiratory tract. These receptors are excited by noxious mechanical stimuli and also by irritant aerosols and gases. Type J receptors are found next to capillaries, and they are innervated chiefly by unmyelinated afferent fibers. J receptors are excited strongly by pulmonary congestion, pulmonary edema, microembolization, and irritants.

27. Recordings in the past have not clearly revealed a class of gastrointestinal nociceptors with properties that are distinctive from those of mechanoreceptors. However, it may be that nociceptors will be found when more appropriate stimuli are used or after prior sensitization, since the gastrointestinal tract is generally insensitive to noxious stimuli that commonly cause pain when applied to the skin. Instead, pain results from irritation or from such mechanical stimuli as distension, contractions or traction on the mesentery. Recently, it has been possible to demonstrate that the biliary system is innervated by nociceptors, as well as mechanoreceptors. The approach used was to produce a controlled amount of distension of the biliary system and to correlate the activity evoked in afferents with the pressure required to increase the systemic blood pressure.

28. Bladder nociceptors, like gastrointestinal nociceptors, have not been identified. However, good evidence is available concerning one type of visceral nociceptor, the polymodal nociceptor that supplies the testis. These afferents behave in a manner similar to cutaneous C polymodal nociceptors. However, the afferent fibers are of both Aδ and C diameter.

Chapter 4

Nociceptive Afferent Input to the Dorsal Horn

Historical Overview

The concept that there is a separation of function of dorsal and ventral spinal roots can be attributed to *Bell* and *Magendie* [see excerpts from their work in *Cranefield,* 1974]. The dorsal root was found to contain sensory axons, whereas the ventral root had a motor function. Pain could be produced in awake, behaving animals by stimulation of a dorsal root, and it was abolished when the dorsal roots were cut [*Magendie,* 1822a]. However, *Magendie* [1822b] noticed that stimulation of ventral roots also produced signs of pain, which he could eliminate by cutting the dorsal roots. He later suggested that this pain was due to sensory fibers that run part way in the ventral root and then loop back into the dorsal root [*Cranefield,* 1974]. A similar 'recurrent sensibility' has been observed in human subjects who experience pain when the ventral root is stimulated [*Frykholm,* 1951; *Frykholm* et al., 1953; *White and Sweet,* 1955]. This pain in human subjects can also be eliminated by sectioning the dorsal root [*Frykholm,* 1951]. *Frykholm* et al. [1953] suggest that the pain produced by ventral root stimulation might be secondary to muscular contractions. *Cranefield* [1974] lists a number of other possible explanations of pain produced by ventral root stimulation. Recent evidence concerning ventral root afferents is reviewed by *Coggeshall* [1980].

As nociceptive afferent fibers approach the spinal cord, they and other fine afferents become segregated within the dorsal root, occupying the lateral part of the root. Then, the fine afferent fibers enter the dorsolateral fasciculus of Lissauer [reviewed by *Kerr,* 1975a]. It has been suggested that the afferents destined for Lissauer's tract can be interrupted by lesions placed in the lateral parts of the dorsal roots, and that such lesions can prevent pain [*Ranson and Billingsley,* 1916; *Spivy and Metcalf,* 1959]. However, the success of such lesions has been suggested to be due instead to interruption of the blood supply of the dorsal horn [*Wall,* 1962].

The composition of Lissauer's tract has been a subject of controversy. The original descriptions indicated that a major component was primary

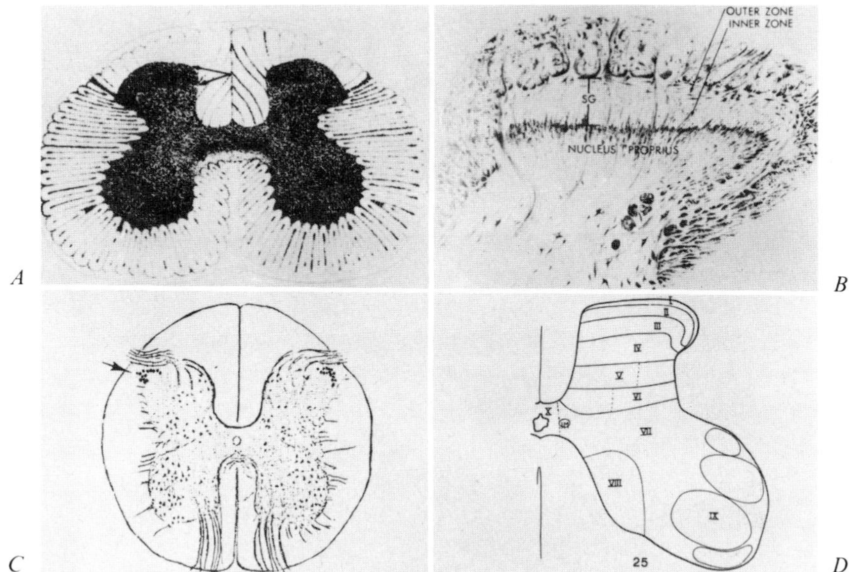

Fig. 4/1. The drawing in *A* shows the 'gelatinous substance' in the superficial dorsal horn. [From *Rolando,* 1824.] *B* is a drawing of the spinal cord showing the marginal layer and the substantia gelatinosa. The border between the latter and the nucleus proprius is well demarcated by the presence of myelinated fibers in the nucleus proprius but not in the substantia gelatinosa. [From *Clarke,* 1859.] The drawing in *C* shows various cell groups in the spinal cord gray matter, including the marginal cells. [Redrawn from *Waldeyer,* 1888; specimen of gorilla spinal cord.] The drawing in *D* shows Rexed's laminae in cat spinal cord. [From *Rexed,* 1954.]

afferent fibers of fine caliber [*Bechterew,* 1887; *Lissauer,* 1885]. However, later investigators denied this [*Earle,* 1952]; they felt instead that Lissauer's tract was formed chiefly of the axons of dorsal horn neurons. More recent studies indicate that Lissauer's tract includes a large component of primary afferents, in addition to propriospinal axons [*Chung and Coggeshall,* 1979; *Chung* et al., 1979; *Coggeshall* et al., 1981]. An important role for Lissauer's tract in pain is suggested by the reports of *Hyndman* [1942] and *Hyndman and Wolkin* [1943] that sectioning Lissauer's tract can supplement the analgesia produced by cordotomy by raising the segmental level of pain loss.

The fine afferents in Lissauer's tract terminate in the upper layers of the dorsal horn. These layers include the marginal zone (fig. 4/1C) [*Waldeyer,*

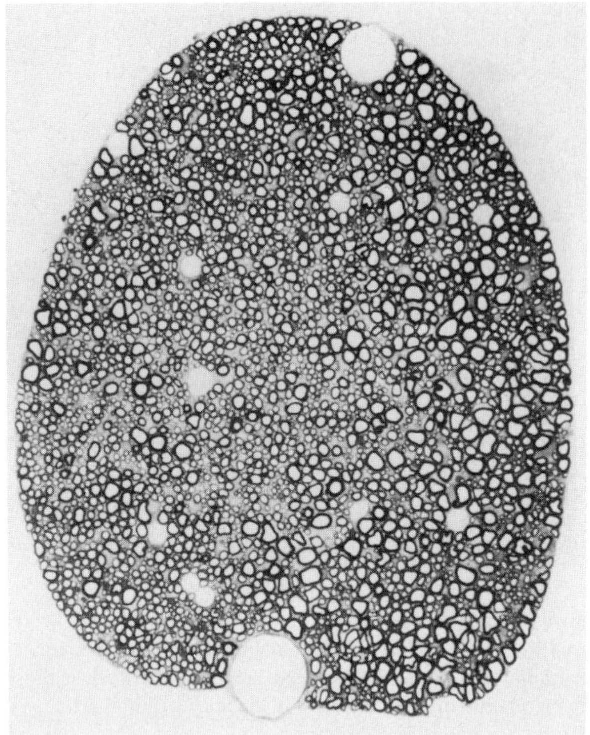

4/2A

1888] and the substantia gelatinosa (fig. 4/1A, B) [*Rolando,* 1824]. Some fine afferents also terminate in the ventral part of the nucleus proprius. The layering pattern of the dorsal horn was codified by *Rexed* [1952, 1954] (fig. 4/1D), based on the appearance of Nissl-stained sections of the cat spinal cord. However, *Rexed* [1964] felt that a similar lamination pattern applies also to other mammals [cf. *McClung and Castro,* [1978]. Most investigators now use Rexed's scheme so that there is uniformity in terminology, if not in understanding, of the relationship of dorsal horn neurons. In this scheme, the dorsal horn includes laminae I–VI. The marginal zone is lamina I and the nucleus proprius is laminae III–V. However, there has been a disagreement as to the identification of the substantia gelatinosa with certain of Rexed's laminae. Some workers have considered laminae II and III to be equivalent to the substantia gelatinosa [*LaMotte,* 1977; *Szentágothai,* 1964]. At present, however, there is general agreement that the substantia gelatinosa corresponds to Rexed's lamina II [*Cervero and Iggo,* 1980].

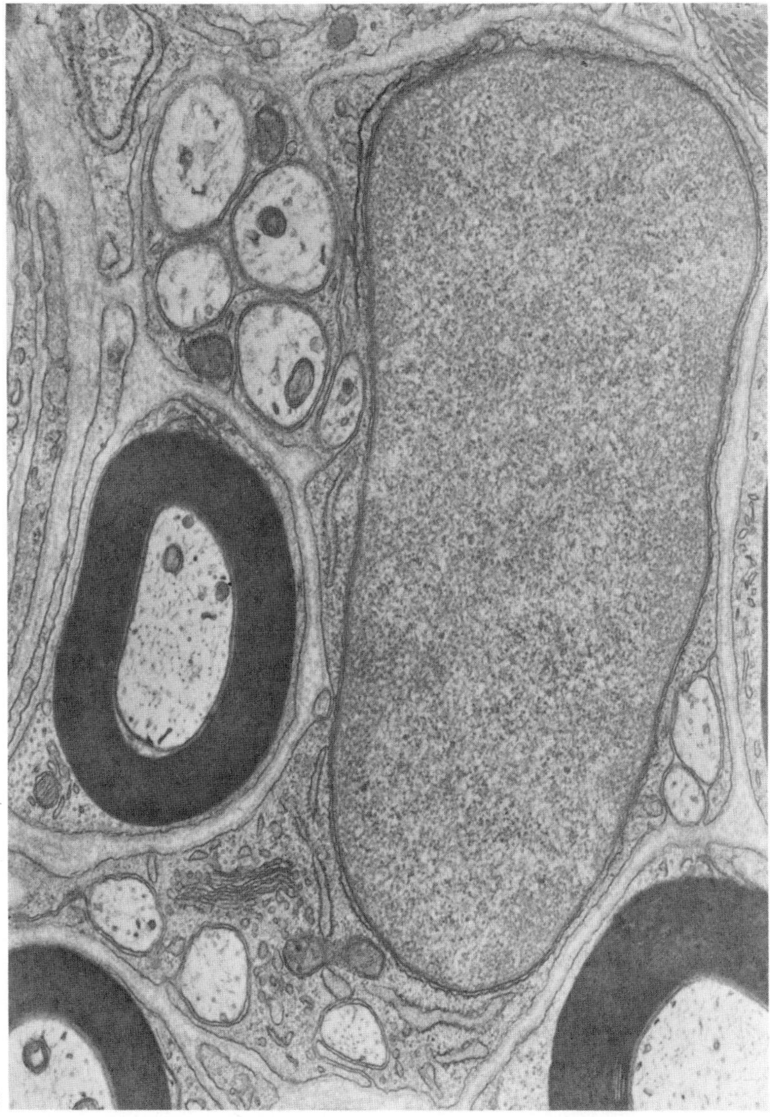

4/2B

Fig. 4/2. A Lumbosacral dorsal root in the rat. At this magnification, the C fibers are not resolved. Myelinated fibers of large, medium and small size are seen, as well as the circular profiles of capillaries. [From *Langford and Coggeshall,* 1979.] *B* Higher magnification electron micrograph showing several small myelinated axons and also several clusters of unmyelinated axons, each of which indents the cytoplasm of a Schwann cell. [From *Langford and Coggeshall,* 1979.]

Several other controversies have also developed concerning the substantia gelatinosa. These include the nature of the specific inputs to particular parts of the upper layers of the dorsal horn, the types of responses of the neurons within the substantia gelatinosa, the relationship of function to morphology of gelatinosa neurons, the pattern of axonal connections of gelatinosa neurons and whether or not they form a closed system, and the association of specific neurotransmitters with cells in the upper layers of the dorsal horn. A further issue is the relationship between neurons in the upper layers of the dorsal horn and cells in the deeper layers. Finally, an unsolved problem is the linkage between primary afferent fibers and nociceptive tract cells, especially the ways in which this linkage is modulated by the activity of neurons in the superficial layers of the dorsal horn. Many of these issues have been reviewed by *Cervero and Iggo* [1980] and by *Willis and Coggeshall* [1978].

Dorsal Root

With the exception of some afferent fibers that enter the spinal cord by way of the ventral roots (see below), all nociceptive afferents terminating in the spinal cord enter through the dorsal roots. Dorsal roots contain afferent fibers having a broad spectrum of sizes, including large, medium and small sized myelinated axons (fig. 4/2A) [*Langford and Coggeshall,* 1979] and unmyelinated axons (fig. 4/2B). Nociceptive fibers within a dorsal root would have either small myelinated axons or unmyelinated axons (fig. 4/2B). However, it is impossible to tell from structure alone which axons belong to nociceptors and which to other kinds of sensory receptors. Thus, the function of the axons shown in figure 4/2B is unknown.

It is generally thought that each dorsal root ganglion cell gives rise to a single central process that enters the spinal cord through a dorsal root and a single peripheral process that supplies one or more receptors, branching only as it nears the receptors. This concept is based in part on counts of dorsal root ganglion cells and of axons in dorsal roots, using silver-stained material for the axonal counts [*Duncan and Keyser,* 1936, 1938; *Holmes and Davenport,* 1940]. However, the counts of axons in the light microscope necessarily underestimate the number of unmyelinated axons, since many of these have sizes at or below the limit of the resolution of the light microscope [*Gasser,* 1955]. A recount using the resolving power of the electron microscope shows that, at least in the rat, there are more dorsal root axons

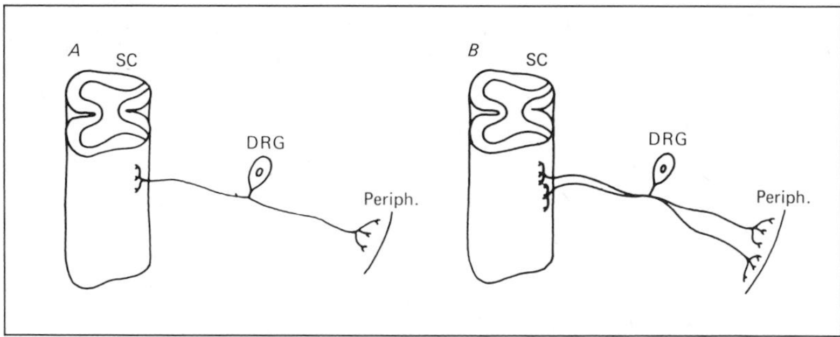

Fig. 4/3. In *A* the drawing shows the traditional concept of the processes of a dorsal root ganglion cell. Although the central and peripheral processes branch, the branching occurs near the terminals. In *B* is a modified concept that may apply to some dorsal root ganglion cells. The central and/or peripheral process branches near the cell body and then again near the terminals. [From *Langford and Coggeshall,* 1981.]

than dorsal root ganglion cells [*Langford and Coggeshall,* 1979; see also *Gasser,* 1955]. For instance, the L6 and S1 dorsal root ganglia average about 7,000 ganglion cells, whereas in the same roots there is an average of more than 10,000 axons, giving an excess of axons over ganglion cells of 43%. Lesion experiments indicate that few or none of the dorsal root axons are efferent [*Langford and Coggeshall,* 1979]. Evidently, dorsal root axons may branch somewhere between the ganglion and the spinal cord. Branching may also occur peripheral to the dorsal root ganglion in the spinal nerve [*Langford and Coggeshall,* 1981]. Thus, some dorsal root ganglion cells may give rise to more than one dorsal root axon and/or more than one peripheral axon (fig. 4/3) [*Langford and Coggeshall,* 1981].

The number of unmyelinated axons exceeds the number of myelinated axons in the rat dorsal root by a ratio of about 2:1 [*Langford and Coggeshall,* 1981]. A 3.7:1 ratio is found for unmyelinated and myelinated axons in the human sural nerve [*Ochoa and Mair,* 1969]. However, some of these unmyelinated axons belong to the sympathetic nervous system. Muscle nerves in the cat hindlimb have a 2:1 ratio of unmyelinated to myelinated axons, provided that the somatic motor and sympathetic fibers are removed by suitable lesions [*Stacey,* 1969]. *Langford* [1983] finds the following percentages of unmyelinated afferent fibers (taking into account the somatic and autonomic motor fibers by removing them surgically) in hindlimb nerves of the cat: medial gastrocnemius, 58%; sural, 61%; medial articular nerve of the knee, 49%; lateral articular nerve, 45%.

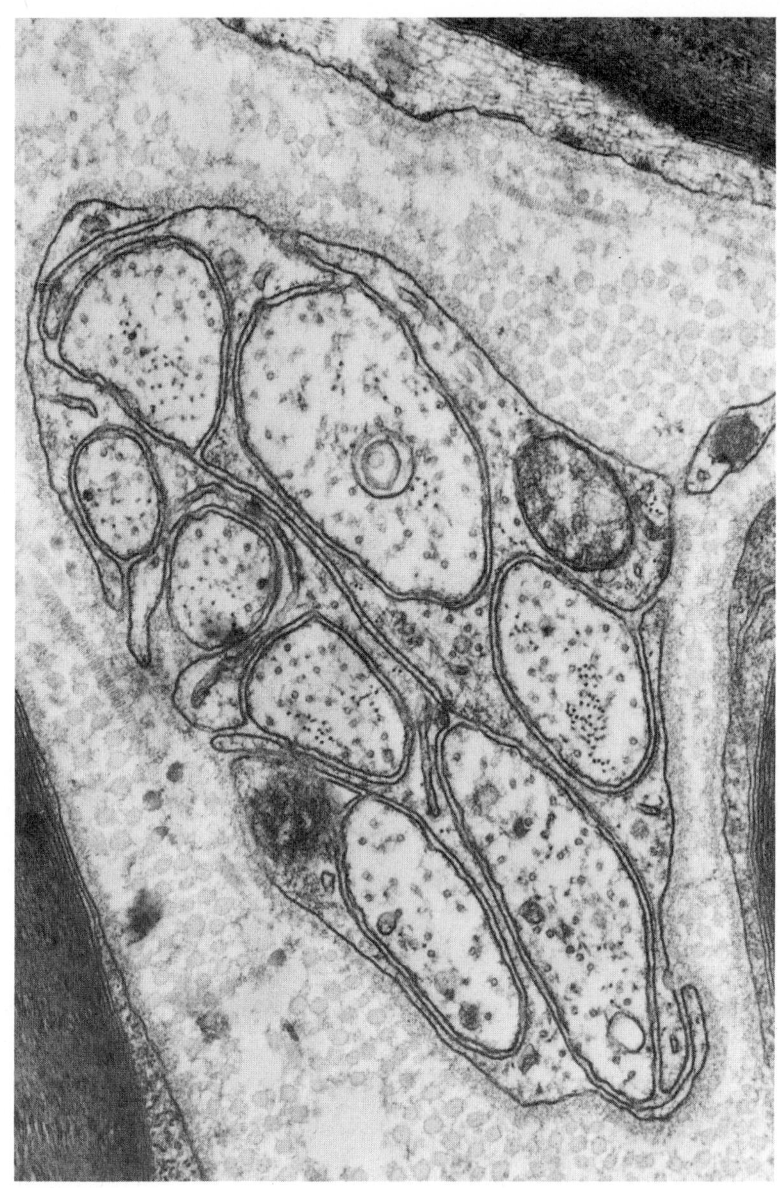

4/4A

Fig. 4/4. A shows unmyelinated axons in a human ventral root. [From *Coggeshall* et al., 1975.] *B* shows an interpretation of the route that might be taken by a ventral root afferent fiber that directly enters the spinal cord over the ventral root. The drawing in *C* shows a possible arrangement of an afferent that has a peripheral process that takes a looping path into the ventral root; alternatively, the central process could make the loop. [From *Coggeshall,* 1979.]

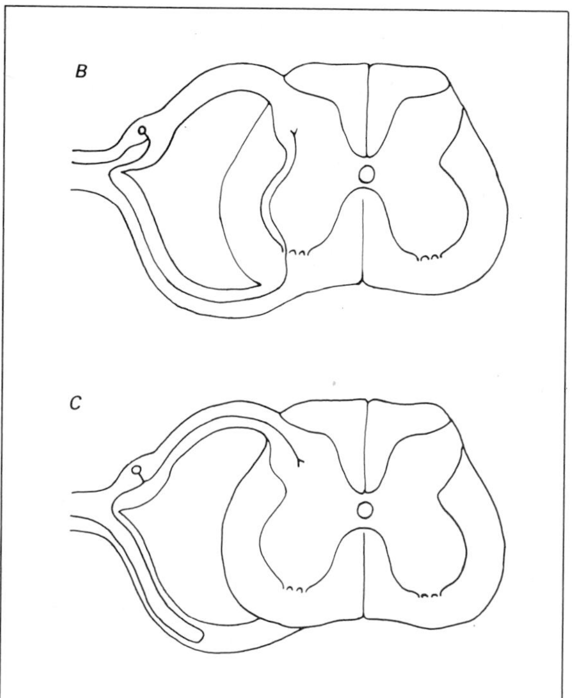

4/4B C

Ventral Root

Ventral roots in mammals (including humans) contain a large proportion of unmyelinated axons, about 30% of the total fiber count, even in segments not contributing to the main autonomic outflows (fig. 4/4A) [*Applebaum* et al., 1976; *Clifton* et al., 1976; *Coggeshall* et al., 1974, 1975; *Davenport and Ranson,* 1931; *Duncan,* 1932; *Holmes and Davenport,* 1940; *Matthews,* 1968]. In the cat, most of these ventral root unmyelinated fibers have been shown to originate from dorsal root ganglion cells [*Coggeshall* et al., 1974], indicating that they have a sensory function. Receptive fields in peripheral tissue have been described for unmyelinated ventral root afferent fibers [*Clifton* et al., 1976; *Coggeshall and Ito,* 1977; *Floyd* et al., 1976b]. Ventral roots also contain a few sensory fibers of myelinated size [*Dimsdale and Kemp,* 1966; *Floyd* et al., 1976b; *Kato and Hirata,* 1968; *Kato and Tanji,* 1971; *Loeb,* 1976; *Ryall and Piercey,* 1970; *Sherrington,* 1894; *Webber and Wemett,* 1966; *Windle,* 1931].

At least some ventral root afferent fibers appear to enter the spinal cord directly through the ventral root (fig. 4/4B) [*Light and Metz,* 1978; *Mawe* et al., 1984; *Maynard* et al., 1977; *Mikeladze,* 1965; *Yamamota* et al., 1977]. However, the number of ventral root unmyelinated fibers decreases in sections taken progressively nearer the spinal cord, suggesting that many of the fibers do not enter the cord directly [*Risling and Hildebrand,* 1982]. Instead, the majority of the ventral root afferents may loop back into the dorsal root (fig. 4/4C) [see discussion in *Cranefield,* 1974]. Alternatively, some ventral root afferents may be peripheral processes of receptors with terminals in the meninges [*Dalsgaard* et al., 1982; cf. *Clark,* 1931].

The observations that many of the ventral root unmyelinated afferents are nociceptors [*Clifton* et al., 1976; *Coggeshall and Ito,* 1977] makes it likely that they account for the 'recurrent sensibility' described by *Magendie* and for the pain experienced by patients caused by ventral root stimulation [*Frykholm,* 1951; *Frykholm* et al., 1953; *Magendie,* 1822b; *White and Sweet,* 1955]. Furthermore, it is possible that ventral root afferent fibers account in part for the poor success of dorsal rhizotomies in eliminating segmental pain [*Coggeshall,* 1979; *Coggeshall* et al., 1975]. There is evidence that ventral root afferents entering the cord directly produce autonomic reflex changes [*Longhurst* et al., 1980]. There is also preliminary evidence for a weak flexion reflex action on tenuissimus motoneurons [*Voorhoeve and Nauta,* 1983]. Looping ventral root fibers may explain the activation of spinal interneurons by stimulation of the distal stump of a cut ventral root [*Chung* et al., 1983].

Segregation of Large and Small Afferent Fibers in the Dorsal Root

The large and small axons in a dorsal root are randomly dispersed throughout most of the length of the root (cf. fig. 4/2A). However, near the entry of the dorsal root into the spinal cord, the large fibers tend to become segregated from the small ones, with the large myelinated axons becoming medial and the fine myelinated and unmyelinated afferent fibers lateral in the dorsal rootlets (fig. 4/5) [*Kerr,* 1975a]. This arrangement was first noted by *Lissauer* [1885] and by *Bechterew* [1887] and has frequently been confirmed [*Cajal,* 1909; *Ingvar,* 1927; *O'Leary* et al., 1932; *Ranson,* 1913, 1914a, b; *Ranson and Billingsley,* 1916; *Sindou* et al., 1974b]. However, several authors have provided contrary evidence [*Earle,* 1952; *Snyder,* 1977; *Wall,* 1962], at least for the cat. It appears that the segregation of fine

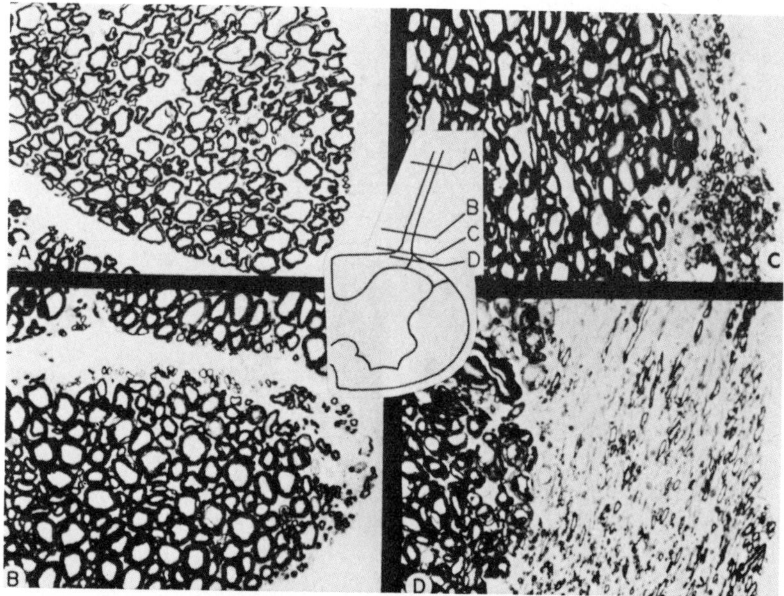

Fig. 4/5. Segregation of large and small fibers in a dorsal rootlet. The electron micrographs show the L7 dorsal root in the cat at different distances from the dorsal root entry zone: *A* is at 5 mm; *B* at 1 mm; *C* very near the entry zone, and *D* as the root fibers merge with the tract of Lissauer. [From *Kerr,* 1975a.]

afferents in the lateral parts of dorsal rootlets does not occur in the cat but does in the macaque monkey [*Kerr,* 1975a; *Snyder,* 1977].

In man, the majority of the small fibers assume a ventrolateral position in dorsal rootlets as they enter the spinal cord, but some aggregate dorsomedially (fig. 4/6). Of the latter, some cross the entering bundle of large myelinated fibers, following an oblique course within the spinal cord, whereas a few enter the dorsal horn directly [*Sindou* et al. 1974b]. The fact that the segregation of fine afferent fibers in the monkey occurs on a rootlet by rootlet basis makes it difficult to interrupt the fine fibers selectively by a surgical approach. However, *Light and Perl* [1979a] provide evidence that this can be done in the monkey, and *Sindou* et al. [1974a] were able to use this approach clinically to relieve pain. It is unclear to what extent there was interruption of the vascular supply of the dorsal horn in the procedure used by *Sindou* et al. in 1974a [also cf. *Wall,* 1962]. The lesions reported by

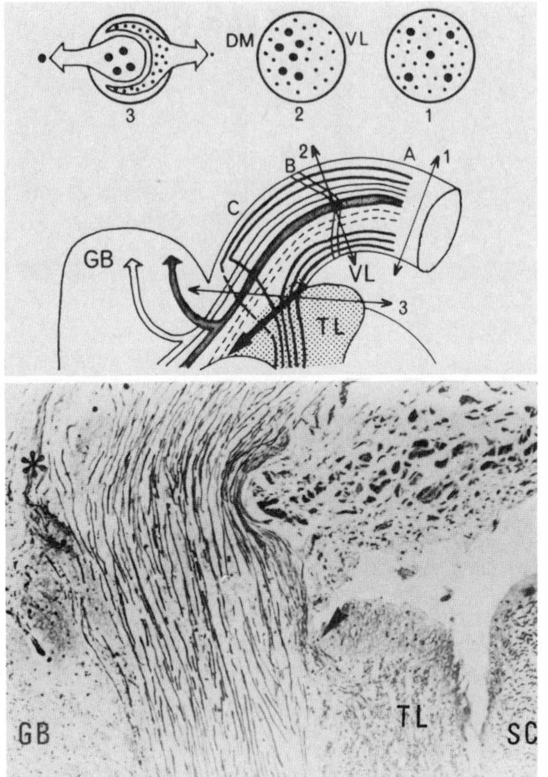
DM
VL
3
2
1
A
B
C
GB
VL
3
TL
GB
TL
SC

4/6

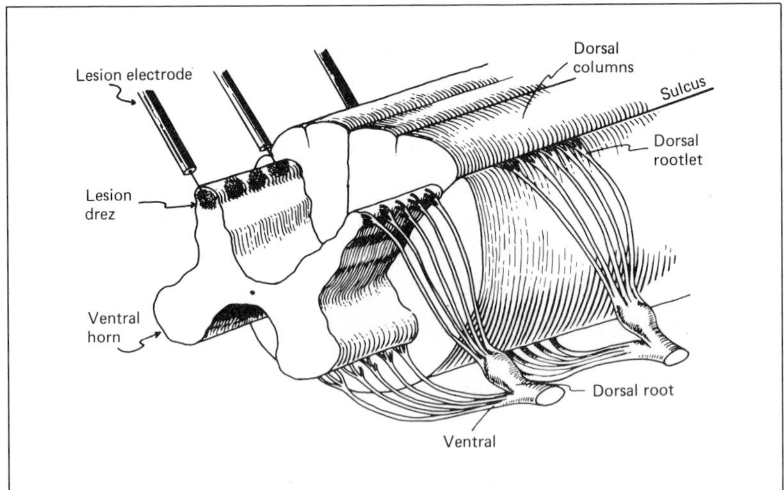
Lesion electrode
Dorsal columns
Sulcus
Dorsal rootlet
Lesion drez
Ventral horn
Dorsal root
Ventral

4/7

Nashold and Ostdahl [1979] in human subjects for interrupting nociceptive input to the spinal cord are not restricted to the dorsal roots, but rather involve a deliberate intrusion into the dorsal horn (fig. 4/7).

Lissauer's Tract

The original view of Lissauer's tract was that it consists chiefly of small sized primary afferent fibers en route to their synaptic terminals in the dorsal horn [*Lissauer,* 1885; see also *Ranson,* 1913, 1914a, b]. However, this view was challenged by investigators who thought that Lissauer's tract consists largely of propriospinal axons [*Earle,* 1952; *Nathan and Smith,* 1959; *Szentágothai,* 1964]. What few primary afferent axons that are present were thought to be located in the medial part of the tract [*Kerr,* 1975a; *Nathan and Smith,* 1959; *Ranson,* 1914b; *Szentágothai,* 1964]. However, *LaMotte* [1977] found that about half of the axons in Lissauer's tract in the monkey are primary afferent fibers. Her results were based on radioautography, following an injection of labeled amino acid into the dorsal root ganglion. She agreed that few primary afferents could be seen when she used a degeneration technique and attributed this to inadequate staining of small processes. *Proshansky and Egger* [1977], *Light and Perl* [1977, 1979a], *Beattie* et al. [1978] and *Morgan* et al. [1981], using anterograde transport of horseradish peroxidase to label primary afferent fibers, also found numerous afferent fibers in Lissauer's tract.

The composition of Lissauer's tract has recently been reassessed by *Coggeshall* and his associates, using electron microscopic counts of axonal numbers in the tract on the side of a dorsal rhizotomy by comparison with the intact side. This group has found that more than two-thirds of the axons in Lissauer's tract in the rat are primary afferent fibers [*Chung* et al., 1979].

Fig. 4/6. Segregation of fine afferent fibers in the human dorsal root entry zone. The fine afferent fibers collect in a ventrolateral position (arrowhead) and also to some extent dorsomedially (asterisk) at the dorsal root entry zone. This is shown schematically in the diagram and also in a light micrograph of a section cut through the dorsal root entry zone. Many of the dorsomedial group of fine axons cross the large myelinated fiber bundle, following an oblique course to Lissauer's tract. [From *Sindou* et al., 1974b.]

Fig. 4/7. Dorsal root entry zone lesions for the relief of pain. The procedure for placing a series of lesions in the area of Lissauer's tract and the substantia gelatinosa is shown. [From *Nashold and Ostdahl,* 1979.]

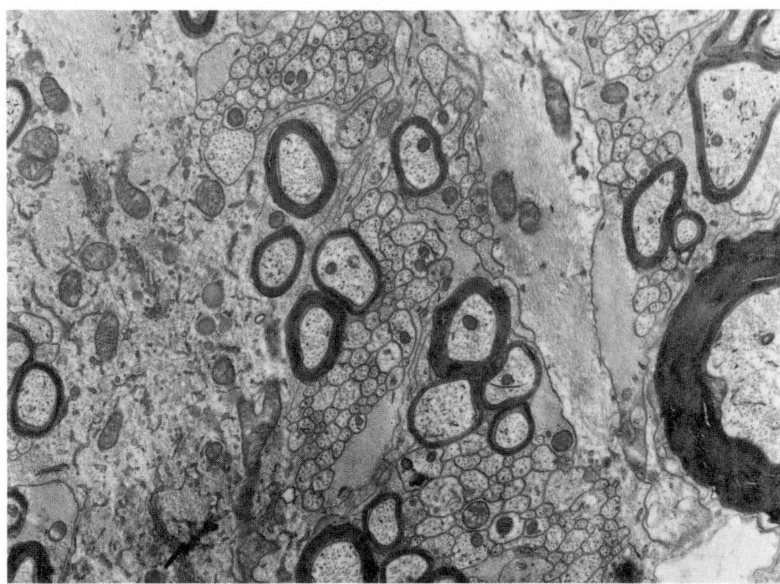

Fig. 4/8. Fine myelinated and unmyelinated axons in the tract of Lissauer in the rat. Astrocytic processes filled with filaments are also seen. [From *Chung* et al., 1979.]

The tract consists primarily of small myelinated and unmyelinated axons (fig. 4/8). The ratio of unmyelinated to myelinated fibers within Lissauer's tract in the rat is 3.8 to 1. When the results of multiple rhizotomies were compared to those of single rhizotomies, it became clear that primary afferent fibers ascend several segments in the lumbosacral cord but terminate within a single segment in the thoracic cord. There were slightly larger numbers of primary afferent fibers in the medial than in the lateral part of Lissauer's tract in the thoracic cord but not in the lumbosacral cord [*Chung* et al., 1979].

These results in the rat were then compared with findings in the cat [*Chung and Coggeshall,* 1979] and the monkey [*Coggeshall* et al., 1981]. The major differences in Lissauer's tract in these animals, as compared with the rat, were that there is a smaller proportion of primary afferent fibers in Lissauer's tract in the cat (about 50%) and a larger proportion in the monkey (80%) than in the rat, and that there are more primary afferents in the medial than in the lateral part of the tract at lumbosacral levels in the cat. Another difference in the monkey is the presence of bundles of small unmyelinated axons (2,000 to 10,000 axons in a bundle), as shown in figure 4/9

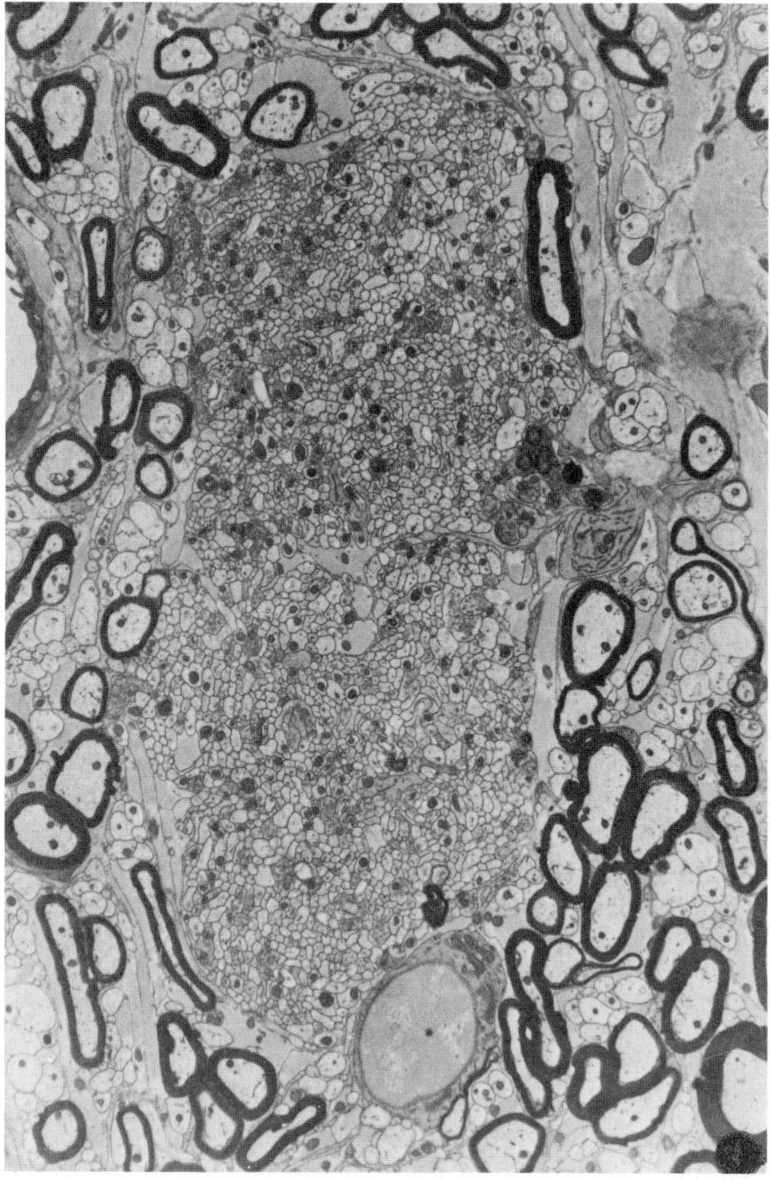

Fig. 4/9. Fine myelinated and unmyelinated axons in the tract of Lissauer in the monkey. The central part of the field of view contains an axonal bundle, which consists of more than 2,000 small unmyelinated axons. [From *Coggeshall* et al., 1981.]

[*Coggeshall* et al., 1981]. These axonal bundles could be important for pain mechanisms in primates, since primates appear to have fewer C mechanoreceptors than do cats [*Kumazawa and Perl,* 1976, 1977], and so a larger proportion of unmyelinated primary afferents must be nociceptive in the primate. However, it is not known if the axons in axonal bundles come from cutaneous receptors or if they are nociceptive.

Lesions restricted to Lissauer's tract produced degeneration in lamina I and the outer part of lamina II [*LaMotte,* 1977]. The degeneration extended only a few millimeters rostral and caudal to the lesion. The primary afferent fibers were located throughout the tract at the level of their entry, but they shifted medially at more rostral and more caudal levels [*LaMotte,* 1977]. Thus, primary afferent fibers were found by radioautography in both the lateral and the medial parts of Lissauer's tract, a finding consistent with the axon counts of the *Coggeshall* group [*Chung* et al., 1979; *Chung and Coggeshall,* 1979; *Coggeshall* et al., 1981].

Terminations of Fine Afferent Fibers in the Dorsal Horn

Many of the studies of the distribution of primary afferent fibers in the spinal cord gray matter have contributed chiefly to our understanding of the projections of the large myelinated fibers, since the large fibers are more readily demonstrated by commonly employed anatomical methods, such as the silver techniques for staining degenerating axons [e.g. *Carpenter* et al., 1968; *Imai and Kusama,* 1969; *Shriver* et al., 1968; *Sprague and Ha,* 1964; *Sterling and Kuypers,* 1967], than are the fine fibers. The general pattern of afferent connections to the spinal cord has been reviewed by *Réthelyi and Szentágothai* [1973].

Cajal [1909] illustrated the projection of small fibers that he regarded as primary afferents into the superficial layers of the dorsal horn, using the Golgi technique (fig. 4/10). However, *Scheibel and Scheibel* [1968] thought that these fine axons originate from neurons of the substantia gelatinosa. *Szentágothai* [1964] proposed that they came from both sources, the dorsal root and the substantia gelatinosa. This controversy undoubtedly stems from one of the major difficulties with the Golgi technique, the problem of following an axon through several sections, especially if the axon changes course abruptly [see discussion in *Willis and Coggeshall,* 1978].

Cajal [1909], *Scheibel and Scheibel* [1968] and *Beal* [1979a] also described what they regarded as primary afferent terminals entering the

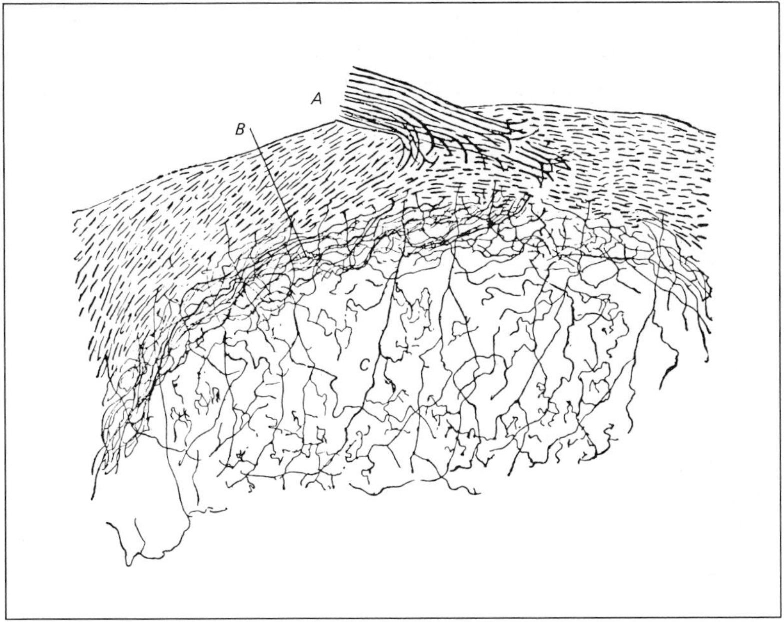

Fig. 4/10. Transverse section of part of the dorsal column and substantia gelatinosa of a newborn cat stained by the Golgi method. The dorsal root is shown at *A*, and the marginal plexus in *B*. The fine collaterals in *C* enter the substantia gelatinosa from its dorsal aspect. [From *Cajal*, 1909.]

inner part of the substantia gelatinosa from its ventral aspect. These terminals represent the dorsalmost extent of the 'flame-shaped arbors' [*Scheibel and Scheibel*, 1968] that have recently been equated with the endings of hair follicle afferents [*Brown* et al., 1977b]. In addition, some Aδ down hair afferents appear to make some synaptic contacts in the inner part of lamina II [*Light and Perl*, 1979b]. These, of course, are the terminals of tactile fibers and so would not be part of the nociceptive pathways, but they might contribute to the modulation of nociceptive transmission.

Another controversy has developed recently. Although the protagonists all accept that many of the fine axons entering the superficial layers of the dorsal horn, as seen in Golgi material, are primary afferent fibers, they dispute the target zones of the small myelinated fibers and the unmyelinated fibers. *Gobel*, studying mainly the subnucleus caudalis of the spinal trigeminal nucleus, has suggested that C fibers project to Rexed's lamina I and that Aδ fibers terminate in the external and internal parts of the substantia gelatinosa [*Gobel,* 1979; *Gobel and Binck,* 1977; see also *Gobel and Falls,* 1979]. *Réthelyi* [1977], on the other hand, proposed that unmyelinated primary afferent fibers give rise to an extensive system of fine terminal arborizations with large synaptic endings in the substantia gelatinosa and that C fibers are the only primary afferent fibers to enter the substantia gelatinosa. *Beal and Fox* [1976] observed fibers with a diameter of 2 μm in Golgi material distributing to lamina I and a system of fine axons with large endings in lamina II. They regarded these as primary afferents. Other terminal systems with small endings they believed to be intrinsic connections. *Beal and Bicknell* [1981] conclude that the most substantial input to the marginal zone is made by Aδ fibers and that the input to the substantia gelatinosa is by C fibers; however, they could not rule out an additional input to the marginal zone by C fibers.

The problem of tracing axons in Golgi material is being addressed in several ways. *Réthelyi and Capowski* [1977] have reconstructed the course of individual fibers using a three-dimensional computer graphics system. Another approach is a comparison between the appearance of Golgi-stained axons and of primary afferent fibers labeled with horseradish peroxidase [*Beal and Bicknell,* 1981].

Additional useful information learned about the primary afferent fiber projection to the superficial layers of the dorsal horn in Golgi material is the orientation of the axonal plexuses. *Beal and Bicknell* [1981], using monkey spinal cord, observed that the marginal zone has a more superficial zone in which the afferent fibers are oriented chiefly transversely (fig. 4/11; see also fig. 4/10), paralleling the orientation of the more superficial dendrites of neurons in that layer. Deep to this transverse marginal plexus is a sheet of axons and dendrites that are oriented longitudinally within lamina I (fig. 4/12). *Scheibel and Scheibel* [1968] emphasized the longitudinal orientation of the neuropil of the substantia gelatinosa, and both *Beal and Fox* [1976] and *Réthelyi* [1977] find that the fine afferent fibers to the substantia gelatinosa run chiefly longitudinally in this layer.

A number of studies have been done using silver impregnation techniques to demonstrate degenerating primary afferent fibers following dorsal rhizotomy [*Carpenter* et al., 1968; *Imai and Kusama,* 1969; *Shriver* et al., 1968; *Sprague and Ha,* 1964; *Sterling and Kuypers,* 1967]. As with the Golgi technique, there were methodological problems that led to controversy. Fine afferent fibers degenerate at a different time than do coarse afferent fibers. Thus, in some studies little degeneration was seen in the substantia gelatinosa at times following dorsal rhizotomy when there was abundant

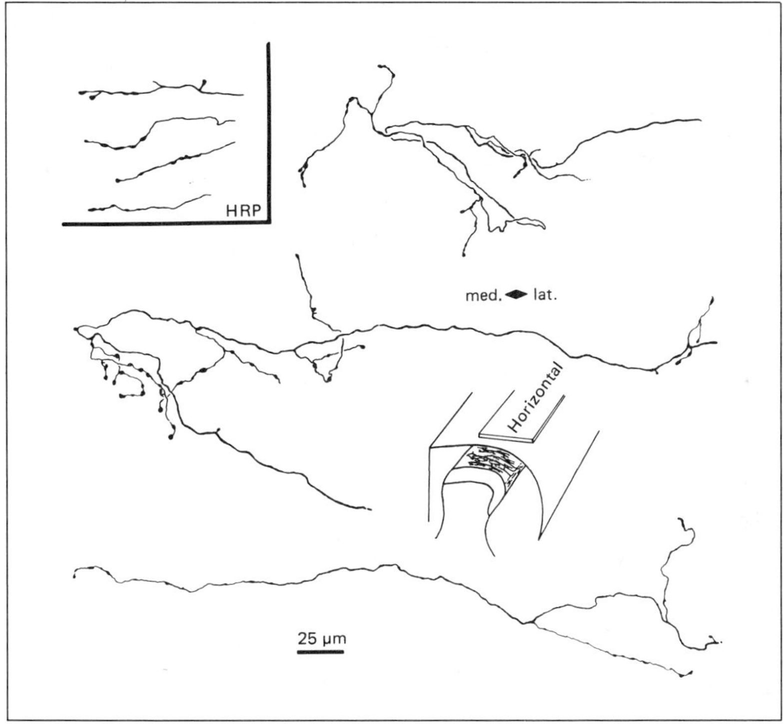

Fig. 4/11. The diagram shows the arrangement of axons in a transversely coursing sheet in the superficial part of the marginal zone. The drawings show representative axons as visualized with the Golgi technique, using monkey spinal cord. The inset shows similar axons stained with horseradish peroxidase transported from the dorsal root, proving that these axons are primary afferent fibers. [From *Beal and Bicknell,* 1981.]

degeneration in more ventral layers of the dorsal horn [e.g. *Anderson,* 1960; *Petras,* 1965; *Sterling and Kuypers,* 1967]. A similar problem occurred when degeneration was sought in the electron microscope [*Ralston,* 1965, 1968b]. However, other investigators found considerable degeneration in the substantia gelatinosa after sectioning the dorsal root [*Sprague and Ha,* 1964; *Szentágothai,* 1964]. *Heimer and Wall* [1968] and *Petras* [1968], using improved techniques and various survival times, confirmed that there can be an abundance of degeneration in lamina II under the right experimental conditions.

LaMotte [1977] used both a silver stain to demonstrate degenerating axons after dorsal rhizotomy and anterograde labeling with radioactive

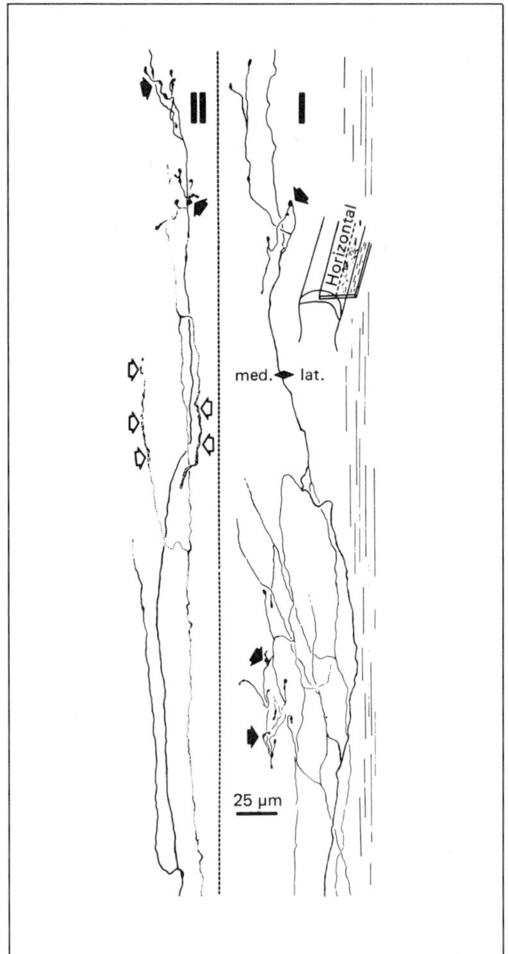

Fig. 4/12. Drawing of longitudinal plexus of axons found in the deep part of the marginal zone and in the substantia gelatinosa. Representative axons from these layers are also drawn. [From *Beal and Bicknell,* 1981.]

amino acids injected into the dorsal root ganglion to trace the distribution of small primary afferent fibers into the dorsal horn (fig. 4/13). She compared her results with these approaches to the effects of lesions of Lissauer's tract and of the dorsal column. Her conclusion was that fine primary afferent fibers project to laminae I, II and III by way of Lissauer's tract. She

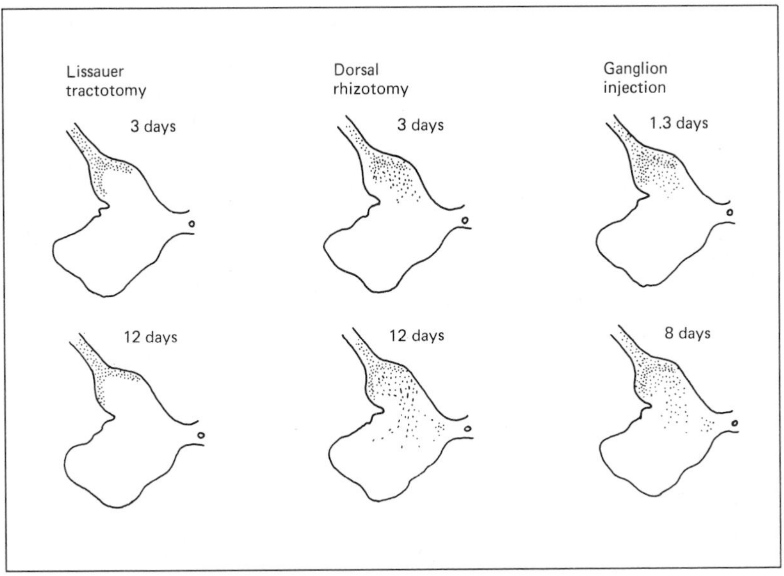

Fig. 4/13. Distribution of axons following interruption of Lissauer's tract or dorsal roots and following injection of tritiated amino acids into a dorsal root ganglion in the monkey. The patterns of projection seen after short and longer survival times are indicated. [From *LaMotte,* 1977.]

could distinguish two populations of axons on the basis of the times required for them to degenerate. One population took about 3–5 days for optimum degeneration, and that population projected to laminae II and III. She thought that these axons might be C fibers. The other population could be demonstrated best after 12 days of degeneration, and it projected to lamina I and the outer part of lamina II. *LaMotte* [1977] proposed that these axons might be Aδ fibers.

Réthelyi et al. [1979] also used the anterograde transport of tritiated amino acids injected into dorsal root ganglia to demonstrate projections of primary afferent fibers into the superficial layers of the dorsal horn. They were able to observe not only a dense projection to lamina II but also to show that in different animals the density of projections to laminae I, II and III–IV could vary, indicating independence of the projections to these different zones. They noted a prominent crossed projection to the contralateral marginal zone and lamina V in their preparations of spinal cord taken from sacral and coccygeal segments.

Several groups have shown the practicality of labeling dorsal root or even peripheral axons with horseradish peroxidase and then tracing the projections of these axons into the spinal cord [*Beattie* et al., 1978; *Proshansky and Egger,* 1977; *Sugimoto and Gobel,* 1982; *Ygge and Grant,* 1983]. Horseradish peroxidase transported anterograde demonstrates abundant projections of dorsal root axons to the substantia gelatinosa.

Light and Perl [1977, 1979a] made use of this technique, combined with selective lesions of either the lateral or the medial aspect of dorsal rootlets in the monkey, to trace the distribution of fine afferent fibers into the dorsal horn. In preparations without lesions, numerous horseradish peroxidase-labeled axons were seen projecting to laminae I and II, as well as to deeper layers of the gray matter and also to the contralateral side. Labeled axons were found both in the dorsal column and in Lissauer's tract. The latter indicated that there is a substantial population of primary afferent fibers in Lissauer's tract. Labeled fibers were found in both the lateral and the medial parts of Lissauer's tract (see previous section on Lissauer's tract). Labeled axons in lamina II were oriented longitudinally and had a small diameter; those in lamina I included both very fine fibers and thicker ones. When the lateral aspect of a dorsal rootlet was cut, labeling in the dorsal horn was restricted to axons that projected chiefly to the nucleus proprius, whereas when the medial aspect of a rootlet was interrupted, there was abundant label in Lissauer's tract and laminae I and II, but only a small amount of label in the ventral nucleus proprius (fig. 4/14). Small diameter fibers judged to be myelinated appeared to terminate both in the marginal zone and in the outer layer of the substantia gelatinosa. Within both the outer and inner zones of the substantia gelatinosa were very fine fibers thought to be the terminals of unmyelinated afferents. A few coarse fibers sent terminals into the deep layer of the substantia gelatinosa by way of lamina III.

More detailed information about the projections of at least myelinated fibers has come from the injection of horseradish peroxidase into functionally identified axons through an intracellular microelectrode. Much of this information pertains to large fibers, as would be expected, since these are more readily impaled than are fine myelinated axons [*Brown,* 1981; *Brown* et al., 1977b, 1978, 1980a, 1981]. For example, the 'flame-shaped arbors' [*Scheibel and Scheibel,* 1968] formed by large myelinated afferents that descend into the deep layers of the dorsal horn and then turn dorsally to end in laminae III, IV and the ventral part of II [*Cajal,* 1909] can now be identified with the projections of hair follicle afferents [*Brown* et al., 1977b].

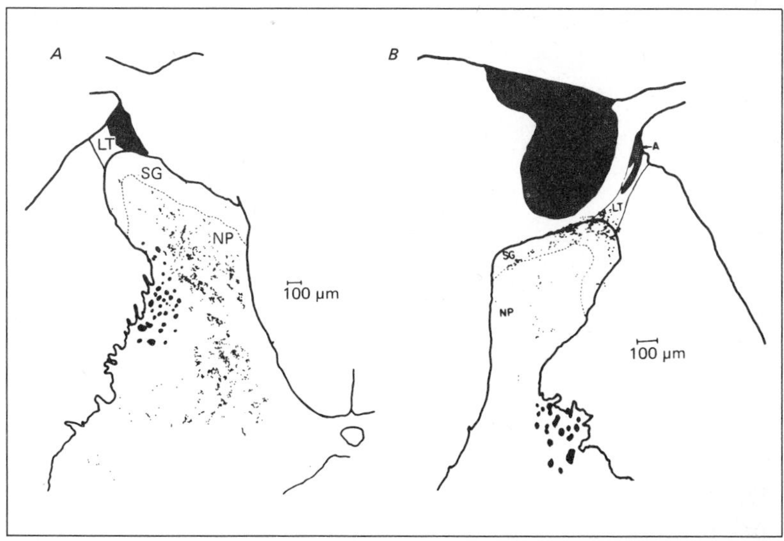

Fig. 4/14. Patterns of projection of large and small primary afferent fibers as demonstrated by anterograde transport of horseradish peroxidase in dorsal rootlets. In *A* the lateral aspect of the dorsal root had been sectioned, and so the intact axons were of large caliber. In *B* the medial aspect of the rootlet had been interrupted, and so the intact axons were of small caliber. Note that in this case there are labeled axons in Lissauer's tract. [From *Light and Perl,* 1979a.]

However, *Light and Perl* [1979b] have been able to label functionally identified Aδ fibers and to trace their projections into the dorsal horn. The Aδ fibers supplying down hair receptors (Aδ mechanoreceptors) were found to project to the inner layer of the substantia gelatinosa and to the outer nucleus proprius (fig. 4/15C), whereas Aδ mechanical nociceptors project largely to the marginal zone and to the ventral nucleus proprius (fig. 4/15A, B). Some of the Aδ fibers projected to the contralateral side of the cord. This observation is consistent with the findings of other laboratories [e.g. *Matsushita and Tanami,* 1983; *Morgan* et al., 1981; *Sprague and Ha,* 1964].

Jancsó and Király [1980] observed degeneration in both laminae I and II following administration of capsaicin to neonatal rats. Similarly, intrathecally administered capsaicin produces terminal degeneration in synaptic endings of laminae I and II in adult rats [*Palermo* et al., 1981]. If the degeneration were restricted to the C polymodal nociceptive afferents, this would be evidence that C fibers project to both laminae.

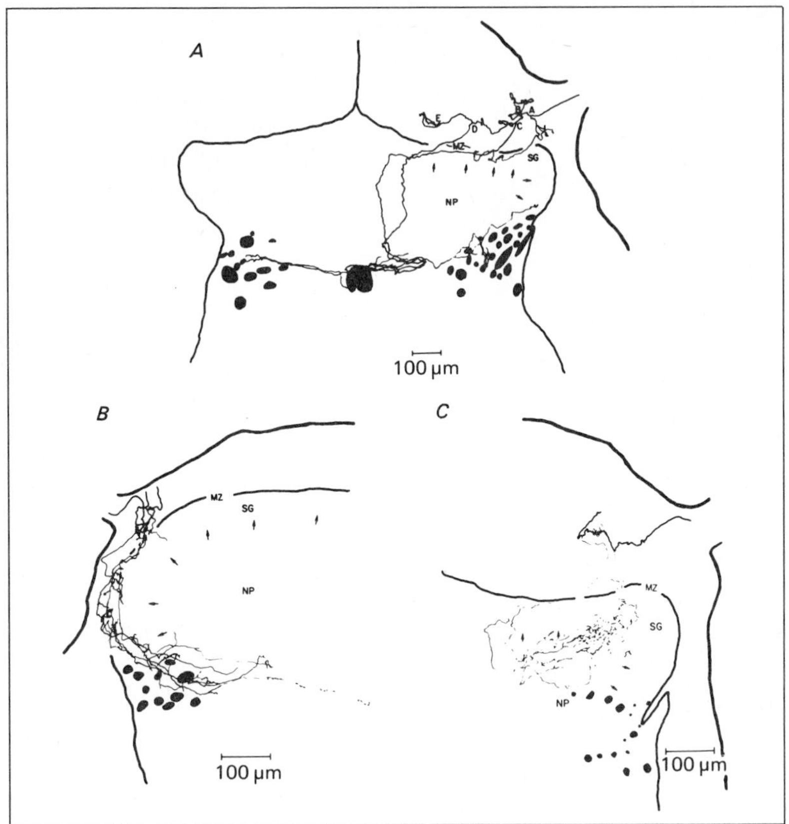

Fig. 4/15. Course followed by Aδ fibers within the spinal cord. The axons were injected intracellularly with horseradish peroxidase after functional identification. The axons in *A* and *B* were Aδ mechanical nociceptors and that in *C* was a D-hair follicle receptor. *A* was from monkey and *B* and *C* were from cat. [From *Light and Perl,* 1979b.]

Synaptic Endings of Primary Afferent Fibers in the Dorsal Horn

Although useful information about the distribution of primary afferent fibers within the dorsal horn has been obtained in studies at the light microscopic level, it is necessary that sections of the dorsal horn be examined at the electron microscopic level to determine the types of synapses made by primary afferents and their specific arrangements.

Primary afferent fibers make several types of synaptic contact within the dorsal horn. One general class of ending is often called the 'central terminal', named for its position in complex synaptic arrays called 'glomeruli' that are commonly observed in the dorsal horn (fig. 4/16) [*Beattie* et al., 1978; *Coimbra* et al., 1974; *Gobel,* 1974, 1976; *Kerr,* 1975a; *Knyihár and Gerebtzoff,* 1973; *Knyihár* et al., 1974; *Ralston,* 1968b; *Réthelyi and Szentágothai,* 1969]. Several opinions about the synaptic arrangements to be found in the glomeruli are reviewed by *Willis and Coggeshall* [1978].

Recent papers by *Ralston and Ralston* [1979, 1982], *Snyder* [1982], *Ribeiro-da-Silva and Coimbra* [1982], and by *Knyihár-Csillik* et al. [1982] provide detailed descriptions of the types of synaptic endings of primary afferent fibers to be found in the dorsal horn. *Ralston and Ralston* [1979, 1982; see also *Beattie* et al., 1978] feel that there are three basic types of synapses: the central terminals, large granular vesicle terminals and round vesicle terminals. The central terminals are in glomeruli, and there are several subtypes of these endings.

Many of the terminals that contain large granular vesicles are concentrated in lamina I and in the outer part of lamina II (fig. 4/16A). *Ralston and Ralston* [1979] draw attention to the similarity in the distribution of these terminals and of substance P (see next section) and propose that large granular vesicle endings contain substance P. *Knyihár-Csillik* et al. [1982] agree that at least some of these might be substance P endings. These endings often develop an electron lucent form of degeneration when dorsal roots are cut [*Ralston and Ralston,* 1979]. *Duncan and Morales* [1973] and *Snyder* [1982] also find dense core vesicle endings of primary afferent fibers in laminae I and II of the cat.

Most of the central terminals located in the substantia gelatinosa (fig. 4/16B) undergo an electron-dense form of degeneration when the dorsal root is cut, in contrast to many of those found in Rexed's laminae III–VI (and in the inner part of lamina II), which develop a neurofilamentous type of degeneration. *Ralston and Ralston* [1979] argue that the terminals in the substantia gelatinosa that develop an electron-dense type of degeneration are from unmyelinated primary afferent fibers, whereas those in deeper layers developing a neurofilamentous type of degeneration are the endings of Aβ fibers. *Knyihár-Csillik* et al. [1982] suggest that the central endings in the substantia gelatinosa may be analogous to the terminals that contain fluoride-resistant acid phosphatase in rodents [*Coimbra* et al., 1974; *Knyihár and Gerebtzoff,* 1973; *Knyihár* et al., 1974]. Comparable terminals were described by *Snyder* [1982] in the cat.

4/16A

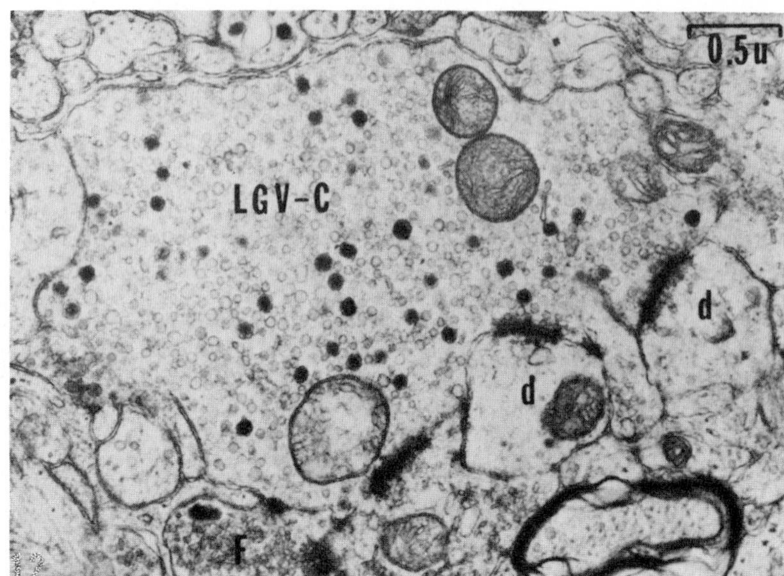

Fig. 4/16. A Central terminal formed by a large granulated vesicle ending (LGV-C) in relation to several dendritic spine heads (d) and receiving a synaptic contact from a terminal containing flattened vesicles (F). From the outer part of lamina II. [*Ralston,* 1979.] *B* Two 'dense sinusoid' type of central terminals (DSA) in relation to dendritic profiles, located in the substantia gelatinosa. [From *Knyihár-Csillik* et al., 1982.] *C* Central terminal containing regular synaptic vesicles (RSV), located in the inner part of lamina II. Note relationship to several dendrites (D), including presynaptic dendrites that contain vesicles (PD). Triadic relationship indicated with two of these by arrows. [From *Knyihár-Csillik* et al., 1982.]

The third type of ending described by *Ralston and Ralston* [1979, 1982] contains round vesicles (fig. 4/16C). These terminals are found throughout the dorsal horn, and they undergo a neurofilamentous type of degeneration after dorsal rhizotomy. Only some of these belong to dorsal root afferent fibers. Similar terminals were found in the cat by *Snyder* [1982].

Knyihár-Csillik et al. [1982] also distinguish between three types of central terminal in the superficial dorsal horn: large dense-core vesicle endings; dense sinusoid endings; and regular synaptic vesicle endings (fig. 4/16). The large dense-core vesicle endings are comparable to *Ralston and Ralston's* large granular vesicle endings in laminae I and outer II. The dense sinusoid endings are equivalent to the central terminals described by *Ral-*

/16B

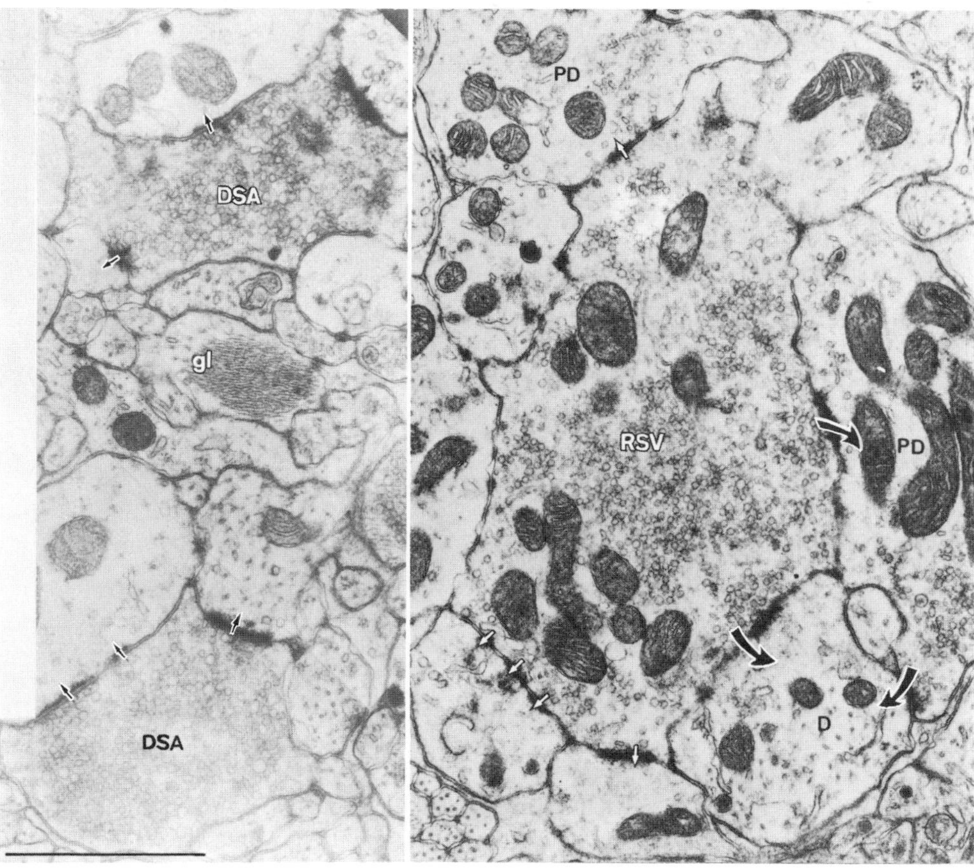

4/16C

ston and Ralston [1979] that end in the substantia gelatinosa and that show an electron-dense type of degeneration after dorsal rhizotomy. The regular synaptic vesicle endings presumably are the population of central terminals that undergo neurofilamentous degeneration described by *Ralston and Ralston* [1979, 1982].

A large fraction of the central terminals observed by *Ralston and Ralston* [1979, 1982] could either be shown to degenerate following dorsal rhizotomy or to label with tritiated amino acid injected into a dorsal root ganglion. Presumably all or nearly all would be shown to have an origin from dorsal root ganglia if several adjacent dorsal roots were cut or ganglia injected with label. Many of the large granular vesicle and round vesicle endings studied by *Ralston and Ralston* [1979, 1982] were not of dorsal root

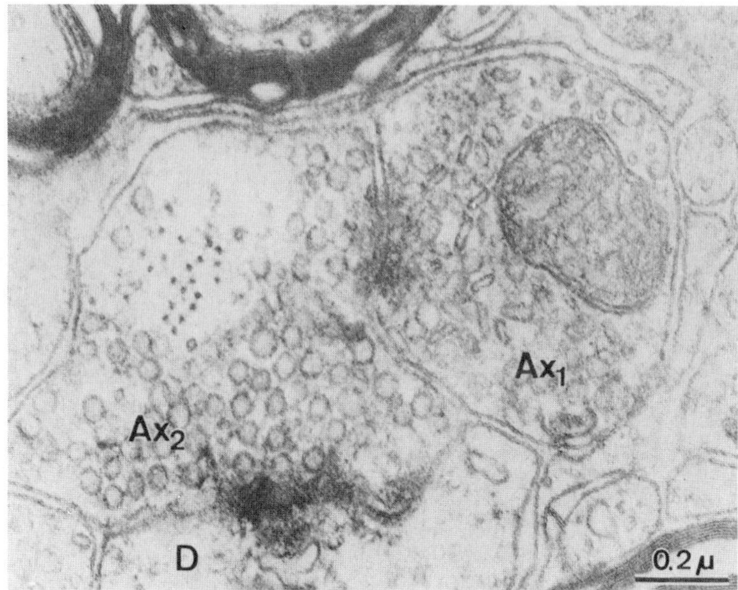

Fig. 4/17. Axoaxonic synapses in the substantia gelatinosa of the rat. The presynaptic element contains flattened vesicles (Ax_1) and the postsynaptic element round vesicles (Ax_2); the latter synapses with a dendrite, D. [From *Zhu* et al., 1981.]

origin, so there is a substantial overlap in the structures representing primary afferent terminals in their experiments and in those of *Knyihár-Csillik* et al. [1982]. However, the latter authors may not have included the round vesicle terminals of *Ralston and Ralston* in their study.

Ribeiro-da-Silva and Coimbra [1982] found few central terminals in lamina I of the rat spinal cord. They observed chiefly dark central terminals in lamina II and neurofilament-containing lucent central terminals in lamina III, although some of each type were found in both laminae.

Axoaxonic and dendrodendritic synapses have been observed in the dorsal horn, in addition to the much more abundant axodendritic synapses [*Ralston,* 1965, 1979; *Ralston and Ralston,* 1979, 1982]. Several types of synaptic arrangements have been described in which the presynaptic elements contain either round or flattened vesicles or in which either the pre- or postsynaptic elements have large granular vesicles [*Zhu* et al., 1981]. One major technical problem in most of the studies of such 'unconventional' synaptic arrangements is that presynaptic dendrites cannot readily be distinguished from axonal terminals unless ribosomes or rough endoplasmic reticulum are recognized in the same element as synaptic vesicles. Such recognition generally depends upon reconstructions of the terminals, either in serial sections [*Knyihár-Csillik* et al., 1982] or in sections containing labeled neuronal processes [*Gobel* et al., 1980].

In the study by *Knyihár-Csillik* et al. [1982], no axoaxonal synapses were observed in the superficial layers of the dorsal horn on dense sinusoidal or large dense core vesicle endings, although axoaxonic complexes were seen on regular synaptic vesicle endings. The presynaptic elements in the latter cases contained flattened vesicles (fig. 4/17) [*Zhu* et al., 1981]. However, elaborate triadic and other arrangements involving presynaptic dendrites were common (fig. 4/16C).

The function of presynaptic dendrites is unclear, but they undoubtedly are important elements in the local circuits of the dorsal horn. *Gobel* et al. [1980] suggest that they make inhibitory postsynaptic and also presynaptic contacts. Axoaxonic synapses are thought to be the structural basis for presynaptic inhibition [*Eccles,* 1964]. However, dendroaxonic synapses would serve as well.

Transmitters Associated with Fine Afferent Fibers

The large dorsal root ganglion cells are thought to employ an amino acid neurotransmitter, such as glutamic or aspartic acid [*Curtis* et al., 1960; *Curtis and Watkins,* 1960; *Graham* et al., 1967; *Puil,* 1981; *Watkins and Evans,* 1981; *Zieglgänsberger and Puil,* 1973]. The cells can be identified histologically by immunocytochemical staining using an antibody against neurofilament protein (fig. 4/18) [*Dodd* et al., 1984].

More pertinent to the nociceptive transmission system are the small dorsal root ganglion cells. One population of these has been found to contain substance P (fig. 4/18) [*Chan-Palay and Palay,* 1977a, b; *Hökfelt* et al., 1975, 1976]. In the rat, the substance P-containing neurons represent some 15–20% of the total population of dorsal root ganglion cells (table 4/I). Substance P is found in the dorsal root, but there is little in the ventral root [*Takahashi and Otsuka,* 1975; see *Dalsgaard* et al., 1982, for evidence concerning ventral root afferents that contain substance P]. Substance P is concentrated in Lissauer's tract and in lamina I and the outer part of lamina II in the spinal cord dorsal horn (fig. 4/19) [*Barber* et al., 1979; *Barbut* et al., 1981; *Chan-Palay and Palay,* 1977a; *deLanerolle and LaMotte,* 1983; *Di-*

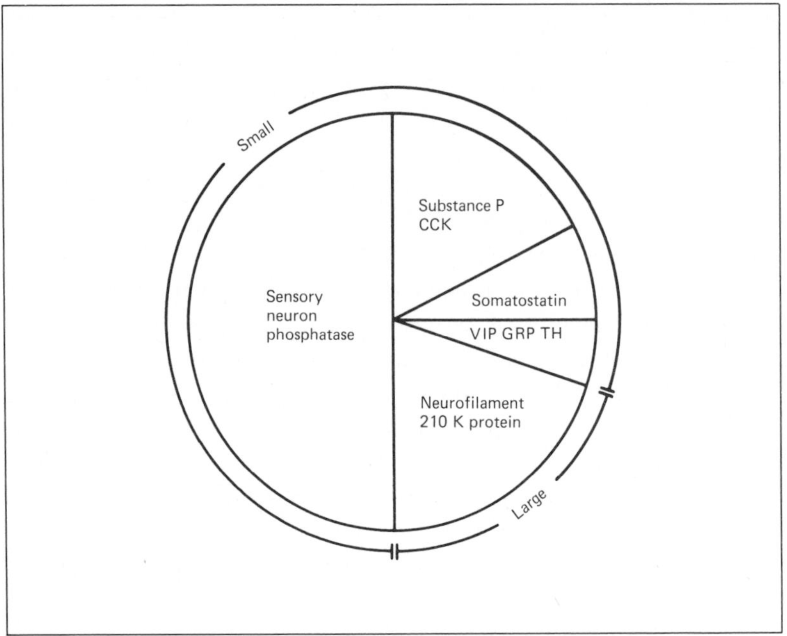

Fig. 4/18. Neuronal markers for various subpopulations of dorsal root ganglion cells in the rat. [From *Dodd* et al., 1984.]

Table 4/I. Neuropeptides present within subpopulations of rat dorsal root ganglion neurons [from *Dodd* et al., 1984]

Peptide	Percentage of neurons	Spinal termination
Substance P	15–20	laminae I–III, V
Somatostatin	5–10	laminae I–III
Cholecystokinin	15–20	laminae I–III
Vasoactive intestinal polypeptide	<5	lamina I
Angiotensin II	not known	lamina II
Gastrin-releasing peptide	5	laminae I, II
Dynorphin	not known	not known
Enkephalin	<1	not known

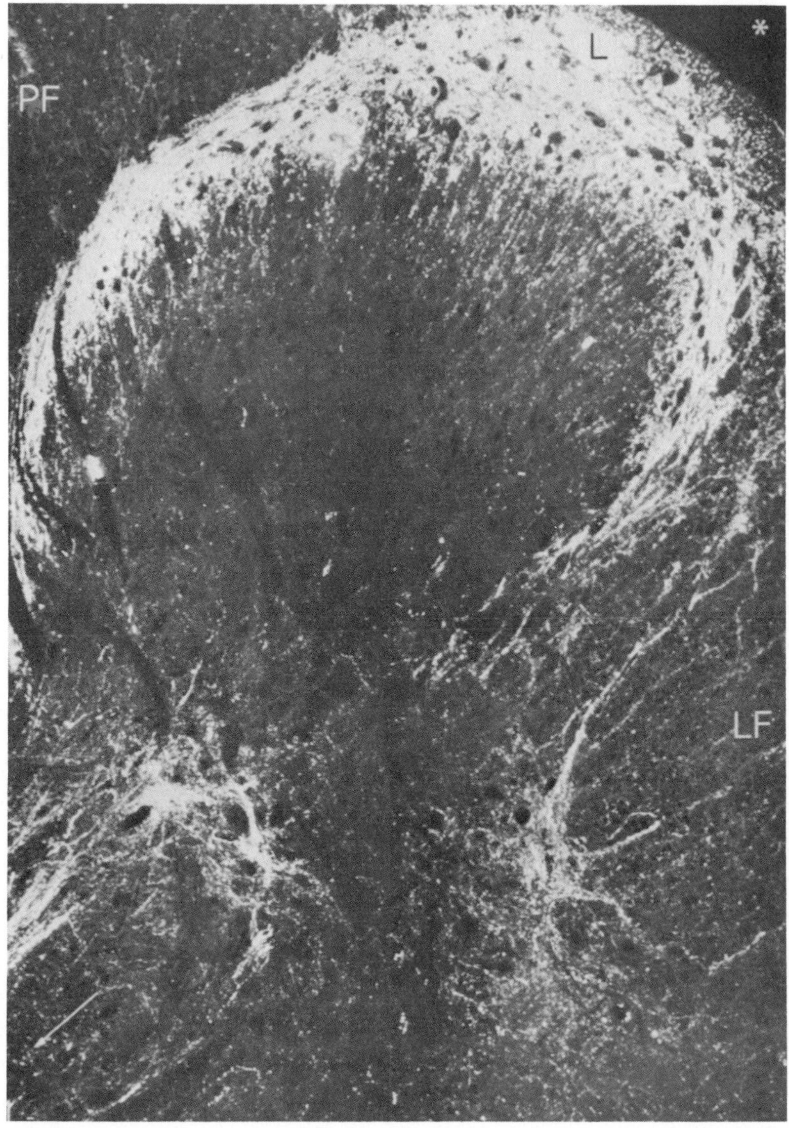

Fig. 4/19. The distribution of substance P immunoreactivity in the dorsal horn of the cat. PF = posterior funiculus; L = Lissauer's tract; LF = lateral funiculus. [From *Hökfelt* et al., 1975.]

Figlia et al., 1982; *Ditirro* et al., 1981; *Gibson* et al., 1981; *Hökfelt* et al., 1975, 1976; *Hunt* et al., 1981; *Jessell* et al., 1979; *Lembeck and Donnerer,* 1981; *Nagy* et al., 1981; *Pickel* et al., 1977; *Seybold and Elde,* 1980]. Many of the substance P-containing axons in Lissauer's tract in the monkey are unmyelinated, but some are small myelinated fibers (fig. 4/20) [*deLanerolle and LaMotte,* 1983; *DiFiglia* et al., 1982]. It is not known if the latter are primary afferent fibers. Substance P endings are also found in the nucleus proprius and in the ventral horn [*deLanerolle* et al., 1983; *DiFiglia* et al., 1982; *Hökfelt* et al., 1975].

The substance P in the dorsal horn originates largely from primary afferent fibers, since most of it disappears following dorsal rhizotomy. However, some of the substance P is contained in axons arising from spinal cord interneurons or from brain stem neurons that project to the cord [*Barber* et al., 1979; *Hökfelt* et al., 1975; *Ljungdahl* et al., 1978; *Takahashi and Otsuka,* 1975]. Substance P in the ventral horn is not obviously affected by dorsal rhizotomy [*Hökfelt* et al., 1975; *Takahashi and Otsuka,* 1975].

Substance P can be depleted from the afferent terminals in the dorsal horn by several maneuvers, in addition to dorsal rhizotomy, including peripheral nerve section [*Barbut* et al., 1981; *Jessell* et al., 1979] and administration of capsaicin [*Ainsworth* et al., 1981; *Cuello* et al., 1981; *Gamse* et al., 1980; *Jessell* et al., 1978; *Lembeck and Donnerer,* 1981; *Nagy* et al., 1980, 1981]. Interestingly, substance P immunoreactivity reappears after its initial reduction by dorsal rhizotomy, presumably because of the sprouting of terminals made by intrinsic or projection neurons containing substance P [*Tessler* et al., 1980, 1981]. This return of substance P requires about 30 days and may explain the failure of *Duncan and Morales* [1973] to see a reduction in synaptic endings containing dense cored vesicles after dorsal rhizotomy, since they sacrificed their animals 55–71 days after rhizotomy.

Studies using immunocytochemistry at the electron microscopic level have demonstrated that substance P terminals do not form typical central terminals in glomerular complexes in the rat, but rather occur as isolated boutons [*Pickel* et al., 1977]. By contrast, in the monkey there are many substance P endings associated with glomeruli as central terminals [*deLanerolle and LaMotte,* 1983; *DiFiglia* et al., 1982]. Substance P is found in dense core vesicle endings, although the same endings also contain small clear vesicles [*deLanerolle and LaMotte,* 1983; *DiFiglia* et al., 1982; *Pickel* et al., 1977]. Most of the substance P endings form axodendritic synapses, but some make axosomatic endings on neurons in the marginal zone, and

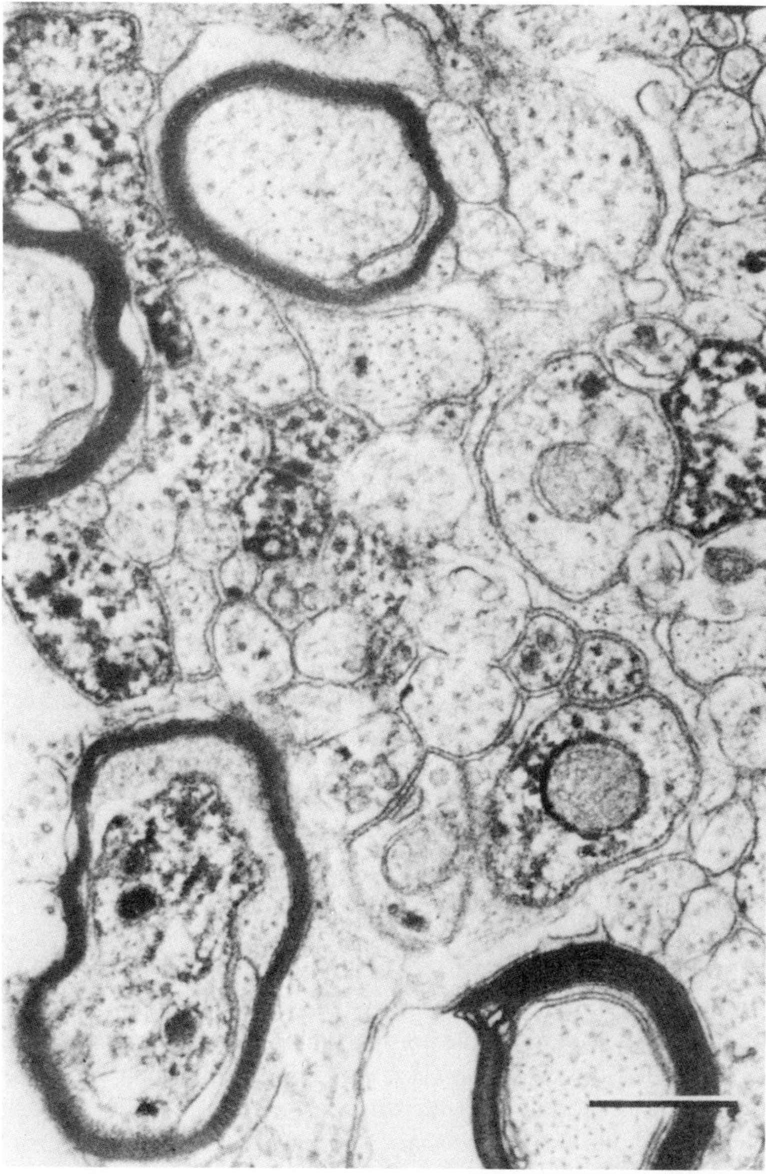

Fig. 4/20. Axons of Lissauer's tract containing substance P immunoreactivity. Note that several unmyelinated axons are labeled, as well as one small myelinated fiber in the lower left. [From *DiFiglia* et al., 1982.]

there are occasional axoaxonic contacts on substance P endings [*deLanerolle and LaMotte,* 1983; *DiFiglia* et al., 1982].

Peptides other than substance P are also found in some dorsal root ganglion cells and in synaptic terminals in the superficial layers of the dorsal horn. These include somatostatin, a peptide resembling cholecystokinin and vasoactive intestinal polypeptide (table 4/I) [*Burnweit and Forssmann,* 1979; *Forssmann,* 1978; *Gibson* et al., 1981; *Hökfelt* et al., 1976; *Kawatani* et al., 1983; *Larsson and Rehfeld,* 1979; *Maderdrut* et al., 1982; *Schultzberg* et al., 1982; *Seybold and Elde,* 1980; however, cf. *Marley* et al., 1982]. The somatostatin-containing dorsal root ganglion cells are a different population than that which contains substance P [*Hökfelt* et al., 1976]. As with substance P, there are intrinsic somatostatin-containing neurons in the dorsal horn, as well as somatostatinergic axons [*Burnweit and Forssmann,* 1979; *Forssmann,* 1978; *Hunt* et al., 1981].

Another group of small dorsal root ganglion cells, at least in rodents (rats and mice), contain an enzyme known as fluoride-resistant acid phosphatase [*Knyihár,* 1971; *Knyihár and Csillik,* 1976; *Knyihár* et al., 1974]. This enzyme is present in synaptic terminals in the substantia gelatinosa and disappears after dorsal rhizotomy [*Coimbra* et al., 1970, 1974; *Knyihár* et al., 1974]. The function of this enzyme is not known with certainty, although it is thought not to relate to lysosomal activity [*Knyihár and Gerebtzoff,* 1973]. Transection of the sciatic nerve causes a depletion of this enzyme from the substantia gelatinosa at the appropriate segmental levels [*Knyihár and Csillik,* 1976]. Recent evidence suggests a relationship to nucleotide metabolism, and *Jessell* and his colleagues are examining the possibility that these dorsal root ganglion cells utilize a nucleotide transmitter, perhaps ATP [*Dodd* et al., 1984]. It is possible to localize dorsal root ganglion cells containing acid phosphatase activity histochemically (fig. 4/21) [*Dodd* et al., 1984].

It is difficult to be certain which type of nociceptor utilizes what neurotransmitter. While it is known that capsaicin affects transmission in C polymodal nociceptors, the fact that this toxin depletes not only substance P but also other substances contained in primary afferent fibers, including somatostatin and fluoride-resistant acid phosphatase, and perhaps cholecystokinin, makes it premature to identify the transmitter in C polymodal nociceptors as substance P [*Gamse* et al., 1981; *Jancsó* et al., 1981; *Jancsó and Knyihár,* 1979], although this seems to be a good working hypothesis. The relationship between other receptors and their transmitters is speculative at this time.

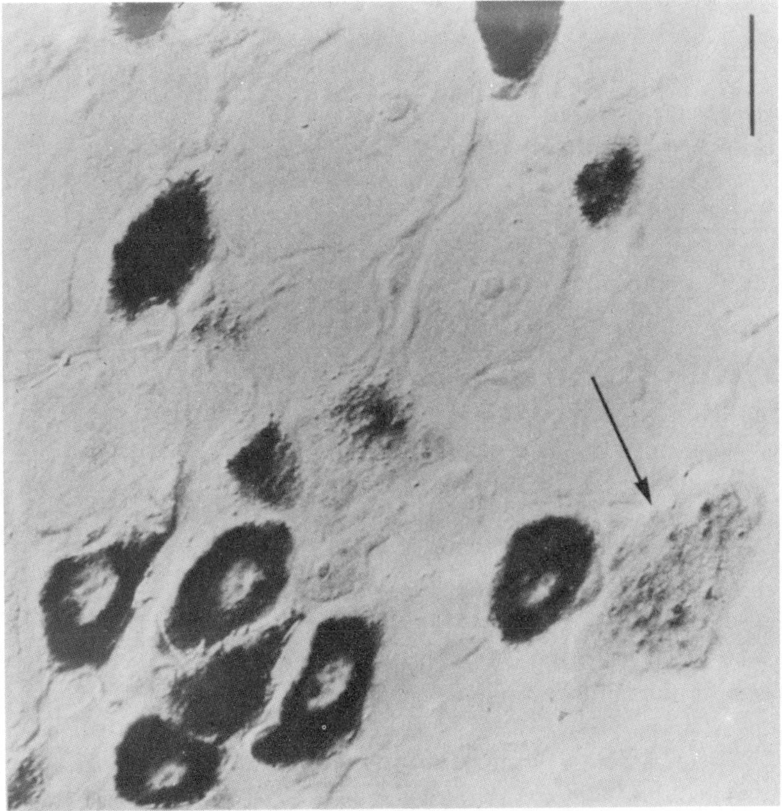

Fig. 4/21. Dorsal root ganglion cells containing acid phosphatase activity as demonstrated histochemically. A number of unstained cells are also seen, as well as cells containing diffuse granular reaction product (arrow). [From *Dodd* et al., 1984.]

Primary afferent fibers appear to contain opiate receptors on their membranes. This is suggested by the reduction in naloxone binding in the dorsal horn that follows dorsal rhizotomy, although the possibility that this change might reflect transsynaptic degeneration could not be ruled out [*La-Motte* et al., 1976]. A number of studies are consistent with the idea that opiates might exert part of their action on presynaptic terminals [*Jessell and Iversen,* 1977; *MacDonald and Nelson,* 1978; *Mudge* et al., 1979].

Cell Types in the Dorsal Horn

Rexed [1952, 1954] described the distribution of neurons of various sizes in the dorsal horn, using this distribution and packing density as criteria for delineating the laminae (fig. 4/1D). Complementary to *Rexed's* description are drawings made from Golgi studies that show the cell types that are typical of the different laminae. Recently, attempts have been made to relate functional properties to particular cell types by combining intracellular recordings and the injection of a marker followed by histological reconstruction of the functionally characterized cells.

Lamina I

The most remarkable cells of the marginal zone or Rexed's lamina I are scattered, large flattened neurons often known as 'marginal cells' or as 'Waldeyer cells' (fig. 4/22, 4/23A). These cells were first noticed by *Clarke* [1859], but they were emphasized by *Waldeyer* [1888] in his comparative study of the gorilla spinal cord (fig. 4/1C). *Cajal* [1909] illustrates these cells chiefly in transverse sections, where they appear to be oriented parallel to the surface of the lamina. In addition to the large cells, lamina I contains many smaller neurons [*Ralston,* 1968a; *Rexed,* 1952].

The ultrastructure of marginal neurons has been described by *Narotzky and Kerr* [1978]. In a study emphasizing longitudinal sections, *Gobel* [1978a] was able to distinguish between 4 different types of cells in the marginal zone of the cat trigeminal caudal spinal nucleus. These are smooth and spiny pyramidal cells (fig. 4/22B) and two types of multipolar neurons (with compact and loose dendritic branching patterns). *Gobel* emphasizes that the dendrites of the marginal neurons tend to be restricted to the marginal zone, although some dendritic intrusion was noted into the white

---

Fig. 4/22. A shows the appearance of a marginal cell in the caudal spinal nucleus of the trigeminal nerve of the cat. The inset shows the orientation of the section to be transverse. In *B* a cell of the marginal layer is seen to have a pyramidal shape when viewed in a section taken tangentially to the nucleus. The large number of dendritic spines suggested the name 'spiny pyramid' for this type of cell. [From *Gobel,* 1978a.]

Fig. 4/23. A shows a marginal neuron in the monkey spinal cord that sends a major dendritic projection through lamina II and into lamina III. *B* shows several marginal cells, including a cluster with dendrites extending into Lissauer's tract (A–D) and a medially located cell with a dendrite descending into the substantia gelatinosa (E). [From *Beal,* 1979b.]

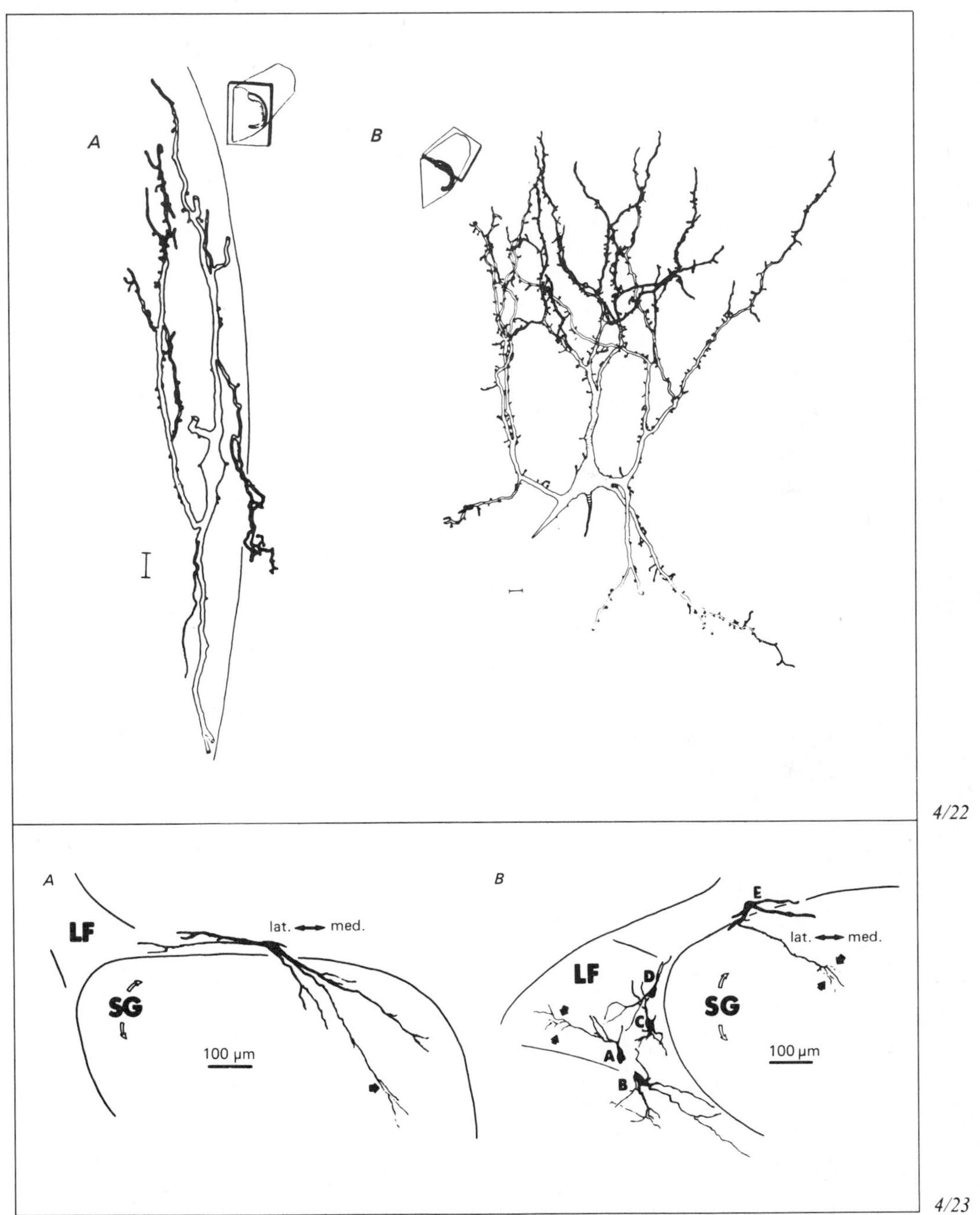

4/22

4/23

Table 4/II. Major properties of primate marginal cell groups [from *Beal* et al., 1981]

	Group I Aspiny cells w/thick blunt dendrites	Group II Large to medium spiny cells		Group III Aspiny cells w/thin, tapering dendrites		Group IV Small spiny cells	
		IIA: longitudinal, spiny cells	IIB: fan-shaped cells	IIIA: aspiny fusiform cells with tortuous dendrites	IIIB: aspiny neurons with confined dendrites	IVA: in lamina I proper	IVB: in dorsolateral fasciculus
Size	large, medium, and small; 70 × 35 to 18 × 10 μm	large and medium; 60 × 35 to 34 × 22 μm	large; 50 × 40 μm	medium and small; 30 × 20 to 25 × 10 μm	medium and small; 43 × 17 to 25 × 10 μm	small; 26 × 16 to 16 × 9 μm	small[1]
Shape	polygonal, pyramidal, oval, or fusiform	polygonal, pyramidal, of fusiform	oval or fusiform	oval or fusiform	fusiform, pyramidal, or polygonal	fusiform, oval, or pyramidal	fusiform, oval, or pyramidal
Location	superficial and deep	superficial and deep	superficial	superficial especially near the dorsolateral fasciculus	superficial and deep	mostly deep	dorsolateral fasciculus
Horizontal dendritic arbor	longitudinal, transverse, or radial w/blunt terminal dendrites	mainly longitudinal	fan-shaped, some w/blunt terminal dendrites	oval, w/tortuous tapering dendrites	mainly longitudinal, w/tapering dendrites	longitudinal	mainly longitudinal
Horizontal dendritic spread	300–1,500 μm	up to 2,000 μm longitudinally; 500 μm or less transversely	up to 1,400 μm longitudinally; 500 μm transversely	300–800 μm longitudinally; 225–500 μm transversely	up to 700 μm longitudinally; 50–250 μm transversely	up to 900 μm longitudinally; less than 150 μm transversely	–[1]
Dendritic spines	few	moderate to numerous	numerous on proximal dendrites	few	few	moderate	moderate
Interstitial dendrites	several	several	several	numerous	rare	rare	most in the dorsolateral fasciculus
Recurrent dendrites	rare	few	rare	few	prominent	few	–[1]
Axons	myelinated	myelinated; some with unmyelinated intrasegmental collaterals	myelinated	probably finely myelinated	probably finely myelinated	probably finely myelinated	unmyelinated

[1] Insufficient number of cells for determination of all variables.

matter overlying the lamina. Most of the cells are of the Golgi type I and so can be regarded as projection cells.

If virtually all of the dendrites of marginal cells are confined to lamina I, then these cells might receive direct synaptic inputs from only a restricted population of primary afferent fibers [*Gobel,* 1978a]. It would be feasible for relayed activity to reach the marginal cells by way of interneurons in deeper laminae [*Gobel,* 1978b; *Price* et al., 1979]. However, the Golgi and horseradish peroxidase study of *Beal* [1979b] of marginal cells in the monkey spinal cord reveals that many of the cells, particularly those located in the medial part of the marginal layer, have dendrites that project ventrally into the substantia gelatinosa or even to lamina III (fig. 4/23). Thus, at least some marginal neurons could receive synaptic inputs from primary afferent fibers that terminate exclusively either in the marginal zone or in the substantia gelatinosa. Of course, relayed activity would also be feasible. Marginal cells located in the lateral part of lamina I often have dendrites that extend into Lissauer's fasciculus, and in some cases the cell body is in Lissauer's tract (fig. 4/23B). Presumably, the synaptic input to these cells could originate from any of the axonal types within Lissauer's tract.

Beal et al. [1981] provide a more detailed description of the marginal cells in the monkey spinal cord. Some of the cell types that they describe are similar to those reported by *Gobel* [1978a] but others are distinctly different. The large marginal cells can be aspinous or spiny, and there are smaller cells that may or may not have dendritic spines. Table 4/II shows the features of the various cell types. Most marginal cells appear to have myelinated axons, but a group of small spiny neurons located in Lissauer's tract seem to give rise to unmyelinated axons. Furthermore, some of the myelinated axons from larger marginal cells give off unmyelinated collaterals that ramify within the marginal layer. Collaterals of the axons of marginal cells studied by intracellular recording and then histologically reconstructed after intracellular injection of horseradish peroxidase have been found to terminate in laminae I within the dendritic territory of the labeled cell [*Bennett* et al., 1981].

A number of studies have confirmed the suggestion from the Golgi studies that cells in the marginal zone include projection cells. For example, cells in the marginal layer undergo chromatolysis after cordotomy [*Foerster and Gagel,* 1932; *Kerr,* 1975a; *Kuru,* 1949; *Morin* et al., 1951]. Many cells in the marginal zone of the spinal cord or caudal spinal trigeminal nucleus can be labeled with horseradish peroxidase injected into the thalamus or midbrain [*Albe-Fessard* et al., 1975; *Carstens and Trevino,* 1978a; *Hayes and Rustioni,* 1980; *Hockfield and Gobel,* 1978; *Kruger* et al., 1977; *Mené-*

trey et al., 1982; *Trevino,* 1976; *Trevino and Carstens,* 1975; *Willis* et al., 1979]. These labeled cells include small neurons located in the vicinity of Lissauer's tract [*Willis* et al., 1979]. Some marginal cells can also be activated antidromically from stimuli applied within the thalamus or midbrain [*Craig and Kniffki,* 1982; *Price* et al., 1976, 1978; *Trevino* et al., 1973; *Willis* et al., 1974] or from the spinal cord white matter at more rostral levels [*Cervero* et al., 1979a; *Kumazawa* et al., 1975]. In addition to these ascending tract cells, some marginal neurons appear to be propriospinal cells, since they can be labeled retrogradely with horseradish peroxidase injected into more caudal levels of the spinal cord [*Burton and Loewy,* 1976; *Matsushita* et al., 1979; *Molenaar and Kuypers,* 1978; *Skinner* et al., 1979; *Yezierski* et al., 1980].

Some neurons within the marginal layer have been shown to contain enkephalin by immunocytochemistry [*Aronin* et al., 1981; *Glazer and Basbaum,* 1981]. Whether enkephalin-containing marginal cells are ascending tract cells or propriospinal neurons is not yet known. The observation that marginal cells may give off local collaterals [*Bennett* et al., 1981] is consistent with the hypothesis of *Glazer and Basbaum* [1981] that enkephalin may be utilized as a local inhibitory transmitter in lamina I. Cells in lamina I have also been found to contain substance P [*Tessler* et al., 1981].

At an ultrastructural level, lamina I can be distinguished from lamina II by the following features [*Ralston,* 1968a, 1979]: (1) the axonal and dendritic processes in lamina I tend to be oriented horizontally in lamina I, whereas they are more likely to be vertical or longitudinal in lamina II; (2) although many of the neurons in lamina I are small, there are occasional large neurons in the layers, whereas lamina II contains an abundance of small neurons; (3) lamina I contains some myelinated axons, whereas these are scarce in lamina II (apart from the bundles of large myelinated axons that traverse lamina II from the dorsal column en route to deeper layers of the gray matter).

A major advance in the field of pain mechanisms was the discovery of *Christensen and Perl* [1970] that many of the cells in lamina I respond specifically to noxious inputs (fig. 4/24). Prior to this finding, only one electrophysiological study had reported specifically nociceptive neurons [*Kolmodin and Skoglund,* 1960], and these cells were concentrated in deeper layers of the dorsal horn and the medial ventral horn. The observation of *Christensen and Perl* [1970] has been confirmed by several groups, both in experiments in which the axonal projections of the cells were not investigated and in experiments in which the cells were shown by anti-

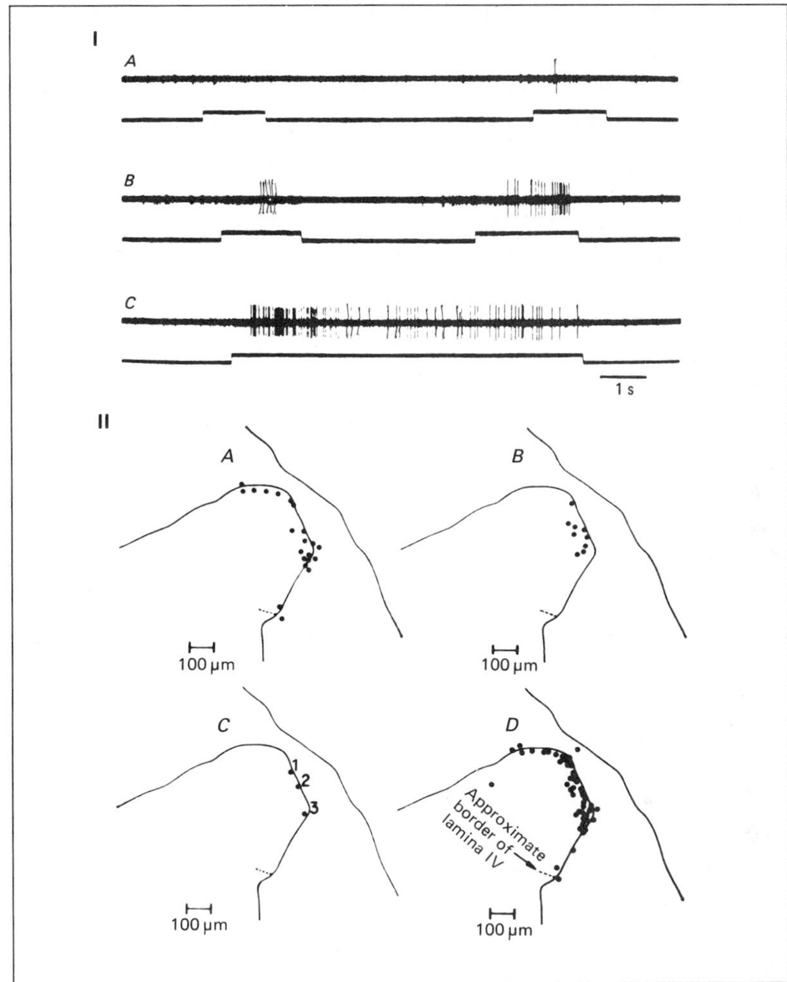

Fig. 4/24. The responses in I are from an individual neuron recorded from the marginal zone. The unit was activated by Aδ fibers. In *A,* stroking with a glass rod had little effect. In *B,* the unit was activated by squeezing the skin with forceps that had a smooth surface. In *C,* the forceps used were serrated. The unit was not excited by cooling or noxious heating. The recording sites in II were marked with an extracellular deposit of dye. Units recorded at the sites shown in *A* were excited only by strong mechanical stimulation. Those in *B* were excited by noxious heat, as well as by strong mechanical stimuli. *C* shows 3 recording sites in one experiment. Unit 1 was activated by cooling, strong mechanical stimuli and noxious heat; 2 by strong mechanical and noxious heat stimuli; 3 only by strong mechanical stimuli. *D* shows all the recording points in the study. [From *Christensen and Perl,* 1970.]

dromic activation to project to more rostral levels, and in the caudal spinal trigeminal nucleus as well as in the spinal cord [*Bennett* et al., 1981; *Cervero* et al., 1976, 1979a; *Chung* et al., 1979; *Fitzgerald and Wall,* 1980; *Kenshalo* et al., 1979; *Kumazawa* et al., 1975; *Kumazawa and Perl,* 1976, 1978; *McMahon and Wall,* 1983; *Menétrey* et al., 1977; *Mosso and Kruger,* 1972, 1973; *Price and Mayer,* 1974; *Price* et al., 1976, 1978, 1979; *Randić and Miletić,* 1977; *Shigenaga* et al., 1976; *Willis* et al., 1974; *Woolf and Fitzgerald,* 1983; *Yokota,* 1975].

However, not all marginal cells are specifically nociceptive. For example, some marginal cells respond to innocuous thermal stimuli [*Christensen and Perl,* 1970; *Hellon and Misra,* 1973; *Iggo and Ramsey,* 1976; *Kumazawa and Perl,* 1978; *Kumazawa* et al., 1975; *Mosso and Kruger,* 1972, 1973; *Poulos and Molt,* 1976]. Other lamina I neurons can be activated by both innocuous and noxious stimulation of the skin [*Bennett* et al., 1981; *Cervero* et al., 1976; *Chung* et al., 1979; *Kenshalo* et al., 1979; *McMahon and Wall,* 1983; *Menétrey* et al., 1977; *Price* et al., 1976, 1978, 1979; *Woolf and Fitzgerald,* 1983]. These cells are often given the name 'wide dynamic range cells' [*Mendell,* 1966], since the thresholds for their inputs range from low to very high. However, many workers in the field believe that a term like 'multireceptive' would be preferable.

In several recent studies, it was possible to obtain intracellular recordings from functionally defined lamina I cells and then to inject horseradish peroxidase or Lucifer yellow into the cells in order to allow a later histological reconstruction of the morphology of the cell, so that an attempt could be made to correlate form with function. *Light* et al. [1979] labeled several neurons with cell bodies in the marginal zone that had nociceptive or thermoreceptive inputs. A better correlation was noted between the location of the major dendritic tree of a neuron than with the location or type of cell body. Neurons with an input from Aδ mechanical nociceptors had a prominent dendritic ramification in the marginal zone; cells with a strong input from unmyelinated nociceptors or from thermoreceptors had a major dendritic arbor in the outer part of the substantia gelatinosa; neurons with nociceptive inputs from both myelinated and unmyelinated fibers had dendrites that spanned both the marginal zone and the outer substantia gelatinosa; cells with an input from sensitive mechanoreceptors and unmyelinated afferents had a principal dendrite in the inner part of the substantia gelatinosa; neurons with a sensitive mechanoreceptor input involving small myelinated afferent fibers had a dendrite that extended through the inner substantia gelatinosa toward the nucleus proprius.

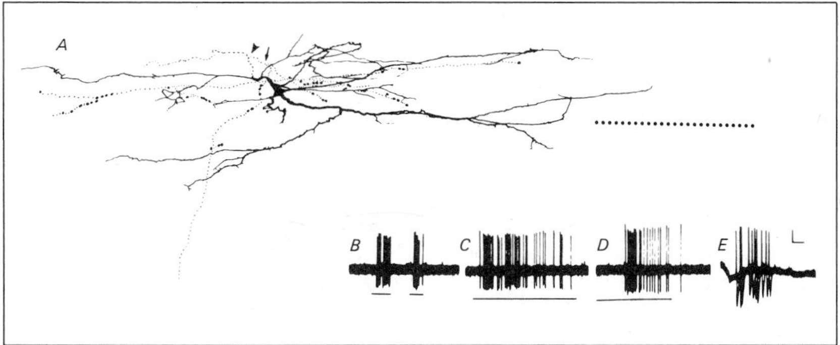

Fig. 4/25. The neuron in *A* was labeled with an intracellular injection of horseradish peroxidase. Before impalement, its receptive field was characterized. The cell responded to brushing *(B)*, pinching the skin with forceps *(C)*, noxious heat *(D)*, and stimulating the skin with electric shocks *(E)*. The latency of the initial discharge indicated that the cell had an input from Aδ fibers. The axon of the cell was found to give off a collateral that had terminals within lamina I (and II). The axon is shown by the dashed line. The arrowhead indicates the origin of the collateral and the arrow the first bifurcation of the collateral. [From *Bennett* et al., 1981.]

In another study that attempted to correlate structure and function of marginal cells, *Bennett* et al. [1981] were able to label 14 lamina I cells with intracellular injection of horseradish peroxidase. 8 of these were classified as wide dynamic range cells (fig. 4/25) and 6 as high threshold cells. The input could be restricted to fine myelinated fibers or could include both fine myelinated and unmyelinated fibers.

Woolf and Fitzgerald [1983] marked cells in lamina I of the rat or in the white matter just superficial to lamina I with horseradish peroxidase or Lucifer yellow. The cells in lamina I could be classified as high threshold or as wide dynamic range (multireceptive) cells. Those in the white matter were either low threshold or wide dynamic range cells.

Lamina II

Cajal [1909] described two major types of cells in the substantia gelatinosa: central cells and limiting cells (fig. 4/26). The central cells appear to have a radial orientation when viewed in transverse section, but a longitudinal orientation in sagittal section (fig. 4/27) [*Scheibel and Scheibel,* 1968]. Central cells are found throughout both the outer and inner part of lamina II. The limiting cells are located in the outer part of the substantia gelatino-

4/26

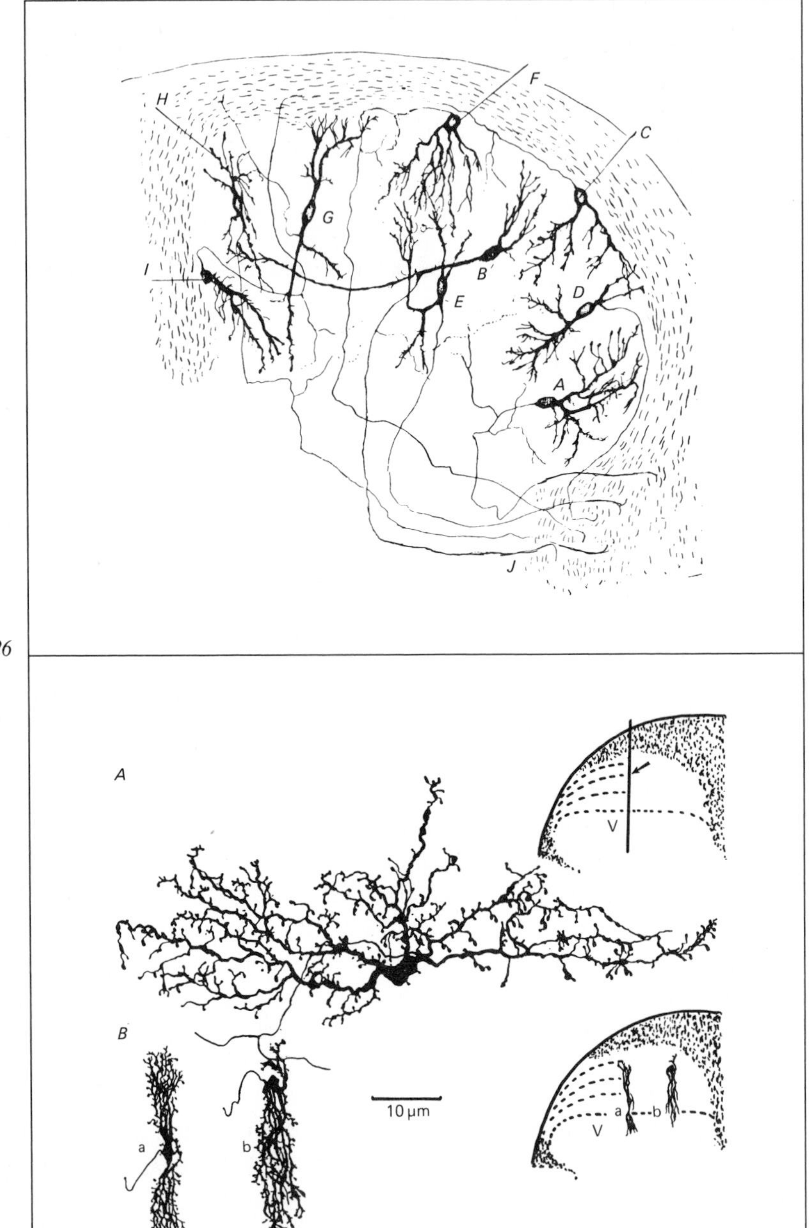

4/27

4/28

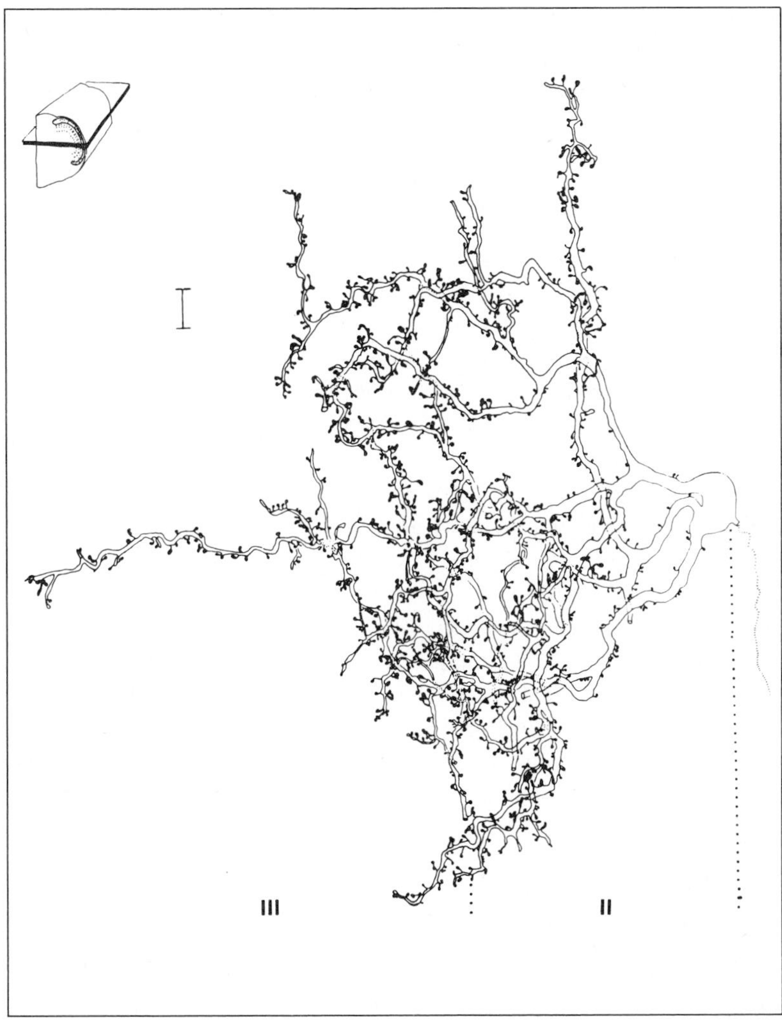

Fig. 4/26. Cells of the substantia gelatinosa. The two main types of cells illustrated (Golgi stain, embryonic chicken) are the limiting cell (C, F) and the central cell (D, H). [From *Cajal,* 1909.]

Fig. 4/27. A shows a neuron of the substantia gelatinosa (60-day-old cat) in sagittal section (plane of section shown in inset at right). *B* shows two neurons of the substantia gelatinosa as viewed in transverse section. Locations of the cells are indicated at right. [From *Scheibel and Scheibel,* 1968.]

Fig. 4/28. A stalked cell of the caudal spinal trigeminal nucleus in the cat. The cell body is in the outer part of lamina II. Dendrites extend into lamina III, and the axon enters lamina I. [From *Gobel,* 1978b.]

sa, near the border with lamina I. Limiting cells are flattened and have a transverse orientation. The opinions of different authors about these cells are discussed by *Cervero and Iggo* [1980] in their review of the substantia gelatinosa.

Gobel [1978b] did a detailed study of the substantia gelatinosa of the caudal spinal trigeminal nucleus, using the Golgi stain. He described 4 types of cells: stalked cells; islet cells; arboreal cells; and II–III border cells. Stalked cells are characterized by numerous stalk-like branches; they also have many dendritic spines on their tertiary and higher order dendrites (fig. 4/28). The dendrites project away from lamina I and into deeper parts of lamina II and generally into lamina III and even IV. The axon usually projects into lamina I, where it ramifies. *Gobel* states that these neurons are equivalent to the limiting cells of *Cajal* [1909]. He further speculates that these cells may be excitatory interneurons. The rationale for this proposal is that lamina I cells often receive an excitatory input from Aδ nociceptors, and yet *Gobel* feels that there are no direct synaptic connections between Aδ nociceptors and marginal cells. Thus, the input might be relayed from lamina II to the marginal layer by way of stalked cells. In support of this hypothesis are the observations of *Price* et al. [1979] that the latency of excitation of pairs of cells in laminae I and II with the same receptive fields is shorter for the lamina II cells than for the marginal cells. Thus, the input, which is in part due to Aδ nociceptors, may excite stalked cells that in turn excite the marginal cells. However, it seems unlikely that stalked cells are universally excitatory, since at least some contain enkephalin, a putative inhibitory transmitter [*Bennett* et al., 1982].

The other major cell type described by *Gobel* [1975, 1978b] he called the islet cell. Islet cells are equivalent to *Cajal's* central cells. These neurons extend rostrocaudally for 500 μm; their dendrites are oriented longitudinally within lamina II, and some recurve back toward the cell body. The axons of islet cells are unmyelinated, and they run for as much as 1 mm within lamina II. Axonal ramifications are extensive, often having recurrent branches and occurring over the full thickness of the lamina. *Gobel* [1978b] suggests that these cells may be inhibitory interneurons of the Golgi type II class.

Arboreal cells and II–III border cells have somewhat more extensive dendritic fields than do islet cells, and their axons might well affect neurons in the adjacent laminae I and III, as well as cells in lamina II. These cell types are also considered Golgi type II inhibitory interneurons by *Gobel* [1978b].

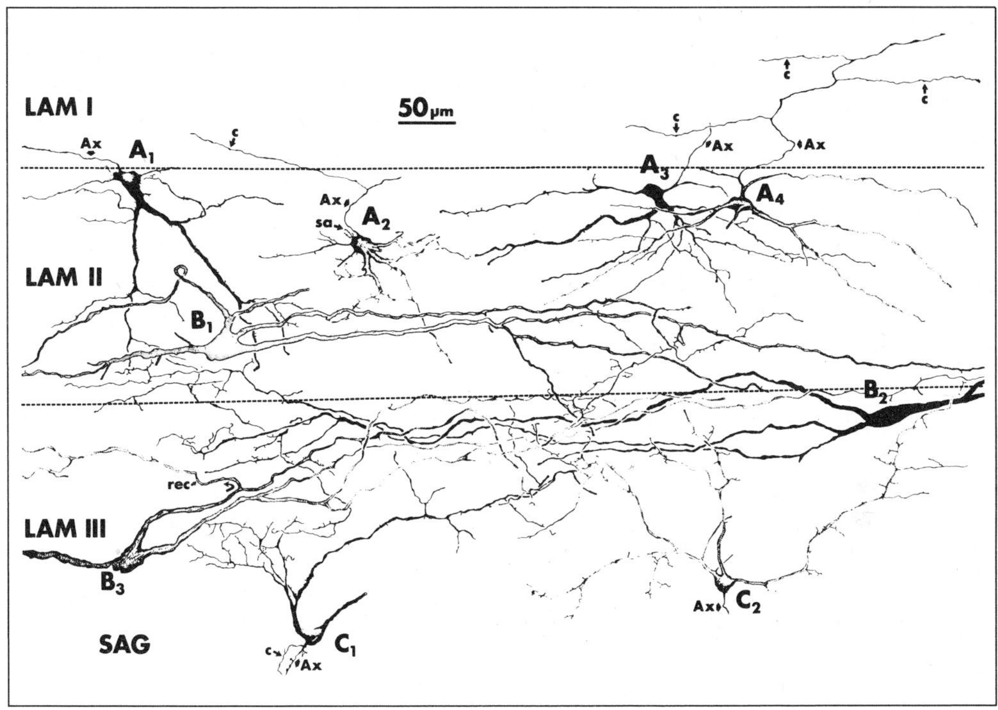

Fig. 4/29. Neurons of the superficial layers of the dorsal horn in sagittal section. Golgi-stained monkey spinal cord. Neuron B_1 is a central cell (or islet cell). Note the recurrent dendrite of cell B_3, a central cell in lamina III. A_1 to A_4 are limiting cells. [From *Beal and Cooper,* 1978.]

Beal and Cooper [1978] studied the cells of the substantia gelatinosa in the adult monkey, using the Golgi technique (fig. 4/29). They found it difficult to classify the cells, stating that cell shape varied over a wide spectrum. Cells in the outer part of lamina II tended to have a polygonal shape, while there were fusiform cells throughout the layer. In sagittal sections, the fusiform cells often give rise to dendrites that have unique recurrent branches (as already mentioned in reference to *Gobel's* analysis). Dendritic spread can be up to 1.5 mm in the rostrocaudal axis in the monkey. The axons of cells in the outer part of lamina II course dorsally (fig. 4/29); they give off collaterals to lamina I and then enter the white matter. These cells are comparable to *Gobel's* 'stalked cells'. The axons of many of the cells of the monkey substantia gelatinosa are relatively thick and may even be myelinated. Many of the neurons have dendritic stalked appendages that may serve to make synaptic contacts with adjacent neurons.

The problem of determining the axonal projections of neurons in the substantia gelatinosa has been discussed by *Willis and Coggeshall* [1978]. In brief, it has been suggested that the substantia gelatinosa is a closed system with its interneurons projecting either locally or to other parts of the substantia gelatinosa either on the same or on the contralateral side of the spinal cord [*Szentágothai,* 1964]. However, several investigators have been able to trace the axons of gelatinosa cells to other laminae [*Beal and Cooper,* 1978; *Gobel,* 1978b], and at least some gelatinosa cells contribute to long ascending tracts [*Giesler* et al., 1978; *Liu,* 1983; *Willis* et al., 1978]. Thus, while it is clear that the cells of the substantia gelatinosa may well serve as local interneurons, in addition at least some of these cells have long axons and may influence neurons in distant parts of the spinal cord and even in the brain. Neurons of the substantia gelatinosa that have long axons appear to develop earlier in the embryo than do the intrinsic neurons [*Beal,* 1983].

Several different putative neurotransmitters have been found in neurons of lamina II (fig. 4/30). Some stalked cells and islet cells contain enkephalin immunoreactivity [*Bennett* et al., 1982; cf. *Aronin* et al., 1981; *Glazer and Basbaum,* 1981; *Hunt* et al., 1981]. Neurons in the outer part of lamina II and others near the lamina II–III border stain immunocytochemically for neurotensin [*Hunt* et al., 1981; *Seybold and Elde,* 1982]. Other cells in lamina II contain substance P, somatostatin, or avian pancreatic polypeptide [*Hunt* et al., 1981; *Tessler* et al., 1981].

At the ultrastructural level, neuronal perikarya in the substantia gelatinosa tend to be small and to have the long axis perpendicular to the lamina [*Ralston,* 1965]. The most characteristic feature of the neuropil is an abundance of unmyelinated axons, mostly running in bundles parallel to the long axis of the spinal cord [*Ralston,* 1965, 1968a, 1979]. Myelinated axons are found medially in the substantia gelatinosa, en route to deeper layers of the gray matter; occasional fine myelinated axons are seen running either vertically or horizontally within lamina II [*Ralston,* 1965]. The border with lamina III is easily recognized because of the presence in this layer of many finely myelinated axons [*Ralston,* 1979].

Gobel et al. [1980] have succeeded in studying several substantia gelatinosa neurons in the spinal cord of the cat at an ultrastructural level after labeling the cells by an intracellular injection of horseradish peroxidase. The label filled the dendrites and at least part of the axon, and so it was

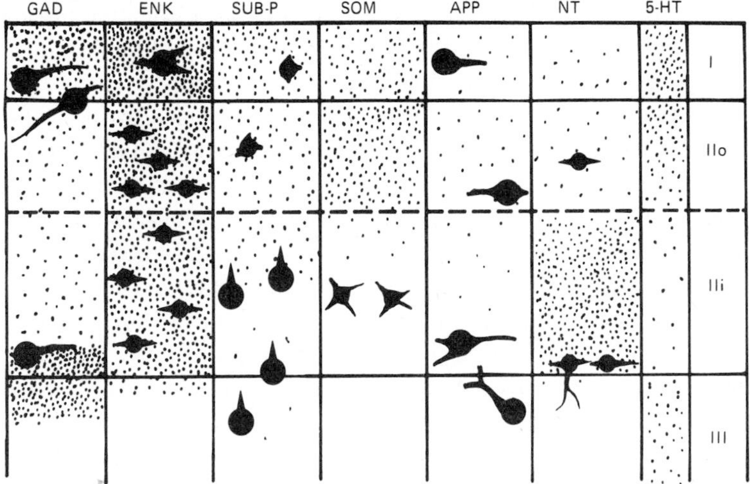

Fig. 4/30. Diagram showing the locations of neurons and of terminal zones that show immunoreactive staining for the following substances: glutamate decarboxylase (GAD); enkephalin (ENK); substance P (SUB-P); somatostatin (SOM); avian pancreatic polypeptide (APP); neurotensin (NT); and 5-hydroxytryptamine (5-HT). [From *Hunt* et al., 1981.]

possible to classify the neurons as 'islet' or 'stalked' cells. Of the two islet cells examined, only a single axosomatic synapse was found on one of them. The islet cell dendrites were clearly participating in the formation of dendrodendritic and dendroaxonic synapses. They contained flattened and oval synaptic vesicles. Dendritic spines entered glomeruli and received a synaptic contact from the central terminal, which can be assumed to originate from a primary afferent fiber. The same dendritic spines also made dendrodendritic synapses with adjacent dendritic processes in the glomeruli. The dendritic spines may also have made dendroaxonic synapses, but they were not postsynaptic to axons other than the primary afferent central endings nor were they postsynaptic to other dendrites. Outside of glomeruli, the dendrites of islet cells synapsed with very fine processes that could be either fine dendrites or unmyelinated axons. *Gobel* et al. [1980] favor the latter explanation. Islet cell dendrites also received axodendritic and dendrodendritic synapses in their course through the neuropil. Some of the axodendritic endings contained flattened vesicles. The axon of the islet cell made synapses on dendritic shafts and spines. The synaptic vesicles were a mixture of round and flattened vesicles. It would be difficult to distinguish between the terminals formed by the axon and those formed by the dendrites of islet cells without tracing the processes serially.

Gobel et al. [1980] also studied a stalked cell labeled by intracellularly injected horseradish peroxidase. It received a few axosomatic synapses. The dendrites did not contain synaptic vesicles, in contrast to the islet cells. Dendritic spines of the stalked cell were found in glomeruli throughout the width of lamina II. Generally, several spines from the stalked cell entered each glomerulus. The stalked cell dendritic spines received an axodendritic synapse from the central terminal of the glomerulus. They also received synapses from axonal endings filled with mixtures of round and flattened vesicles and from dendrites that resembled those of islet cells. The axon of the stalked cell was unmyelinated, and it ascended into lamina I. Several synaptic endings could be observed in lamina I.

Gobel et al. [1980] interpret their electron microscopic observations to be consistent with the view that islet cells are inhibitory interneurons, receiving axonal input from primary afferent fibers and also from some other sources (fig. 4/31). The synaptic output of the dendritic processes and also axonal terminals of the islet cells is to dendrites of other neurons within lamina II, and perhaps also to axonal endings (including unmyelinated axons). Some of the elements presumed to be affected by islet cells include stalked cells and primary afferents. Neurons in laminae III and IV that send dendrites dorsally into lamina II may also be affected. It is suggested that many of the endings that stain immunocytochemically for glutamic acid decarboxylase (and so presumably utilize GABA as a neurotransmitter) may be dendrodendritic and dendroaxonic endings formed by islet cells [cf. *McLaughlin* et al., 1975]. The stalked cells have quite another function, since they do not make dendritic synapses. They have a synaptic input from primary afferents and they synapse within lamina I (fig. 4/31), and so *Gobel* et al. [1980] propose, in keeping with *Gobel's* [1978b] previous suggestion, that these cells are excitatory interneurons.

Extracellular recordings have been made by several groups from neurons thought to be in the substantia gelatinosa [*Cervero* et al., 1977b, 1979b, c; *Dubuisson* et al., 1979; *Hentall,* 1977; *Kumazawa and Perl,* 1976, 1978; *Millar and Armstrong-James,* 1982; *Price* et al., 1979; *Wall* et al., 1979a; *Woolf and Fitzgerald,* 1983]. A variety of approaches have been taken to add confidence to the supposition that the cells were in fact in the substantia gelatinosa. The difficulty, of course, is that extracellular recordings can be made from units whose cell bodies are actually located at some distance

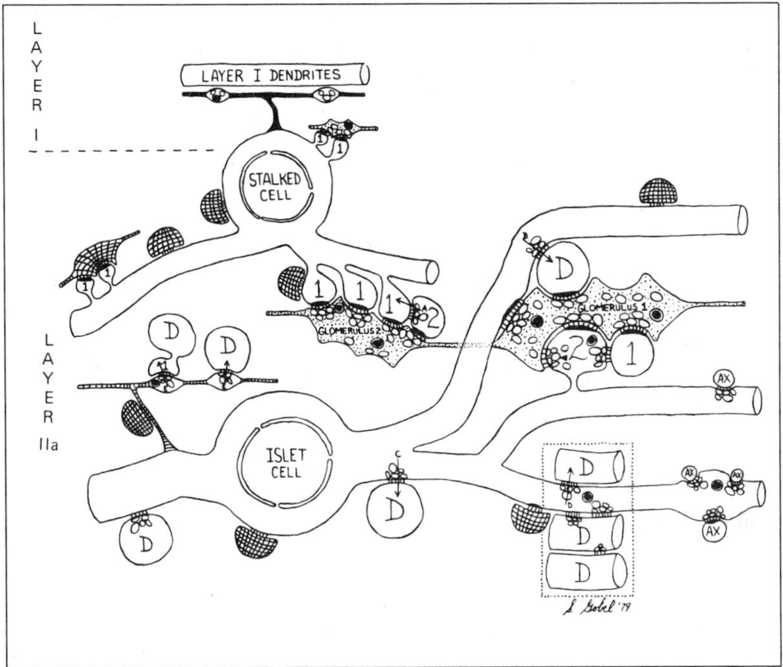

Fig. 4/31. Diagram showing interpretation of synaptic interrelations of an islet cell and a stalked cell in the substantia gelatinosa of the cat spinal cord. Both cell types receive synaptic input from primary afferents in glomeruli. They also receive other inputs (cross-hatched endings). The islet cell makes numerous dendrodendritic synapses both within and outside of glomeruli. It also makes dendroaxonic contacts within and outside of the glomerulus. The axon receives an axoaxonic contact and makes axodendritic synapses. The stalked cell axon projects to lamina I and synapses on dendrites there. [From *Gobel* et al., 1980.]

from the microelectrode tip, and so the site of a dye mark is only an approximation to the location of the recorded neuronal soma. One approach, taken by the Edinburgh group, was to exclude cells that had responses typical of neurons in the adjacent laminae of the dorsal horn [*Cervero* et al., 1977b, 1979b]. However, this tactic may exclude many cells that belong to the substantia gelatinosa and may therefore bias the sample toward cells with unusual properties. Another approach was to use stimulation of the tract of Lissauer to activate neurons in the substantia gelatinosa either antidromically or orthodromically [*Cervero* et al., 1979b, c; *Wall* et al., 1979a]. This technique was based on the assumption that all of the axons in Lissauer's tract arise from cells of the substantia gelatinosa. However, many of the orthodromic responses caused by such stimulation could as well be due to activation of primary afferent fibers as to activation of axons arising from neurons of the substantia gelatinosa [cf. *Chung and Coggeshall,* 1979]. Still another strategy has been to record first from an antidromically activated spinothalamic tract cell in lamina I and then to record the activity of a cell in the subjacent gray matter within a distance that indicates a location in lamina II [*Price* et al., 1979]. Finally, *Millar and Armstrong-James* [1982] restricted their sample to neurons that could be excited by iontophoretic release of glutamate.

These studies of extracellularly recorded activity in the substantia gelatinosa provide a rather diverse picture. On the one hand, there are several reports of quite unusual response types, while on the other hand there are reports that suggest that substantia gelatinosa cells are much like neurons in adjacent laminae. A major limitation of these investigations is the impossibility of correlating structure with function when the recordings are extracellular.

Kumazawa and Perl [1976, 1978] were able to record from units in the marginal zone and in the substantia gelatinosa. They were struck by the observation that the cells could be activated by fine afferent fibers and by specific classes of receptors. Some of the neurons were nociceptive, others thermoreceptive and still others mechanoreceptive. Aδ fibers appeared to contact neurons in the marginal zone, whereas C fibers activated cells in the substantia gelatinosa.

Hentall [1977], using metal microelectrodes, observed units in the substantia gelatinosa that were activated after a delay by gentle mechanical stimulation. Discharge lasted 5–20 s. Repeated stimuli were not consistently effective, and habituation could often be demonstrated. Tactile stimuli produced an inhibition when there was a background discharge. Noxious stimuli could activate the cells, but this depended upon the presence of background activity, which in turn was augmented by prior noxious stimulation.

Cervero et al. [1979b, c] recorded from units in the substantia gelatinosa that were antidromically or orthodromically activated by stimulation of Lissauer's tract and that had responses that were distinct from those of cells in adjacent laminae. The electrodes were fine glass micropipettes, and some of the recordings were extracellular and some intracellular. Most of the neurons could be classified by a system that is inverse to that usually used for other neurons of the dorsal horn. The cells commonly had a background discharge, and they could be inhibited by peripheral stimulation. Some cells were inhibited by innocuous mechanical stimuli but excited by noxious inputs. Others were inhibited by both innocuous and noxious stimuli. Still others were inhibited by noxious stimuli, but excited by innocuous stimulation.

Wall et al. [1979a] recorded from units in the substantia gelatinosa with metal microelectrodes. They also utilized stimulation of Lissauer's tract to activate cells of the substantia gelatinosa antidromically. No distinction was made between cells located in lamina II or III. Many of the cells had restricted receptive fields, and most could be activated by innocuous, by both innocuous and noxious or by noxious stimulation of the skin. Electrical stimulation indicated that none of the cells were exclusively activated by C fibers. Habituation was pronounced for some of the cells. Many cells could be excited for long periods of time, seconds to minutes, by the appropriate stimulus, either brushing or pinching, depending on the cell. These prolonged discharges were eliminated by barbiturate.

Dubuisson et al. [1979] mapped the receptive fields of neurons in the superficial laminae of the dorsal horn (laminae I–III) repeatedly in decerebrate or spinalized animals and found that the receptive fields were unstable, expanding or contracting over a period of time ('ameboid' receptive fields). This instability was not seen in recordings from deeper layers of the dorsal horn, and it was eliminated by barbiturate anesthesia. Similar results were reported by *Woolf and Fitzgerald* [1983].

Price et al. [1979] were interested in examining *Gobel's* hypothesis that stalked cells are excitatory interneurons that may provide a route for Aδ mechanoreceptor input to marginal cells. They recorded from pairs of neurons, the more superficial of a pair usually being a spinothalamic tract cell in lamina I and the deeper being an interneuron in lamina II. The cells of lamina II could be classified by the same system that applies to other neurons of the dorsal horn: low threshold, wide dynamic range (multireceptive) and nociceptive specific neurons. A few cells with unusual responses were observed. For instance, several cells could only be inhibited by peripheral stimulation. Other cells were excited exclusively by C fibers; these cells were nociceptive specific. It was argued from the fact that the cells in lamina II had shorter latencies to peripheral stimuli than did cells in lamina I that the lamina II cells might relay information to lamina I, in keeping with *Gobel's* hypothesis that stalked cells may be excitatory interneurons.

Fitzgerald [1981] recorded the responses of neurons in laminae I and II of the cat spinal cord. She found that the cells could be activated by innocuous mechanical stimuli (10%), both innocuous and noxious stimuli (64%), or just noxious stimuli (26%). The nociceptive-specific cells tended to be more superficial than the other cells. Many of the cells had convergent inputs from A and C fibers. Some cells showed long-lasting after-discharges, whereas others were characterized by habituation. The responses to bradykinin were not inhibited by stimulation of Aβ fibers.

Millar and Armstrong-James [1982] recorded unitary activity from neurons in the superficial layers of the dorsal horn, using carbon fiber microelectrodes. They accepted data only from units that could be excited by iontophoretically released glutamate ions. The recording electrode apparently had to be quite near the cell soma, since small displacements greatly increased the threshold for glutamate responses. The electrode position was usually much more critical for glutamate responses than for spike amplitude. All but 1 of the 62 units could be activated by tactile stimuli. However, most showed an increased discharge when pressure was applied to the skin; 1 unit was proprioceptive. Unfortunately, the locations of the units were not determined.

In a few studies, cells have been demonstrated to belong to the substantia gelatinosa by intracellular marking with horseradish peroxidase or Lucifer yellow [*Bennett* et al., 1979, 1980; *Light* et al., 1979; *Molony* et al., 1981; *Woolf and Fitzgerald,* 1983]. As already discussed, *Light* et al. [1979] were better able to correlate the functional properties of the neurons in their sample with the location(s) of the principal dendritic field(s) of the cells than with the location of the cell body. Neurons with an input chiefly from Aδ mechanical nociceptors had a dendritic ramification in the marginal zone; cells with inputs from C fibers or from thermoreceptors had a major dendritic field in the outer part of lamina II, whereas those with an input from C fibers and sensitive mechanoreceptors had dendrites in the inner part of lamina II; finally, cells with an input from myelinated afferents from sensitive mechanoreceptors were characterized by having important dendritic branches in the inner substantia gelatinosa and extending into the nucleus proprius.

Several reports from *Dubner's* group confirm the functional classes of neurons they found using extracellular recordings in the substantia gelatinosa. Recordings from stalked cells were reported by *Bennett* et al. [1979], and the activity of both stalked cells and islet cells was described by *Bennett* et al. [1980]. The neurons that were located in the outer part of lamina II (stalked cells and some of the islet cells) were either nociceptive-specific (fig. 4/32) or wide dynamic range (multireceptive) in nature (fig. 4/33), whereas

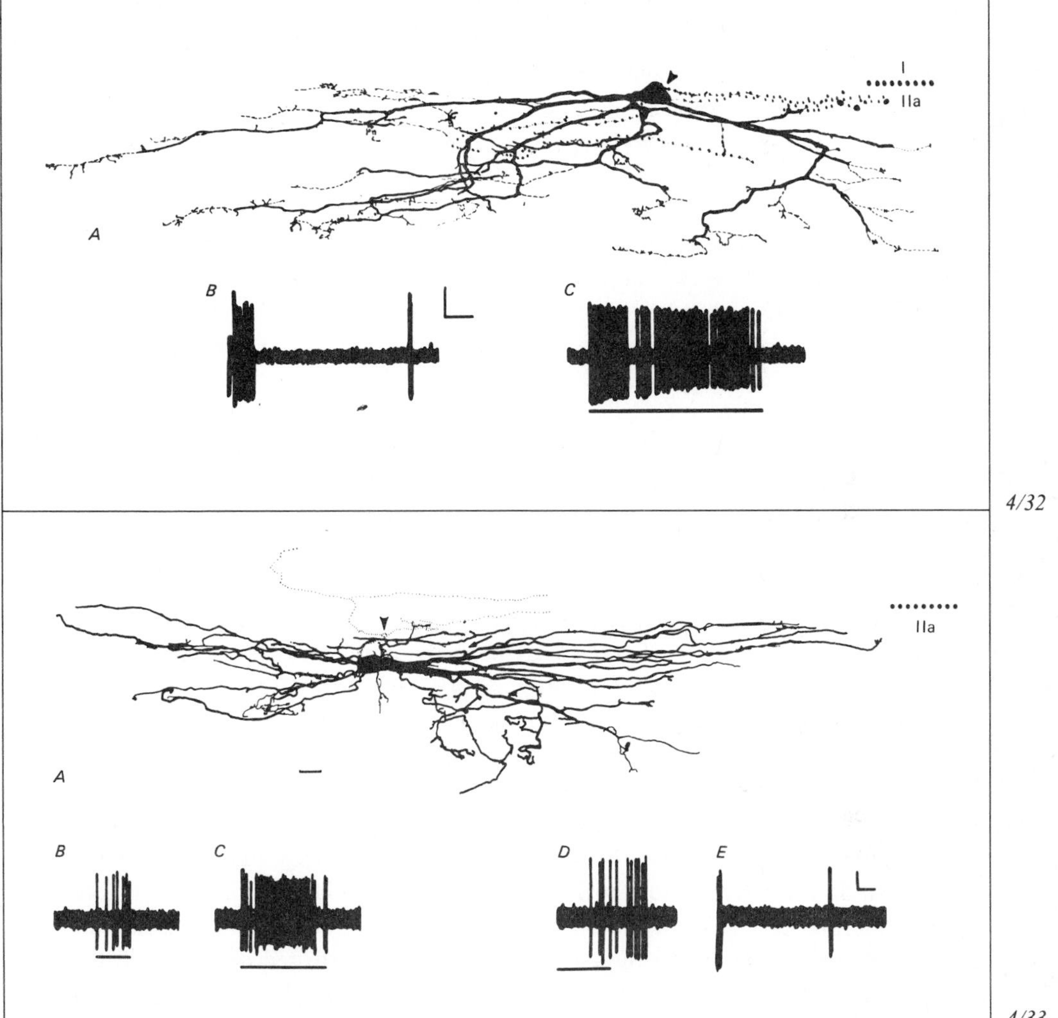

Fig. 4/32. Nociceptive-specific stalked cell. The neuron was labeled with intracellular injection of horseradish peroxidase, as shown in *A*. The responses in *B* are to electrical shocks applied to the receptive field and indicate a convergent input from Aδ and C fibers. The response in *C* is to pinch with toothed forceps. [From *Bennett* et al., 1980a.]

Fig. 4/33. Wide dynamic range or multireceptive islet cell in the outer part of lamina II. The horseradish peroxidase labeled neuron is shown in *A*. The responses are to brushing *(B)*, pinching with toothed forceps *(C)*, noxious heat *(D)*, and an electrical shock applied to the skin *(E)*. The input was from myelinated fibers conducting up to 41 m/s and from C fibers. [From *Bennett* et al., 1980a.]

those found in the inner part of the substantia gelatinosa could be activated only by innocuous stimuli. The nociceptive inputs could be conveyed by Aδ fibers or by both Aδ and C fibers. The tactile inputs could be mediated by Aδ down hair receptors or Aβ guard hair receptors. It was unclear to what extent the Aβ input was direct, for example by way of the 'flame-shaped arbors', and to what extent it may have been relayed through more ventral laminae.

Molony et al. [1981] marked 7 central cells in the cat substantia gelatinosa, after studying their functional properties. All of these cells had responses that could be categorized according to the Edinburgh 'inverse' terminology (see previous discussion). They also marked four nociceptive specific cells that proved to be in lamina I.

Laminae III and IV

As mentioned earlier, lamina III can be distinguished readily from lamina II on the basis of the larger numbers of myelinated axons in lamina III (fig. 4/1B). It is more difficult to separate laminae III and IV, except by the presence of larger neurons in lamina IV [*Rexed,* 1952].

Cajal [1909] illustrated neurons in what would now be considered laminae III and IV of the dorsal horn that projected dendrites dorsally into the substantia gelatinosa (fig. 4/34). These cells were referred to as 'antenna type neurons' by *Réthelyi and Szentágothai* [1973], who thought that they could be identified with the cells of origin of the spinocervical tract. *Brown* and his colleagues have since labeled many spinocervical tract cells intracellularly with procion yellow [*Brown* et al., 1976] or with horseradish peroxidase [*Brown* et al., 1977a], and they find that actually the dorsal dendrites of these neurons tend to deviate horizontally before entering lamina II (fig. 5/35) [*Brown,* 1981]. However, some spinothalamic tract cells in the monkey do send dorsally directed dendrites all the way through lamina II and into lamina I [*Willis* et al., 1979], and so at least a portion of the population of these cells belong to ascending tracts. In addition to the dorsally projecting dendrites, these antenna cells have horizontally running dendrites that are likely to receive a different spectrum of synaptic input than do the dorsal dendrites (fig. 4/34) [*Scheibel and Scheibel,* 1968]. Lamina III also contains many central or islet cells that are similar to the neurons found in lamina II [*Beal and Cooper,* 1978; *Szentágothai,* 1964].

The synaptic architecture and dorsal root projections of lamina IV of the monkey are described by *Ralston* [1982] and *Ralston and Ralston*

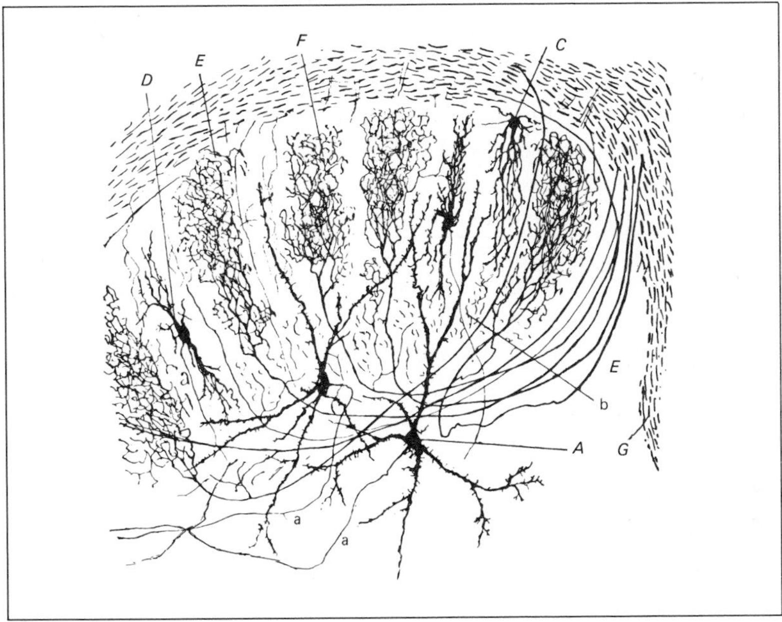

Fig. 4/34. Transverse section of the dorsal horn of the spinal cord of a newborn cat, stained with the Golgi technique. Two large neurons in laminae III and IV send dendritic projections dorsally into the substantia gelatinosa. These cells have been termed 'antenna cells' by *Réthelyi and Szentágothai* [1973]. Also seen is a limiting cell and two central cells, as well as a number of 'flame-shaped arbors'. Note that the dorsal extent of the flame-shaped arbors is less in the adult. [From *Cajal,* 1909.]

[1982]. One conclusion is that C fibers do not appear to project to the deeper layers of the dorsal horn.

Recordings from neurons in laminae III and IV frequently reveal that the neurons respond specifically to innocuous mechanical stimulation of the skin [*Menétrey* et al., 1977; *Pomeranz* et al., 1968; *Price and Browe,* 1973; *Price and Mayer,* 1974; *Wall,* 1967]. Few cells in lamina IV are excited by intra-arterial injection of bradykinin into the limb [*Besson* et al., 1972]. However, some neurons in lamina IV receive an additional noxious input [*Handwerker* et al., 1975; *Price and Browe,* 1973; *Price and Mayer,* 1974] and are thus wide dynamic or multireceptive cells. Details concerning the response properties of cells contributing to particular ascending tracts, including the spinocervical, postsynaptic dorsal column, and spinothalamic tracts, will be discussed in Chapter 5.

Lamina V

This lamina forms the neck of the dorsal horn, and it is characterized by the presence of a reticulated area occupying its lateral margin [*Rexed,* 1952]. Apart from this reticulated zone, which contains longitudinally oriented bundles of axons, the neuropil of lamina V is characterized by a 'verticotransverse' orientation, in contrast to the sagittal arrangement of laminae II–IV [*Scheibel and Scheibel,* 1968]. *Rexed* [1952] divided lamina V into medial and lateral zones (fig. 4/1D). The lateral zone includes the reticulated area and about a third of the gray matter of the lamina. There are a number of transversely oriented fibers in the lateral zone, and the cells tend to be larger than in the medial zone.

The dorsal root projections and synaptic arrangements in lamina V of the monkey are described by *Ralston* [1982] and by *Ralston and Ralston* [1982]. Cells in lamina V contribute to several ascending tracts, including the spinothalamic and spinoreticular tracts (see Chapter 5).

Recordings from neurons in lamina V show that the responses typical of cells in this layer are of the wide dynamic range or multireceptive type [*Handwerker* et al., 1975; *Hillman and Wall,* 1969; *Menétrey* et al., 1977; *Pomeranz* et al., 1968; *Price and Browe,* 1973; *Price and Mayer,* 1974]. However, some of the neurons in this layer can have other response properties, including tactile or nociceptive-specific responses [*Menétrey* et al., 1977; *Price and Mayer,* 1974]. Many of the cells in lamina V respond to the intra-arterial injection of bradykinin into the hindlimb [*Besson* et al., 1972].

Lamina VI

According to *Rexed* [1954], lamina VI exists only at certain levels of the spinal cord. It is present in the upper cervical segments and in the cervical and lumbosacral enlargements, but not in most of the thoracic cord or in caudal segments. Cells in the medial third are more tightly packed, and so this zone can be distinguished from a lateral zone (fig. 4/1D).

Many neurons in lamina VI respond to proprioceptive stimuli [*Menétrey* et al., 1977; *Pomeranz* et al., 1968; *Wall,* 1967]. However, other cells have been reported to be of the tactile, wide dynamic range (multireceptive) or nociceptive-specific type [*Menétrey* et al., 1977; *Price and Browe,* 1973; *Price and Mayer,* 1974].

Intermediate Region and Ventral Horn

Details about the arrangement of the spinal cord gray matter in laminae VII–X can be found in the papers of *Rexed* [1952, 1954]. Although much of the emphasis of workers investigating nociceptive transmission in the spinal cord has been on cells of the dorsal horn, it should be noted that there are cells in the intermediate region and ventral horn that contribute to ascending tracts presumed to be important in nociception, such as the spinothalamic and spinoreticular tracts (see Chapter 5), and recordings have been made of responses of cells in this part of the cord that suggest a nociceptive function [*Menétrey* et al., 1977; *Meyers and Snow,* 1982a; *Molinari,* 1982; *Price and Browe,* 1973; *Price and Mayer,* 1974]. As assessment of the contribution of these more ventrally placed neurons to pain mechanisms is needed.

Lamina X, the area immediately surrounding the central canal, may be particularly concerned in nociception. In a recent study, *Nahin* et al. [1983] have recorded from cells in this area of the rat spinal cord and have found the cells to be uniformly nociceptive-specific and to have small receptive fields [cf. *Honda and Perl,* 1981]. They also were able to label many of the cells in this area by retrograde transport of horseradish peroxidase from the brain stem reticular formation. This region resembles the superficial dorsal horn in its peptide content [*Gibson* et al., 1981].

Responses of Dorsal Horn Interneurons to Noxious Inputs Other than from Skin

The response properties that have been described for dorsal horn interneurons in the previous section have been to input from cutaneous receptors. However, of similar interest are responses to activation of nociceptors in other tissues, such as muscle, joints and viscera. Unfortunately, less is known about the responses to receptors in these structures, especially responses to nociceptors.

Pomeranz et al. [1968] described the excitation of dorsal horn neurons following stimulation of fine muscle afferent fibers and also visceral afferents in the splanchnic nerve. The cells that were responsive to these inputs were also classified as wide dynamic range (multireceptive) cells with respect to cutaneous stimulation. The cells were in lamina V.

Fine muscle afferents have been shown by *Kniffki* et al. [1977] to activate some spinocervical tract cells and by *Foreman* et al. [1977, 1979a, b] to excite many spinothalamic tract cells. As for the interneurons described by *Pomeranz* et al. [1968], these tract cells could be classified as being of the 'wide dynamic range' (multireceptive) type.

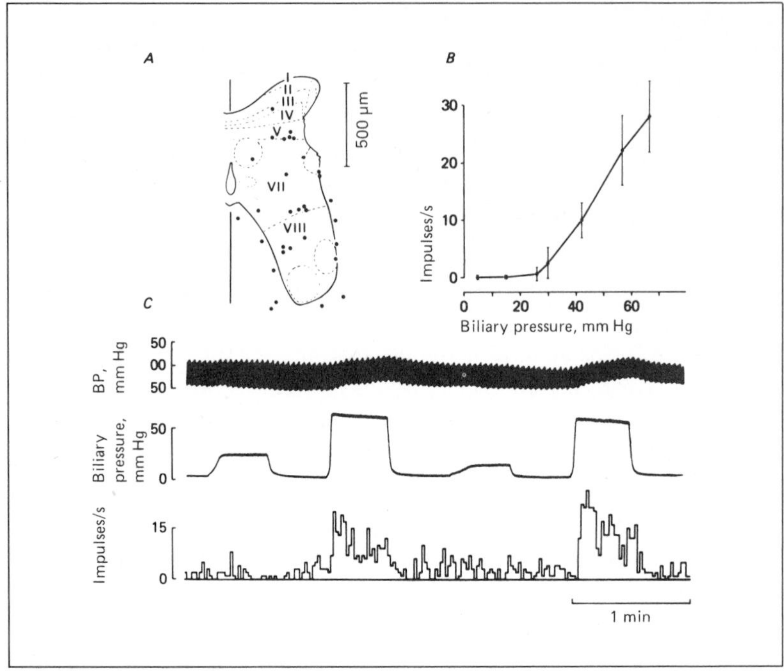

Fig. 4/35. A shows the recording sites of neurons in the T9 segment that could be activated by distension of the biliary system in the cat. The graph in *B* shows the stimulus-response curve for one of the units. Threshold was in the noxious range. The activity of another unit is shown in *C,* bottom trace, along with blood pressure and biliary pressure. The unit discharged only when the biliary distension was great enough to cause an increase in blood pressure. [From *Cervero,* 1983a.]

Several groups have examined the effects of visceral afferents on dorsal horn interneurons [*Ammons and Foreman,* 1984; *Cervero,* 1982b, 1983a, c; *Cervero and Iggo,* 1978; *Fields* et al., 1970a, b; *Foreman,* 1977; *Gokin* et al., 1977; *Guilbaud* et al., 1977a; *Hancock* et al., 1973; *McMahon and Morrison,* 1982a, b; *Selzer and Spencer,* 1969; *Takahashi and Yokota,* 1983] and on spinothalamic neurons [*Blair* et al., 1981; *Foreman* et al., 1981; *Foreman and Weber,* 1980; *Hancock* et al., 1975; *Milne* et al., 1981; *Rucker and Holloway,* 1982]. The cells receiving visceral input are concentrated in laminae I and V–VIII [*Cervero,* 1983c]. However, some are in lamina IV as well [*Ammons and Foreman,* 1984]. The visceral afferent inputs were either evoked by stimulating the splanchnic nerve or by some form of natural-like stimulation. An example of the latter is occlusion of a coronary artery while recording from a spinal neuron responsive to receptors in the heart [*Foreman and Ohata,* 1980].

A very favorable system is that being developed by *Cervero* [1982b, 1983a] in which afferents to the biliary system are stimulated by controlled distension. It is feasible to distinguish between innocuous and noxious stimuli by referring stimulus intensity to the appearance of pseudoaffective reactions in the animal. In a sample of neurons in the thoracic cord receiving a convergent input from afferents in the splanchnic nerve and others supplying the skin, *Cervero* [1983a] found that most of the cells were of the wide dynamic range or multireceptive type or were high threshold. Many could be activated antidromically from the contralateral ventrolateral funiculus. Some of the cells could be excited by distension of the biliary system (fig. 4/35), provided that the distension reached a noxious intensity (defined as causing a blood pressure increase).

To date, there have been few cases observed of spinal cord neurons that respond selectively to visceral stimulation [*Cervero,* 1983b, c; *Gokin* et al., 1977]. In most instances, neurons that have a visceral input also have cutaneous receptive fields. This observation is of relevance to the problem of referred pain. It has further been suggested that allodynia (pain due to innocuous stimulation of normal skin) in such conditions as angina is explainable if visceral nociceptive input makes wide dynamic range (multireceptive) central neurons more responsive to tactile stimuli [*Takahashi and Yokota,* 1983].

Pharmacological Responses of Nociceptive Dorsal Horn Neurons

Neurons that are activated either selectively or preferentially by nociceptors have been studied by a number of groups to see if they respond to the iontophoretic application of neurotransmitter substances that are contained in primary afferent fibers, interneurons of the dorsal horn, or axons descending from the brain. For example, nociceptive cells can be excited by substance P [*Henry,* 1976; *Henry* et al., 1975, 1980; *Randić and Miletić,* 1977; *Sastry,* 1979; *Willcockson* et al., 1984b; *Zieglgänsberger and Tulloch,* 1979a] and by glutamate [*Henry,* 1976; *Henry* et al., 1980; *Jordan* et al., 1978; *Willcockson* et al., 1984a, b; *Zieglgänsberger and Tulloch,* 1979a], amongst the putative primary afferent transmitters. Interestingly, some nociceptive cells can be inhibited by substance P when this agent is applied at certain locations near the cell [*Willcockson* et al., 1984b; see also *Davies and Dray,* 1980], and somatostatin appears to be inhibitory [*Randić and Miletić,* 1978]. Several amino acids (including glycine and GABA), amines (such as serotonin, norepinephrine, and dopamine), and peptides (enkephalin, cholecystokinin) have been reported to have inhibitory effects on nociceptive cells [*Duggan* et al., 1977b, 1979, 1981; *Jordan* et al., 1978; *Randić and Miletić,* 1978; *Randić and Yu,* 1976; *Willcockson* et al., 1984a, b; *Zieglgänsberger and Tulloch,* 1979b] as has morphine [*Belcher and Ryall,* 1978; *Calvillo* et al., 1974; *Dostrovsky and Pomeranz,* 1976; *Duggan* et al., 1977a, 1981; *Zieglgänsberger and Bayerl,* 1976]. On the other hand, there are reports that cholecystokinin and also neurotensin are excitatory when applied to neurons in the dorsal horn [*Jeftinija* et al., 1981; *Miletić and Randić,* 1979; *Stanzione and Zieglgänsberger,* 1983].

Conclusions

1. The discovery that dorsal roots have a sensory function and ventral roots a motor function can be attributed to *Bell* and *Magendie.*

2. *Magendie* also noticed that stimulation of the ventral root causes pain that can be abolished by sectioning the dorsal root. A number of explanations have been developed for this 'recurrent sensibility', including the presence in the ventral root of afferent fibers.

3. As sensory fibers approach the spinal cord through a dorsal root, the fine afferent fibers tend to become segregated in the lateral aspect of the root, entering Lissauer's tract from there (at least in some species). It has been reported that lesions of the lateral aspect of the dorsal root can prevent pain in animals, but an alternative interpretation is that such a procedure interrupts the blood supply to the dorsal horn.

4. Lissauer's tract contains many primary afferent fibers, as well as projections of intrinsic neurons of the dorsal horn.

5. The dorsal horn can be described in terms of a series of laminations, based on its cytoarchitecture. The uppermost layer is lamina I. Lamina II is synonymous with the substantia gelatinosa. The nucleus proprius consists of laminae III to V. The base of the dorsal horn is lamina VI.

6. Nociceptive afferent fibers generally enter the spinal cord through the dorsal roots. It is usually thought that each dorsal root ganglion cell gives rise to a single peripheral and a single central process. This idea is based on light microscopic counts of the numbers of axons in the dorsal

roots and of the numbers of cells in the dorsal root ganglia. However, recent counts done with the electron microscope indicate that there are more peripheral and central processes than dorsal root ganglion cells, indicating that at least some ganglion cell axons branch within the spinal nerve and dorsal root.

7. The unmyelinated afferent fiber population exceeds the myelinated afferent fiber population by a ratio of about 2:1.

8. The ventral roots contain a sizable population of afferent fibers, most of which are unmyelinated. These originate from dorsal root ganglion cells. Many of the ventral root afferent fibers are likely to loop back into the dorsal root, since only some appear to enter the spinal cord directly through the ventral root. A few of these fibers may supply receptor terminals in the meninges. However, many have peripheral receptive fields that are often nociceptive.

9. The segregation of fine afferent fibers within the dorsal root does not occur in the cat, but it does in the monkey and in man. This species difference may have led to some of the controversies concerning occurrence of this segregation.

10. The abundance of primary afferent fibers in Lissauer's tract has been a subject of controversy. Some investigators have observed only a small number of primary afferent fibers entering Lissauer's tract. However, recent evidence from electron microscopic counts indicates that a substantial fraction of the fibers of Lissauer's tract are primary afferents. The numbers of primary afferent fibers are greater in the rat and the monkey than in the cat, however.

11. Primary afferent fibers are found in both the lateral and the medial parts of Lissauer's tract. In the cat, there are more primary afferent fibers medially than laterally.

12. Monkeys have bundles of as many as 10,000 unmyelinated axons in Lissauer's tract. It is not known if these play a special functional role in primates.

13. Lesions of Lissauer's tract produce degeneration of axon endings in lamina I and the outer part of lamina II.

14. Primary afferent fibers enter the substantia gelatinosa from its dorsal aspect. These include fine afferent fibers that originate from nociceptors, as well as those from thermoreceptors and certain mechanoreceptors. Other larger primary afferent fibers approach the ventral aspect of the substantia gelatinosa. These are the terminals of the 'flame-shaped' arbors that are formed by the endings of hair follicle afferents.

15. There are divergent views about the destinations within the superficial layers of the dorsal horn of primary afferent fibers with finely myelinated axons or with unmyelinated axons. One opinion is that C fibers project to the marginal layer and that Aδ fibers project to the substantia gelatinosa. Another opinion is that Aδ fibers terminate in the marginal zone and C fibers in the substantia gelatinosa.

16. Axons within the superficial part of the marginal zone are oriented in transverse sheets, as are many of the dendrites of neurons in this layer. Axons in the deeper part of lamina I and also in the substantia gelatinosa are oriented longitudinally, as are the dendrites of the intrinsic neurons.

17. Although studies with degeneration techniques have often failed to disclose many of the primary afferent terminals within the substantia gelatinosa, it is now clear that there is an abundance of these and that they can be demonstrated with suitable techniques, such as by anterograde transport of tritiated amino acids.

18. Horseradish peroxidase can be used to demonstrate the projections of different sized primary afferent fibers after anterograde transport through a dorsal root filament. The large afferent fibers can be interrupted by lesion of the medial aspect of the rootlet, and when this is done the terminals are in the superficial layers of the dorsal horn, as well as some in the ventral nucleus proprius.

19. Intracellular injection of horseradish peroxidase into Aδ fibers identified by their functional properties has revealed that Aδ mechanical nociceptors project to the marginal layer and to the ventral nucleus proprius.

20. Several types of synaptic terminals are made by primary afferent fibers in the dorsal horn. The most common general class of ending is called the 'central terminal', from its relationship to other processes in the complex synaptic arrays called 'glomeruli'.

21. Many of the central terminals in lamina I and the outer part of lamina II contain large dense core vesicles. When these terminals degenerate, the type of degeneration is often of the electron lucent variety. The large dense core vesicles may contain peptides, such as substance P.

22. A second kind of central terminal tends to have a dark matrix and has been called a 'dense sinusoid ending'. These terminals are found in the substantia gelatinosa, and they undergo an electron dense type of degeneration. It has been suggested that these endings may contain fluoride-resistant acid phosphatase (and perhaps use a nucleotide like ATP as a transmitter).

23. The third type of ending is found in the inner part of lamina II and also in the deeper layers of the dorsal horn. This ending undergoes a neurofilamentous type of degeneration when the dorsal root is cut.

24. Axoaxonic, dendrodendritic and dendroaxonic synapses are found in the superficial layers of the dorsal horn, in addition to abundant axodendritic synapses. Presynaptic dendrites often contain flattened vesicles, and they may form inhibitory connections.

25. Large dorsal root ganglion cells are thought to use an amino acid transmitter, such as glutamate or aspartate.

26. Some small dorsal root ganglion cells are likely to use substance P as a neurotransmitter. Substance P-containing axons in Lissauer's tract are often unmyelinated, but some myelinated axons contain substance P as well.

27. Substance P can be depleted from the terminals of primary afferent fibers in lamina I and the outer part of lamina II by dorsal rhizotomy, peripheral nerve section or capsaicin administration. Substance P content recovers after dorsal rhizotomy, presumably because of sprouting of intrinsic or projection neurons that contain substance P.

28. Other peptides in small dorsal root ganglion cells and fine afferents include somatostatin, a cholecystokinin-like substance, and vasoactive intestinal polypeptide. Other primary afferent neurons contain fluoride-resistant acid phosphatase and may use ATP as a transmitter.

29. A clear identification of particular neurotransmitters with particular classes of receptors is not yet possible.

30. Primary afferent fibers may have membrane receptors for opiates. If so, this suggests that part of the action of opiates could be presynaptic.

31. Lamina I contains large 'Waldeyer' cells and also smaller neurons. Several cell types can be recognized in Golgi-stained preparations. In the caudal spinal trigeminal nucleus, the dendrites of marginal cells tend to be restricted to lamina I. However, in the spinal cord, dendrites from marginal cells often enter Lissauer's tract or project ventrally into the substantia gelatinosa. The issue related to these observations is whether marginal cells necessarily receive all of their input from afferent fibers that terminate in the marginal layer, or whether dendritic contacts are possible in lamina II or in Lissauer's tract.

32. Many or all of the neurons in the marginal layer are projection neurons. Some contribute to long ascending tracts, whereas others are propriospinal cells.

33. Some marginal cells contain enkephalin-like immunoreactivity.

34. Many neurons in the marginal layer respond specifically to noxious stimuli. Others are activated by thermal stimuli or by both innocuous and noxious stimuli. The latter neurons are often called 'wide dynamic range cells', although 'multireceptive' might be a better term.

35. Intracellular injections of horseradish peroxidase make it possible to relate morphology to the functional properties of neurons. One study showed that the responses of marginal neurons depended more on the location of the major dendritic ramifications than on the location of the cell body. Cells with nociceptive inputs from myelinated nociceptive afferents had major dendrites in the marginal zone and from C fibers in the outer part of lamina II; cells with nociceptive inputs from both myelinated and unmyelinated afferents had dendritic processes in both layers.

36. Some marginal neurons could be classified as nociceptive specific, whereas others were of the wide dynamic range (multireceptive) type, according to another study.

37. *Cajal* described two major cell types in the substantia gelatinosa: limiting cells and central cells. Synonyms for these are, respectively, stalked cells and islet cells. Stalked cells are located in the outer part of the layer and send their axons into lamina I. Islet cells are found throughout the lamina, and their axons ramify within lamina II.

38. The neurons of the substantia gelatinosa have a predominantly longitudinal orientation, as do many of the axons in this layer.

39. It has been proposed that stalked cells are excitatory interneurons that relay input from small myelinated nociceptors to marginal cells.

40. It has further been proposed that islet cells are inhibitory interneurons.

41. The axons of some neurons in the substantia gelatinosa project to other regions of the nervous system, rather than or in addition to back to the substantia gelatinosa. The substantia gelatinosa should not be regarded as a closed system, as has been suggested in the past.

42. Some neurons in the substantia gelatinosa contain one or another of the following putative neurotransmitters: GABA; enkephalin; neurotensin; substance P; somatostatin; avian pancreatic polypeptide.

43. Islet cell dendrites form synaptic connections with other dendrites and also with axonal endings. They receive synaptic contacts from primary afferent fibers in glomeruli; they also receive other synaptic contacts outside of glomeruli. Islet cell axons make synaptic contacts that are similar to those made by their dendrites. The vesicles in these cells are often flattened.

44. Stalked cell dendrites do not form synaptic connections on other elements. They receive synaptic contacts from central axon terminals in glomeruli and from presynaptic dendrites, as well as from other axons besides those forming glomeruli. The axons of stalked cells enter lamina I and synapse on the dendrites of marginal cells.

45. Recordings have been made in recent years from neurons of the substantia gelatinosa. Some investigators have found that these cells resemble those of neighboring laminae in their response properties. Other investigators restrict their observations to neurons having unusual responses. Some of the cells in the substantia gelatinosa show prolonged discharges when their receptive fields are stimulated. Others show marked habituation. The receptive fields may be variable or 'ameboid'.

46. One terminology that has developed for describing the responses of substantia gelatinosa neurons has been referred to as the 'inverse' classification. Cells are described by the inputs that inhibit them: innocuous stimuli only; innocuous and noxious stimuli; noxious stimuli only.

47. Some investigators have found that most of the cells of the substantia gelatinosa can be described by the ordinary system of classification by their excitatory responses to innocuous stimuli only, to innocuous and noxious stimuli, or to noxious stimuli only. However, some unusual cells were also encountered by these investigators.

48. Intracellular labeling with horseradish peroxidase has been done on a limited sample of neurons of the substantia gelatinosa. The response properties of these cells were similar to those observed with extracellular recordings.

49. Neurons in laminae III and IV sometimes give off dorsally projecting dendrites, leading to a description of these cells as 'antenna' cells. Other cells in lamina III are typical central or islet cells.

50. Many of the neurons in laminae III and IV respond specifically to innocuous stimuli, although some have an additional noxious input.

51. Neurons in lamina V are often classified as 'wide dynamic range' or multireceptive neurons. However, there are tactile and nociceptive specific cells in this lamina as well.

52. Neurons in lamina VI often respond best to proprioceptive stimuli. However, cells of other response categories can also be demonstrated, including nociceptive-specific and wide dynamic range cells.

53. Nociceptive neurons can also be found in the intermediate region and ventral horn. Their role in nociception is as yet unclear. Several lines of evidence suggest a relationship of cells in lamina X to nociception.

54. The responses of dorsal horn neurons to input from nociceptors in muscle, joint and viscera are so far incompletely characterized. Many of the neurons that have been observed to have inputs from fine muscle afferent fibers or from visceral afferents appear to be of the wide dynamic range (multireceptive) type with respect to cutaneous input. Only rarely have neurons been found that have a visceral input but not a cutaneous input. The common occurrence of convergent inputs may be important for an understanding of referred pain.

55. A number of pharmacological agents have been observed to have excitatory or inhibitory actions on nociceptive dorsal horn neurons. Excitatory actions have been observed for glutamate, substance P, cholecystokinin, and neurotensin, and inhibitory effects for glycine, GABA, serotonin, norepinephrine, dopamine, enkephalin, substance P, somatostatin and cholecystokinin.

Chapter 5

Ascending Nociceptive Tracts

Historical Overview

As discussed in chapter 2, the most important pathways in the human spinal cord transmitting nociceptive information to the brain ascend in the anterolateral white matter on the side contralateral to the source of the noxious input. This is shown by several lines of evidence: (1) interruption of the anterolateral quadrant by cordotomy or by a disease process causes an analgesia on the side of the body contralateral to the lesion [*Brown-Sequard,* 1860; *Foerster and Gagel,* 1932; *Head and Thompson,* 1906; *Spiller,* 1905; *Spiller and Martin,* 1912; *White and Sweet,* 1955, 1969]; (2) interruption of all of the ascending tracts of the spinal cord except those ascending in one anterolateral quadrant does not prevent the transmission of pain information, but rather permits painful stimuli to be felt on the contralateral side of the body below the lesion (and a poorly localized unpleasant feeling on the ipsilateral side) [*Noordenbos and Wall,* 1976]; (3) stimulation of axons in the anterolateral quadrant with a sufficient stimulus strength and frequency elicits pain [*Mayer* et al., 1975].

The ascending pathways known to exist in the anterolateral white matter include the spinothalamic tract, the spinoreticular tract and the spinomesencephalic tract. The first evidence of a direct projection from the spinal cord to the thalamus is attributed to *Edinger* [1889], based on a study of myelogenesis in kittens. The first experimental evidence for the pathway was provided by *Mott* [1895], who followed the degeneration that resulted from a lesion of the anterolateral quadrant in the monkey to the thalamus, using the Marchi stain. In reports of cordotomies done for the relief of pain, the statement is often made that the target of the surgical intervention was the spinothalamic tract. However, there is no way to know from the results of such lesions whether interruption of the spinothalamic tract produces the most significant deficit or if interruption of the spinoreticular or spinomesencephalic tracts is of more consequence. In fact, it would be a reasonable hypothesis that cordotomy is effective because of interruption of all three pathways.

Although some information is available concerning the anatomy of the human spinothalamic tract, this information is incomplete, and there is no direct evidence concerning the functional properties of human spinothalamic tract neurons. Thus, evidence is needed indicating that human spinothalamic tract neurons are involved in pain. It is known that the information transmitted through the anterolateral quadrant contributes to temperature and touch [*Foerster and Gagel,* 1932; *Kuru,* 1949; *Noordenbos and Wall,* 1976; *White and Sweet,* 1955], as well as pain sensation, and so without direct electrophysiological evidence, it would be speculation to attribute any particular sensory function to spinothalamic tract neurons. Until suitable techniques are developed to allow a determination of the functional properties of human spinothalamic tract neurons, it will be necessary to rely on data obtained from animal experiments.

The animal model for investigating an ascending nociceptive transmission system that is most comparable to man is the monkey. Monkeys develop an analgesia on the side contralateral to an anterolateral cordotomy similar to that produced by cordotomy in man [*Kennard,* 1954; *Poirier and Bertrand,* 1955; *Vierck and Luck,* 1979; *Yoss,* 1953]. As is often the case in man [*White and Sweet,* 1969], the analgesia in monkeys decreases over time [*Kennard,* 1954; *Vierck and Luck,* 1979].

There is ample evidence from experiments on monkeys [*Chung* et al., 1979; *Kenshalo* et al., 1979; *Price* et al., 1978; *Willis* et al., 1974], cats [*Craig and Kniffki,* 1982; *Hancock* et al., 1975; *McCreery and Bloedel,* 1975, 1976; *Meyers and Snow,* 1982a] and rats [*Giesler* et al., 1976] to indicate that a large proportion of spinothalamic tract cells do respond to noxious stimuli, either selectively or preferentially. However, spinothalamic tract cells often respond to tactile stimuli in addition to noxious stimuli [*Chung* et al., 1979; *Price* et al., 1978; *Willis* et al., 1974]. It is not known if these tactile responses contribute to a sensation of touch. A few spinothalamic cells respond selectively to tactile stimuli, and these could well contribute to touch. *Craig and Kniffki* [1982] report that at least one of their sample of spinothalamic tract cells that projected to the medial thalamus in the cat responded to innocuous cooling of the skin. This observation is consistent with a thermoreceptive function for some spinothalamic tract cells. It is interesting that studies of spinothalamic tract cells projecting to the ventral posterior lateral nucleus of the thalamus in the monkey have so far not revealed any thermoreceptive neurons although most of the cells are activated by noxious heat [*Chung* et al., 1979; *Kenshalo* et al., 1979; *Price* et al., 1978; *Willis* et al., 1974]. The absence of thermoreceptive cells in the population of spinothalamic tract cells projecting to the lateral thalamus may reflect a sampling error (e.g., thermoreceptive spinothalamic tract cells could be smaller than the cells generally studied and so may have been overlooked), or this may be an indication that such cells either do not exist and another tract is the thermoreceptive pathway, or that they project to other areas of the thalamus. The observation of *Craig and Kniffki* [1982] suggests the possibility that thermoreceptive spinothalamic tract cells have a target in thalamic nuclei more medial than the ventrobasal complex.

There is also evidence from animal experiments that neurons contributing to ascending tracts in the anterolateral quadrant other than the spinothalamic tract may also be nociceptive. For example, neurons contributing to the spinoreticular and spinomesencephalic tracts are often responsive to noxious stimuli [*Fields* et al., 1975, 1977; *Haber* et al., 1982; *Maunz* et al., 1978; *Menétrey* et al., 1980].

In many cases of cordotomies in human subjects, pain returns after a variable interval of months to years [*White and Sweet,* 1969]. Similarly, the responses of monkeys to painful stimulation resume weeks to months after cordotomy has reduced them [*Kennard,* 1954; *Vierck and Luck*, 1979]. Alternative nociceptive pathways include spinothalamic and spinoreticular projections that ascend ipsilaterally to their source of input [cf. *Kevetter* et al., 1982; *Willis* et al., 1979]. However, there may be additional nociceptive pathways in the dorsal half of the spinal cord. In cats, for example, it is known from behavioral experiments that such pathways exist [*Casey* et al., 1981; *Kennard,* 1954; cf. *Ranson and von Hess,* 1915]. Candidate pathways for nociception in the dorsal part of the cord include the spinocervical tract [*Cervero* et al., 1977a; *Kniffki* et al., 1977] and the postsynaptic dorsal column pathway [*Angaut-Petit,* 1975; *Uddenberg,* 1968b]. Alternatively, propriospinal fibers may be able to transmit nociceptive information to the brain by a polysynaptic route [*Basbaum,* 1973; *Breazile and Kitchell,* 1968].

Spinothalamic Tract

Cells of Origin

The locations of the cells of origin of the spinothalamic tract have been investigated in several species, using three different techniques: retrograde chromatolysis, antidromic mapping, and retrograde labeling with substances like horseradish peroxidase.

The human spinothalamic tract obviously cannot be studied using an experimental technique. Thus, the only evidence about the cells of origin of the human spinothalamic tract comes from descriptions of the distribution of cells that show retrograde chromatolysis after lesions, such as those that are produced by cordotomy [*Foerster and Gagel,* 1932; *Kuru,* 1949; *Smith,* 1976]. Chromatolysis has also been studied in the monkey spinal cord following cordotomy [*Kerr,* 1975a; *Morin* et al., 1951]. However, this evidence is difficult to interpret, since cordotomy will cause retrograde chromatolysis not only in the cells of origin of the spinothalamic tract, but also in cells giving rise to other tracts ascending in the anterolateral quadrant. Thus, chromatolytic cells from spinal cords of humans or monkeys who had had recent cordotomies include cells giving rise to the spinothalamic, spinomesencephalic, spinoreticular, ventral spinocerebellar, ventral spino-olivary and other tracts. Nevertheless, examination of a number of cases of cordotomy by the retrograde chromatolysis technique allowed *Kuru* [1949] to suggest a correlation between the extent of analgesia produced by the cordotomy and the presence of chromatolysis in cells of the marginal zone. He further interpreted the function of cells in the neck of the dorsal horn to be tactile.

Specificity has been attained in animal experiments using an electrophysiological approach (antidromic activation) and an anatomical one (retrograde transport of horseradish peroxidase). The antidromic activation technique was first used by *Dilly* et al. [1968] in cats and rats. Later experiments using this approach were reported by *Craig and Kniffki* [1982], *Holloway* et al. [1978], *Levante and Albe-Fessard* [1972], *McCreery and Bloedel* [1975], *Meyers and Snow* [1982b] and *Trevino* et al. [1972] in cats, by *Albe-Fessard* et al. [1974], *Price* et al. [1978] and *Trevino* et al. [1973] in monkeys, and by *Giesler* et al. [1976] in the rat. *Levante and Albe-Fessard* [1972] were able to activate only a few spinothalamic tract cells antidromically in the cat, although they were able to detect spinoreticular neurons by this approach. The distributions of spinothalamic tract cells as determined from antidromic mapping are shown in figure 5/1A–C for the rat, cat and monkey lumbosacral enlargements.

The horseradish peroxidase technique has a major advantage over the antidromic mapping technique, since it may be less susceptible to sampling bias. Antidromic mapping is likely to emphasize the detection of larger neurons, as opposed to smaller neurons, since the sizes of extracellularly recorded action potentials depend upon cell size, whereas retrograde transport of horseradish peroxidase is presumably not so dependent on cell size (although a difference in the numbers of labeled spinothalamic tract cells

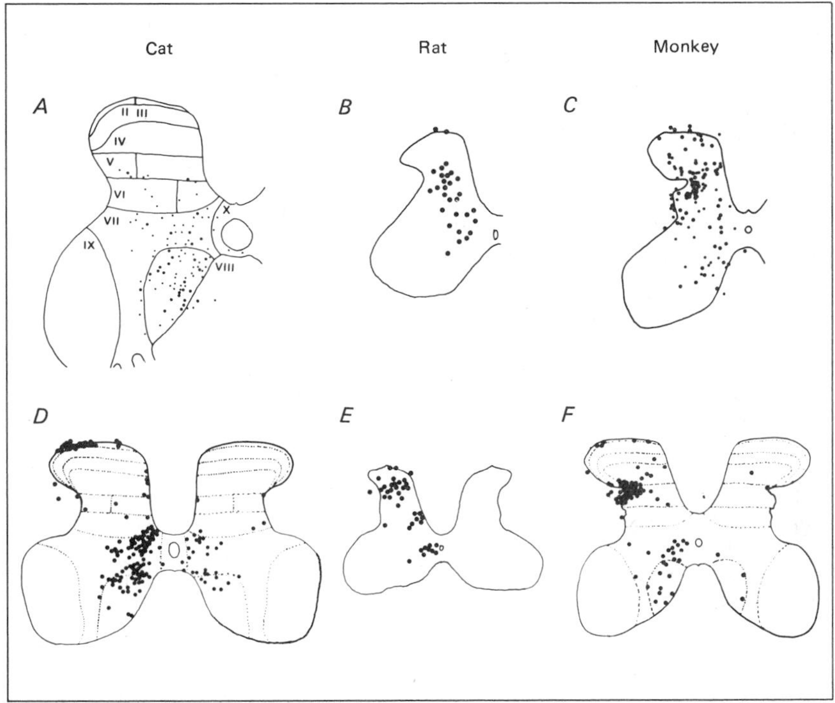

Fig. 5/1. A–C show the recording sites from which antidromic action potentials could be recorded from spinothalamic tract cells in the lumbosacral enlargement of the cat, rat and monkey following stimulation in the contralateral thalamus. [*A* is from *Trevino* et al., 1972; *B* from *Giesler* et al., 1976; *C* from *Trevino* et al., 1973.] *D–F* show the location of spinothalamic tract cells labeled retrogradely following injection of horseradish peroxidase into the thalamus in cat, rat and monkey. [*D* and *F* from *Trevino and Carstens,* 1975; *E* from *Giesler* et al., 1979a.]

has been reported using different chromogens by *Willis* et al. [1979]). Thus, the maps of the distributions of spinothalamic tract cells using horseradish peroxidase, shown in figure 5/1D–F, are probably near the true picture of the source of this pathway.

Figure 5/1A–F shows the distributions of the cells of origin of the spinothalamic tract in the rat, cat and monkey. In all of these species, there are spinothalamic tract cells in the marginal zone, the deeper layers of the dorsal horn, the intermediate region and the ventral horn. However, there is a striking species difference, in that there are fewer spinothalamic tract cells in the nucleus proprius region of the cat lumbosacral enlargement than in

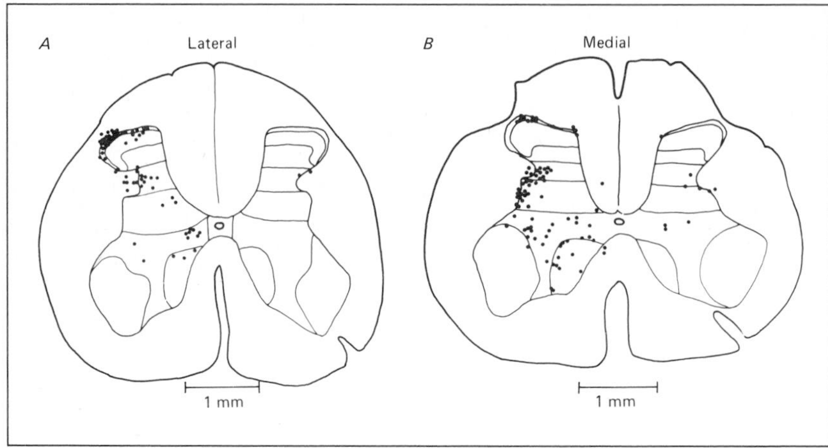

Fig. 5/2. Locations of spinothalamic tract cells labeled with horseradish peroxidase transported retrogradely from the lateral thalamus *(A)* or from the medial thalamus *(B)* to the lumbosacral enlargement in the monkey. Rexed's laminae are indicated on the drawings of cord sections. No boundary is shown between laminae II and III. [From *Willis* et al., 1979.]

the rat or monkey [*Albe-Fessard* et al., 1975; *Carstens and Trevino,* 1978a; *Giesler* et al., 1979a; *Trevino and Carstens,* 1975; *Trevino,* 1976; *Willis* et al., 1979]. Curiously, the distribution of spinothalamic tract cells in the cervical enlargement of the cat spinal cord is very similar to that of comparable cells in the lumbosacral enlargement of the monkey spinal cord [*Carstens and Trevino,* 1978a; *Hayes and Rustioni,* 1980; *Trevino and Carstens,* 1975; *Trevino,* 1976; *Willis* et al., 1979]. This would account for the much greater success that *Dilly* et al. [1968] had in finding antidromically activated spinothalamic tract cells in the dorsal horn of the cervical than in the lumbar cord in cats.

The distribution of the cells of origin of the spinothalamic tract depends not only upon species and level of the spinal cord, but also on the target zone in the thalamus. For example, the population of spinothalamic tract cells projecting to the lateral thalamus (region of the ventrobasal complex) differs from that of the cells projecting to the medial thalamus (region of the central lateral nucleus), with a high proportion of the medially projecting cells being located in the ventral horn, as shown for the monkey in figure 5/2 [*Willis* et al., 1979; cf. *Carstens and Trevino,* 1978a; *Giesler* et al., 1979a]. Some spinothalamic tract cells collateralize and project to both the lateral and medial thalamus [*Craig and Kniffki,* 1982; *Giesler* et al., 1981b; *Kevetter and Willis,* 1983]. Small injections of horseradish peroxidase into the nucleus parafascicularis of the cat label a few neurons in laminae VI–VIII [*Comans and Snow,* 1981]. Neurons that could be backfired from the medial thalamus in the cat were labeled by intracellular injections of horseradish peroxidase by *Meyers and Snow* [1982b]. The cells were large, multipolar neurons with long dendrites that often extended into the white matter. The axons crossed the midline in the ventral white commissure and ascended in the ventral funiculus. No local collaterals were detected.

Most spinothalamic tract cells project to the contralateral thalamus. For example, in the rat, very few spinothalamic tract cells are labeled in the ipsilateral spinal cord after an injection of the thalamus [*Giesler* et al., 1979a, 1981a]. Similarly, in the monkey, 95% of the spinothalamic tract cells labeled in the lumbosacral enlargement from an injection in the region of the ventrobasal complex and 90% of the cells labeled from the region of the intralaminar complex are contralateral to the injection site [*Willis* et al., 1979]. However, in the sacral cord, 26% of the spinothalamic tract cells labeled are ipsilateral to the injection site. Many of these ipsilaterally projecting cells are in Stilling's nucleus [*Willis* et al., 1979] and appear to have a proprioceptive function [*Milne* et al., 1982].

There is another special nucleus of spinothalamic tract cells that is located in Rexed's laminae VII and VIII in the uppermost cervical segments. Cells in this region project either to the contralateral or to the ipsilateral thalamus [*Carstens and Trevino,* 1978a, b]. Spinothalamic tract cells in the thoracic and upper lumbar spinal cord are found throughout Rexed's laminae, but are not seen in Clarke's column [*Foreman* et al., 1981; *Milne* et al., 1981].

Spinothalamic tract axons give off collaterals at several levels of the brain stem. For example, antidromic activation experiments have demonstrated that some spinothalamic tract cells can be activated antidromically both from the ventral lateral nucleus of the thalamus and from the region of the lateral periaqueductal gray in the midbrain [*Price* et al., 1978]. Furthermore, a double labeling study in the rat has shown that spinothalamic tract cells can be labeled from both the contralateral thalamus and from the rhombencephalic reticular formation [*Kevetter and Willis,* 1983]. Antidromic activation from both the thalamus and the medullary reticular formation has also been described in the monkey [*Giesler* et al., 1981b] and from the rostral midbrain and medullary reticular formation in the cat [*Fields* et al., 1977].

A count of the number of labeled spinothalamic tract cells present in the lumbosacral spinal cord after injection of horseradish peroxidase into one side of the thalamus showed a total of 1,215 labeled cells in alternate sections, implying the existence of as many as 2,500 spinothalamic tract cells in this part of the cord and presumably several thousand more at the rostral levels of the cord [*Willis* et al., 1979]. This figure contrasts with the estimate of 1,500 directly projecting spinothalamic axons on one side of the human brain stem by *Glees and Bailey* [1951]. Since the spinothalamic tract is thought to be more prominent in the human than in the monkey, there may be a more substantial number of spinothalamic tract cells in the human.

Organization of the Spinothalamic Tract in the Spinal Cord and Brain Stem

Based on the results of cordotomies and of stimulation within the anterolateral quadrant of the spinal cord, it would be predicted that the spinothalamic tract would have a somatotopic organization within the spinal cord (fig. 5/3A, B) [*Hyndman and Van Epps,* 1939; *Tasker* et al., 1976; *Walker,* 1940; *White and Sweet,* 1969]. The fibers representing the more caudal segments come to lie dorsolaterally to those representing more rostral segments. A somatotopic arrangement is found in the monkey spinothalamic tract that resembles that postulated in humans [*Weaver and Walker,* 1941].

Recordings from the axons of spinothalamic tract cells in the monkey show that the axons have a roughly somatotopic distribution (fig. 5/3C) [*Applebaum* et al., 1975].

The arrangement of spinothalamic axons in the rat spinal cord appears to follow a different principle than that of the primate cord. In the rat, axons arranged along the outer margin of the ventral lateral funiculus project to the ventrobasal complex, whereas axons in the ventral funiculus project to the medial thalamus [*Giesler* et al., 1981a].

The spinothalamic tract has been thought to decussate either within one segment [*Foerster and Gagel,* 1932] or several segments rostrally from the cell bodies [*Horrax,* 1929; *Walker,* 1940]. The reason for the hypothesis that the decussation is several segments rostral to the cell bodies is that the analgesia produced by cordotomy often begins several (as many as five) segments caudal to the level of the cordotomy [*Horrax,* 1929]. However, *Foerster and Gagel* [1932] attributed this to incompleteness of the cordotomy, since they could produce an analgesic level to the dermatome within a

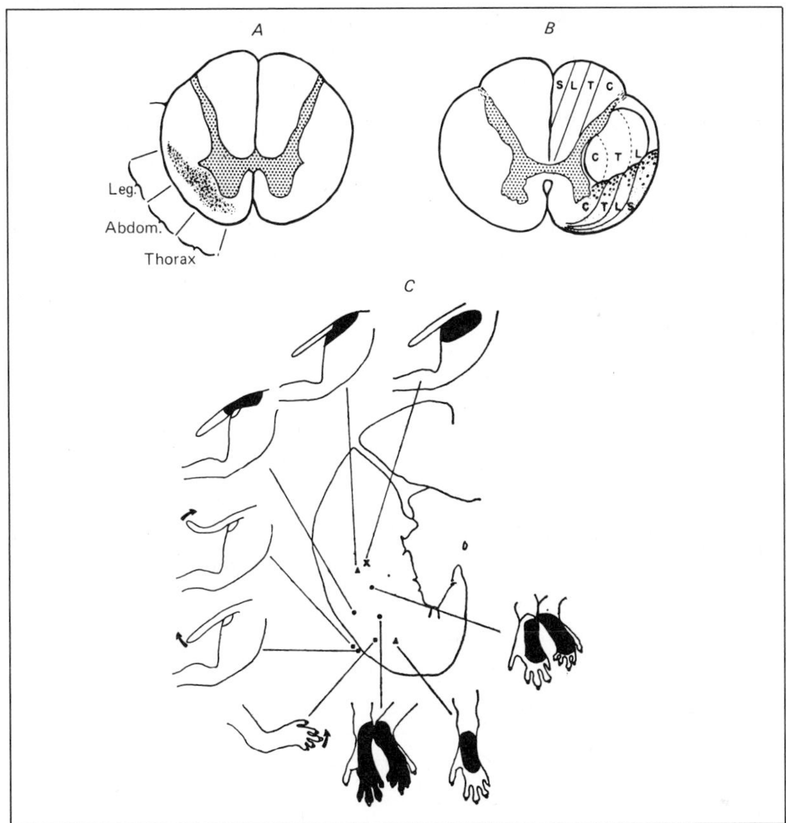

Fig. 5/3. The drawings of the human spinal cord in cross section at the top show the somatotopic organization of the spinothalamic tract, as determined from the results of cordotomies. [From *Hyndman and Van Epps,* 1939 *(A)* and *Walker,* 1940 *(B).*] The drawings below *(C)* indicate the receptive fields of several spinothalamic tract axons recorded in an experiment on the monkey spinal cord, showing the roughly somatotopic organization of the tract. [From *Applebaum* et al., 1975.]

segment of the level of the cordotomy by extending the lesion to near the midline. Another explanation of the discrepancy in the level of analgesia and of the cordotomy was offered by *Hyndman* [1942], who found that interruption of Lissauer's tract could elevate the sensory level of a cordotomy [cf. *Rand,* 1960]. Thus, some nociceptive information might ascend ipsilaterally for several segments in Lissauer's tract to activate spinothalamic tract cells at a more rostral level than that of the sensory input.

Although most investigators have denied that afferent fibers ascend more than a segment or two in Lissauer's tract, it would still be possible for cells of the dorsal horn, for example in the substantia gelatinosa, to transmit information rostrally in Lissauer's tract. In the monkey, at least, some spinothalamic axons decussate within a short distance of the cell body [*Willis* et al., 1979].

Anatomic studies, using anterograde degeneration techniques, demonstrate that the spinothalamic tract ascends through the ventrolateral quadrant of the spinal cord to enter the brain stem anterior to the spinal tract and nucleus of the trigeminal nerve [*Mehler* et al., 1960; *Mehler,* 1962; *Walker,* 1940]. Interruption of the ventral funiculus at a cervical level in the monkey produces a degeneration pattern that for the most part parallels that of degeneration ascending from axons of the ventral part of the lateral funiculus [*Kerr,* 1975b]. The spinothalamic tract ascends in the medulla just dorsal to the inferior olivary nucleus and then through the pons and midbrain in company with the lateral lemniscus [*Rasmussen and Peyton,* 1948; *Walker,* 1940]. This pattern has been seen in a number of species of mammals [see review by *Willis,* 1983]. Stimulation in the midbrain in the area in which the spinothalamic tract courses can cause a sharp pain localized to contralateral extremities [*Nashold* et al., 1969], although commonly there are sensations of cool or warm tingling or burning instead [*Tasker* et al., 1976]. In some cases, the sensory reference is bilateral, but usually it is strictly contralateral [*Tasker* et al., 1976].

Thalamic Nuclei of Termination

The spinothalamic tract projects to several different thalamic nuclei (see also Chapter 6). These include components of the posterior, ventral and intralaminar regions of the thalamus. The main component of the posterior complex that receives a spinothalamic input is the medial part, PO_m [*Berkley,* 1980; *Boivie* 1971, 1979; cf. *Mehler* et al., 1960]. Most current workers include the magnocellular part of the medial geniculate that receives a spinal projection as part of the PO_m nucleus. Some terminals can also be seen in the suprageniculate nucleus [*Burton and Craig,* 1983].

The ventral region involved may depend upon the species. In the human and the monkey, as well as the rat, the spinothalamic tract ends in the ventral posterior lateral nucleus (fig. 5/4, 5/5) [*Berkley,* 1980; *Boivie,* 1979; *Bowsher,* 1957, 1961; *Lund and Webster,* 1967; *Mehler* et al., 1960; *Mehler* 1962; *Pearson and Haines,* 1980; *Peschanski* et al., 1983; *Zemlan* et al., 1978]. The endings in the VPL nucleus in the monkey are in both the

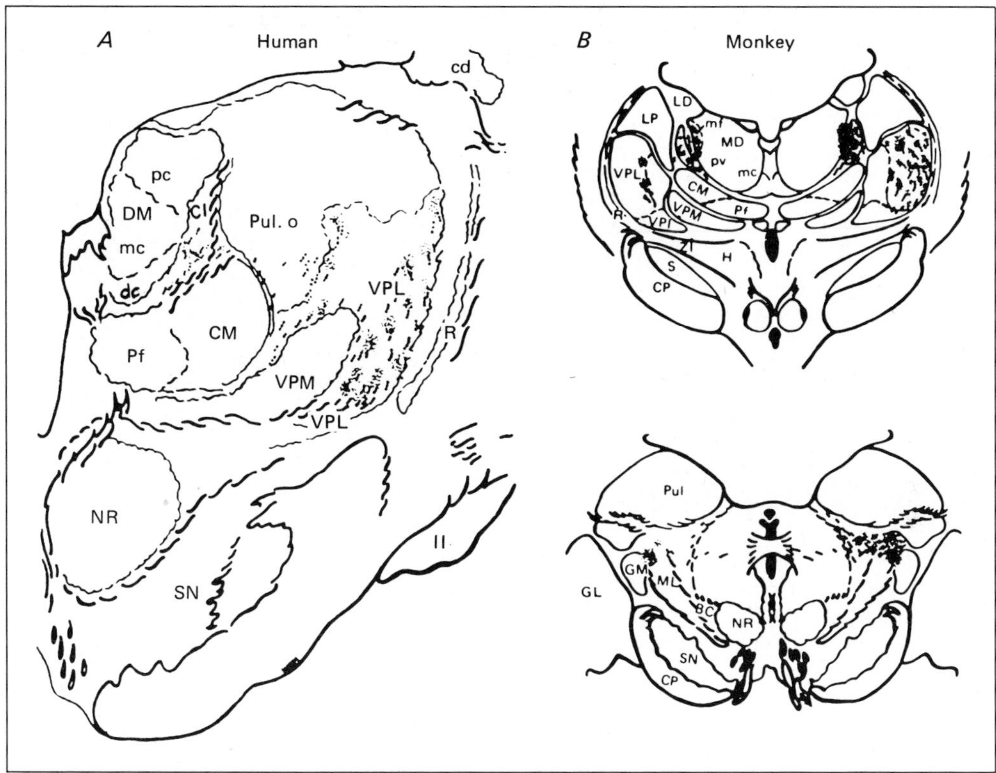

Fig. 5/4. Termination zones of the spinothalamic tract in the human and in the monkey. Terminal areas are shown by fine stippling; fibers of passage by coarse stipple. [*A* is from *Mehler,* 1962; *B* is from *Mehler* et al., 1960.] CL = Central lateral nucleus; CM = centre median nucleus; CP = cerebral peduncle; DM, MD = dorsal medial nucleus; GL, GM = lateral and medial geniculate nuclei; H = field H of Forel; LD, LP = lateral dorsal and lateral posterior nuclei; ML = medial lemniscus; NR = red nucleus, Pf = parafascicular nucleus; Pul = pulvinar; SN = subthalamic nucleus; VPL, VPM, VPI = ventral posterior lateral, medial and inferior nuclei; Zi = zona incerta.

caudal part of VPL (VPL_c of *Olszewski* [1952]), which also receives terminals from the medial lemniscus, and in the oral part (VPL_o). The spinothalamic endings do not occur throughout the VPL_c nucleus, however, but instead form small patch-like zones, called 'bursts' by *Mehler* et al. [1960] (fig. 5/4, 5/5) [cf. *Kerr,* 1975b; *Kerr and Lippman,* 1974; *Ralston,* 1984]. In the cat it has been reported that there are few spinothalamic endings in the ventral posterior lateral nucleus; instead, the tract ends in the border region

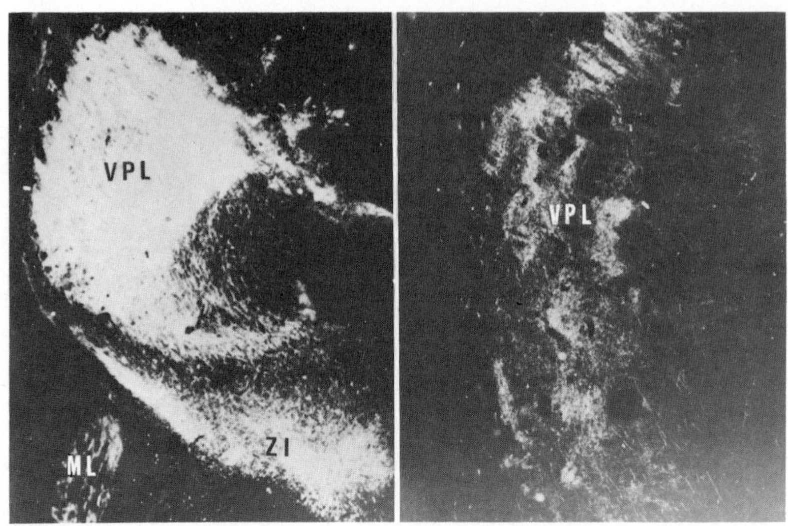

Fig. 5/5. Distribution of label in the thalamus after anterograde transport from the dorsal column nuclei *(A)* and from the anterolateral quadrant at the C3 level of the spinal cord *(B)*. The label was horseradish peroxidase conjugated with lectin. VPL = Ventral posterior lateral nucleus; ZI = zona incerta; ML = medial lemniscus. [From *Ralston,* 1984.]

between the ventral posterior lateral and ventral lateral nuclei and in the posterior complex [*Berkley,* 1980; *Boivie,* 1971; *Jones and Burton,* 1974]. Recent studies employing the very sensitive wheat germ agglutinin-conjugated horseradish peroxidase anterograde technique show a sparse projection of the spinothalamic tract to the VPL nucleus of the cat [*Burton and Craig,* 1983; *Mantyh,* 1983]. The results of an antidromic mapping study are consistent with the anatomical findings that the spinothalamic tract in the cat projects to nuclei other than VPL [*Holloway* et al., 1978].

The part of the intralaminar complex that receives a spinothalamic projection includes chiefly the central lateral nucleus and the adjacent related parts of the medial dorsal nucleus (pars multiformis and pars densocellularis of *Olszewski* [1952]) (fig. 5/4) [cf. *Mehler* et al., 1960; *Mehler,* 1962]. In addition, there are some spinothalamic terminals in parts of the parafascicular, paraventricular, paracentral, central medial and reuniens nuclei [*Boivie,* 1979; *Burton and Craig,* 1983], and there is a dense projection to the nucleus submedius (fig. 5/6A, B) [*Craig and Burton,* 1981]. The spinothalamic tract cells projecting to the nucleus submedius appear to be

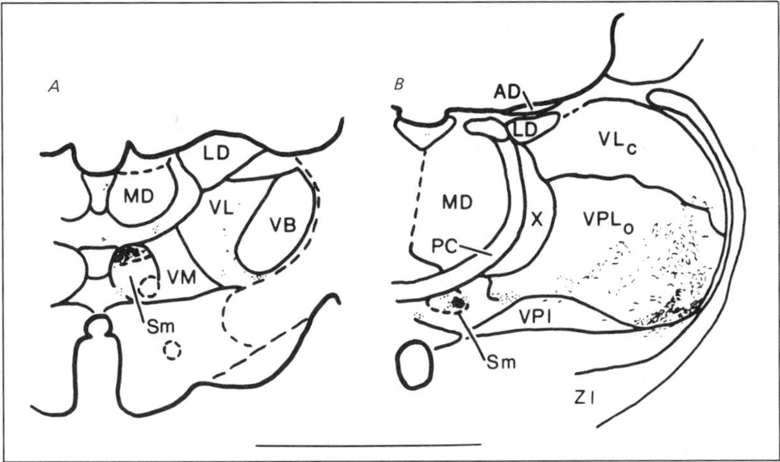

Fig. 5/6. Anterograde labeling that occurs in the submedius nucleus of the thalamus of the rat *(A)* and monkey *(B)* following injection of horseradish peroxidase into the spinal cord, and anterograde transport to the thalamus. [From *Craig and Burton,* 1981.]

located chiefly in lamina I [*Craig and Burton,* 1981]. Several other nuclei have also been reported to receive a spinal input by some authors [*Bowsher,* 1957, 1961; *Mantyh,* 1983].

Spinothalamic terminations in the VPL_c nucleus are somatotopically organized, whereas those in other thalamic nuclei are not [*Boivie,* 1979; *Chang and Ruch,* 1947; *Clark,* 1936; *Lund and Webster,* 1967; *Mehler* et al., 1960; *Peschanski* et al., 1983; *Walker,* 1940].

Identification of Spinothalamic Tract Cells in Physiological Experiments

Recordings can be made from identified spinothalamic tract cells (or their axons) by using the technique of antidromic activation following stimulation within the thalamus or in the spinothalamic tract as it nears the thalamus (fig. 5/7A) [*Albe-Fessard* et al., 1974; *Applebaum* et al., 1975; *Dilly* et al., 1968; *Fox* et al., 1980; *Price* et al., 1978; *Trevino* et al., 1972, 1973]. The criteria for antidromic activation include a constant latency discharge on repeated testing; the capability of following a high frequency train of discharges on a one-for-one basis; and collision of antidromic with

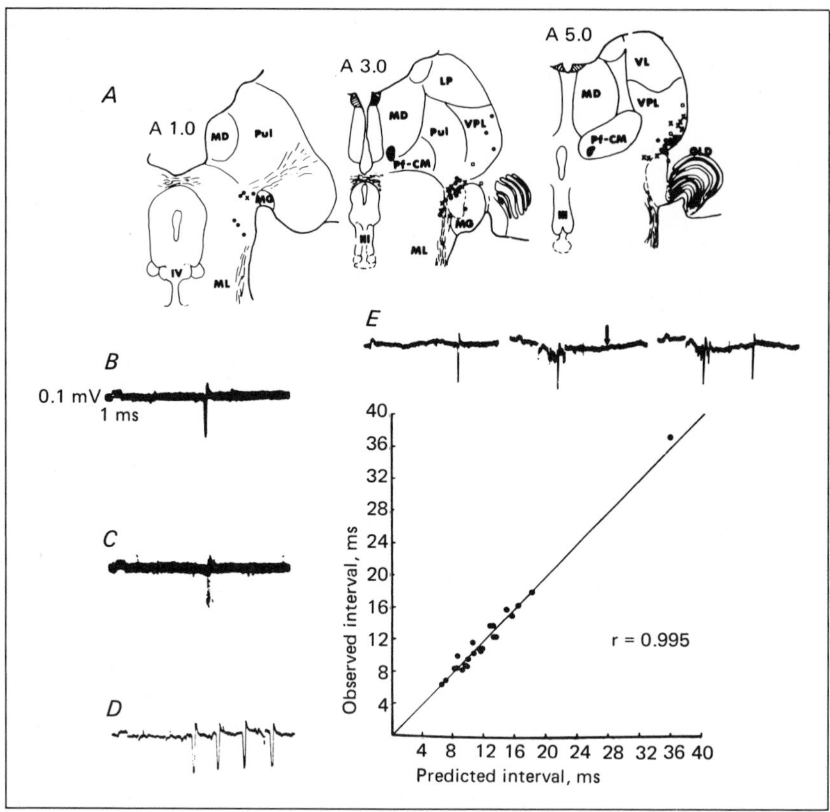

Fig. 5/7. A Sites in the thalamus and in the spinothalamic tract of the midbrain from which spinothalamic tract cells or axons in the spinal cord could be activated antidromically. [From *Applebaum* et al., 1975.] *B* Antidromic action potential in a spinothalamic tract cell. Multiple superimposed sweeps. *C* Antidromic action potential at threshold, showing little jitter in latency. *D* High frequency following (333 Hz). *E* The records at the top show the antidromic spike, collision with an orthodromic spike (antidromic spike would have occurred at arrow), and lack of collision with longer interval. The graph plots the critical intervals for collision in different cells as predicted (abscissa) and as determined experimentally (ordinate). [From *Trevino* et al., 1973.]

orthodromic spikes at the appropriate intervals (fig. 5/7B–E) [*Trevino* et al., 1972, 1973]. Cells that resemble spinothalamic tract cells can be activated antidromically by placing a stimulating electrode in or against the anterolateral quadrant of the spinal cord at an upper cervical level [*Chung* et al., 1984a; *Kumazawa* et al., 1975; *Price and Mayer,* 1974, 1975]. These cells have receptive fields that are indistinguishable from those of spinothalamic tract cells, and they are located in the same part of the dorsal horn and have axons with the same conduction velocities as those of identified spinothalamic neurons [*Chung* et al., 1984a]. However, it is likely that some of the cells investigated in this manner contribute to tracts other than spinothalamic tract. Nevertheless, such incompletely identified neurons are useful in studies that require the absence of analgesia, since identified spinothalamic tract cells would be difficult to investigate in unanesthetized preparations [e.g. *Hori* et al., 1984].

Responses of Spinothalamic Tract Cells to A and C Fiber Volleys

Since nociceptors are connected to Aδ or C fibers, it is of interest to examine the effects of stimulating Aδ and C fibers upon spinothalamic tract cells. In a series of experiments in which graded strengths of electrical stimulation were applied to a peripheral nerve, it was found that many spinothalamic tract cells were excited to discharge a burst of action potentials when the afferent volley was confined to the Aβ fiber group, and they discharged with a second burst when the stimulus was increased to activate the Aδ fibers (fig. 5/8) [*Beall* et al., 1977; *Foreman* et al., 1975; cf. *Albe-Fessard* et al., 1974]. However, in other cases, most of the discharges were produced only when the stimulus intensity was raised enough to activate Aδ fibers (fig. 5/9). By stimulating a peripheral nerve both at a distal and a proximal point along a peripheral nerve, it was possible to demonstrate a shift in the latency of the second burst discharge consistent with the conduction velocities of Aδ fibers [*Beall* et al., 1977]. Furthermore, the Aδ afferent volley continued to be effective, or even became more effective, when the Aβ volley was prevented by anodal blockade (fig. 5/10) [*Beall* et al., 1977].

Volleys in C fibers in a peripheral nerve were also found to excite spinothalamic tract cells (fig. 5/8C, F, 5/11) [*Foreman* et al., 1975; *Chung* et al., 1979]. The late discharge produced by peripheral nerve stimulation could be shown to result from C fibers on the basis of several lines of evidence: the stimulus strength had to be sufficient to produce a C fiber compound action potential in the peripheral nerve; the latency was appropriate for fibers conducting at 0.5–1 m/s; the response could be observed even when conduction was prevented in A fibers by anodal blockade (fig. 5/12); and the response continued to be seen after the spinal cord was transected, ruling out a spinobulbo-spinal relay of activity [*Chung* et al., 1979].

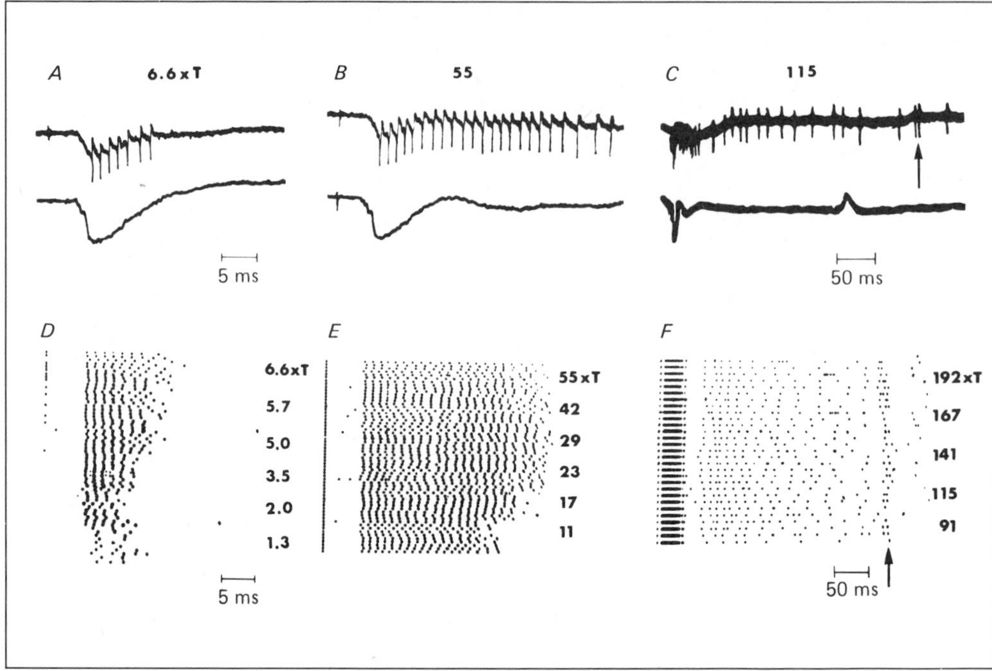

5/8

Fig. 5/8. Responses of a spinothalamic tract cell to graded strengths of electrical stimulation of the sural nerve. The cell was activated weakly by stimuli of Aβ strength (1.3 times threshold or × T), but it was activated progressively more strongly with higher stimulus intensities that would have excited Aδ fibers (up to 23 × T). A few spikes can be attributed to C-fiber volleys (*C* and *F,* arrows). Note the slower time base in *C* and *F,* as compared with *A, B* and *D, E.* The dot raster records in *D–F* show the latencies of spikes on repeated trials at various stimulus intensities; *A–C* show individual records at 3 different intensities. [From *Foreman* et al., 1975.]

Fig. 5/9. Graded excitation of a spinothalamic tract cell by graded stimulation of the sural nerve. The records in *A–D* show the discharges of the cell, the cord dorsum potentials, and the neurogram recorded from the sural nerve following stimulation at the indicated intensities (T = times threshold). *E* and *F* show the sizes of the N_1, N_2 and N_3 waves and the number of discharges of the spinothalamic tract cell in the early and late bursts for different stimulus intensities. Stimuli of 1.35 T and above activated Aδ fibers. Evidently, the cell could be activated to discharge one spike by Aβ fibers, whereas Aδ fibers caused most of the additional discharges as stimulus intensity was increased. [From *Beall* et al., 1977.]

Fig. 5/10. Activation of a spinothalamic tract cell by Aδ fibers of the sural nerve. *A* shows the discharges of a spinothalamic tract cell, the cord dorsum potentials, and a neurogram recorded from the sural nerve. The stimulus strength activated Aβ and Aδ fibers. *B* shows the Aβ volley and the fast part of the Aδ volley have been blocked by anodal current. [From *Beall* et al., 1977.]

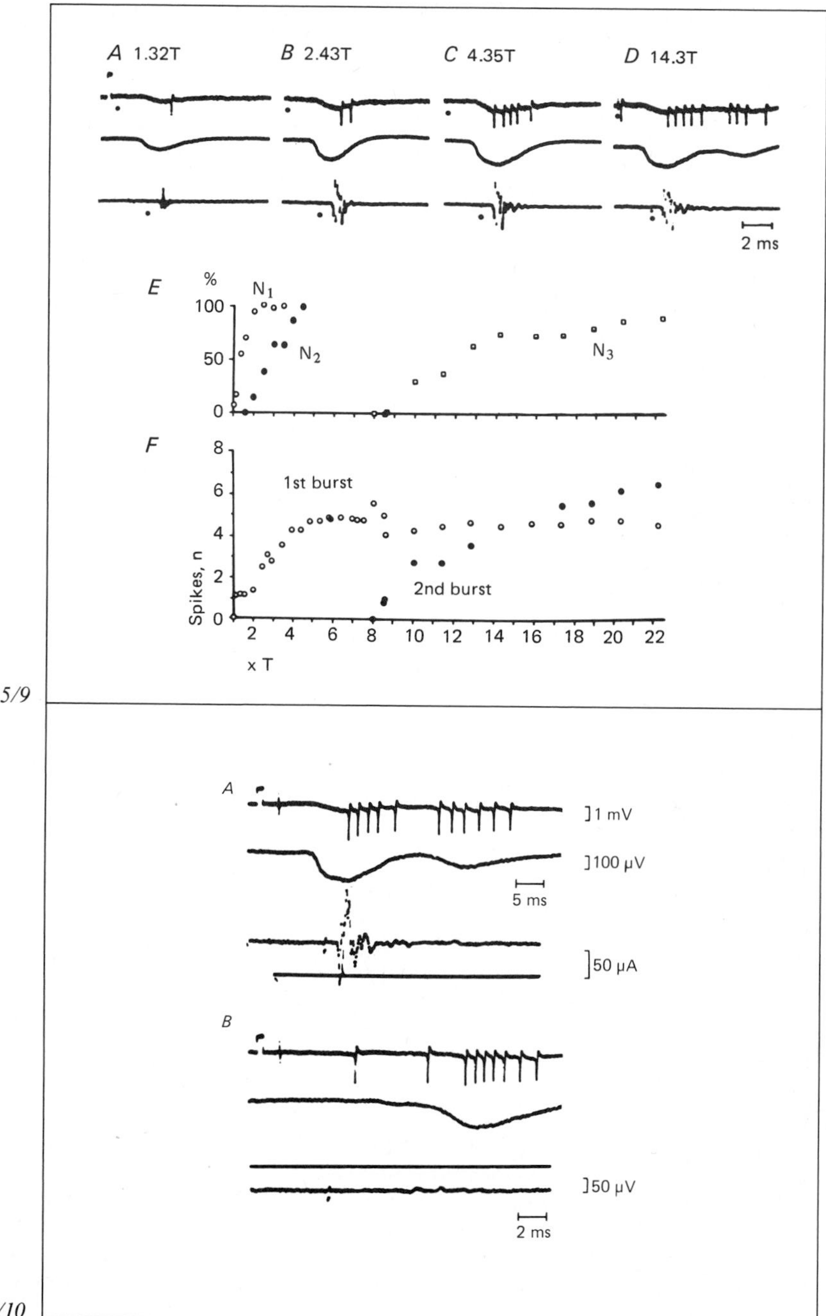
A 1.32T
B 2.43T
C 4.35T
D 14.3T
2 ms
E
%
N_1
N_2
N_3
100
50
0
F
8
6
4
2
0
1st burst
2nd burst
Spikes, n
2 4 6 8 10 12 14 16 18 20 22
x T
A
]1 mV
]100 μV
5 ms
]50 μA
B
]50 μV
2 ms

5/9

5/10

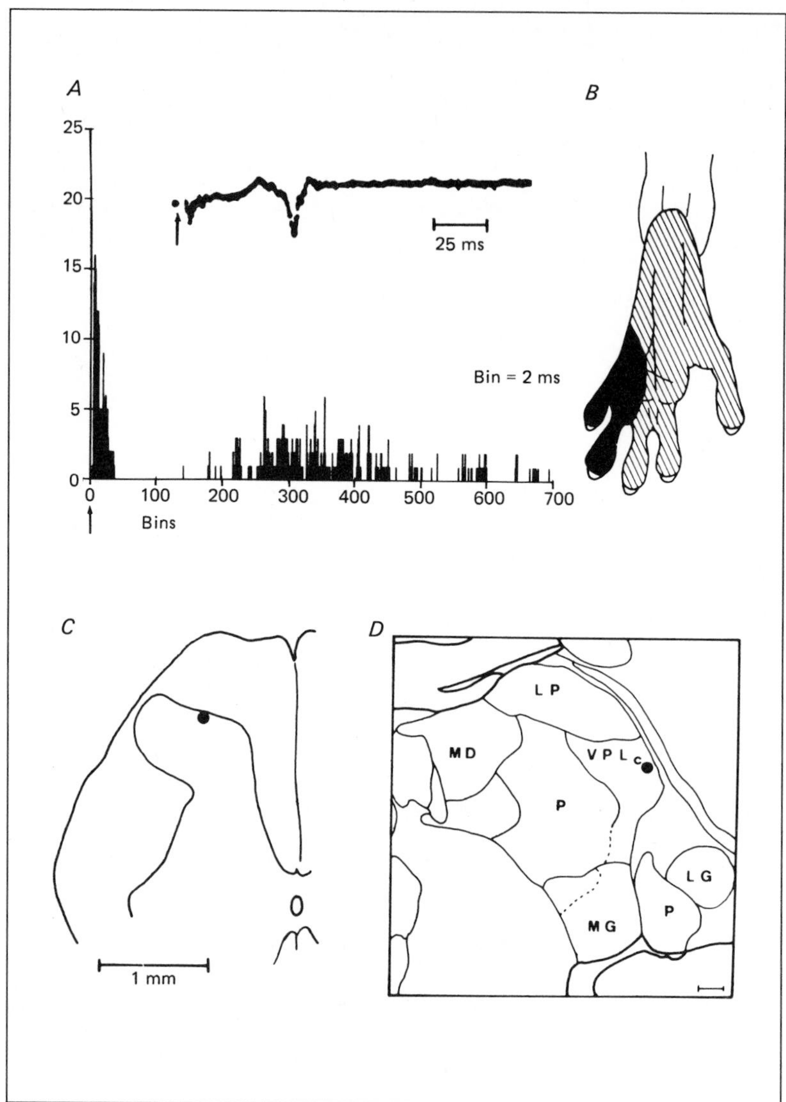

Fig. 5/11. Activation of a spinothalamic tract cell by a volley in A and C fibers in the sural nerve. The histogram in *A* shows an early response due to A fibers and a late response due to C fibers. The inset is the compound action potential recorded from the sural nerve and shows a large wave due to conduction in C fibers. The receptive field of the cell is shown in *B*, with a more sensitive region indicated in black and a higher threshold region that is hatched. The recording site is shown in *C* and the stimulation site for antidromic activation in the thalamus in *D*. [From *Chung* et al., 1979.]

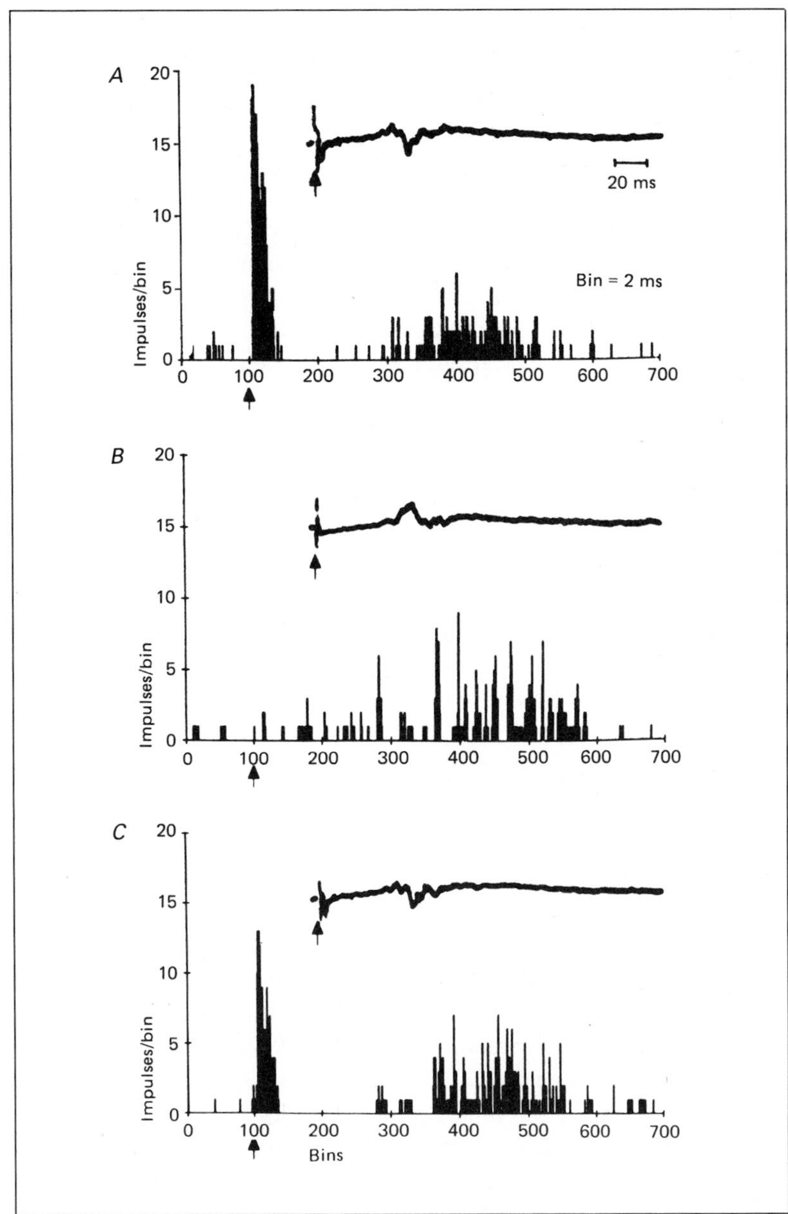

Fig. 5/12. Activation of a spinothalamic tract cell by A and C fiber volleys *(A–C)*. The sural nerve was stimulated at a strength sufficient to excite A and C fibers. In *B*, the A fiber volley was blocked by anodal current. [From *Chung* et al., 1979.]

Interestingly, C fiber volleys were more effective when repetitive stimulation was used, indicating the need for temporal summation, and repeated stimulation at rates of about 0.5 s produced a wind-up of the response to C fiber volleys similar to that observed by others for dorsal horn neurons investigated in unanesthetized preparations [*Mendell,* 1966; *Wagman and Price,* 1969].

Although most of the experiments showing an excitatory input to spinothalamic tract cells from Aδ and C fibers have been done using cutaneous or mixed peripheral nerves, some experiments have instead employed muscle nerves [*Foreman* et al., 1977, 1979a]. Some spinothalamic tract cells could be excited weakly by group I muscle afferents. However, group II and especially group III muscle afferents were more effective. A limited sample of neurons was tested and found to be excited by group IV (unmyelinated) muscle afferent fibers [*Foreman* et al., 1979a].

In a few experiments, it has been shown that Aδ and C fibers in a visceral nerve, the greater splanchnic nerve, can excite at least some spinothalamic tract cells in the midthoracic spinal cord [*Hancock* et al., 1975; *Foreman* et al., 1981]. Furthermore, stimulation of fibers in sympathetic nerves to the heart can activate spinothalamic tract cells in upper thoracic segments [*Blair* et al., 1981; *Foreman and Weber,* 1980].

Responses of Spinothalamic Tract Cells to Natural Forms of Stimulation

Not all of the afferent fibers in the Aδ or C ranges are nociceptors, and so the results of experiments in which electrically evoked volleys are employed, while suggestive, do not necessarily reflect the effects of nociceptive inputs. For example, Aδ fibers include mechanoreceptors supplying down hairs and some thermoreceptors, and C fibers include other thermoreceptors and mechanoreceptors [*Burgess and Perl,* 1973].

When graded intensities of mechanical stimuli are applied to the skin, a few spinothalamic tract cells respond selectively to innocuous stimuli; many respond to innocuous stimuli, but are activated better by noxious stimuli; and others are discharged only by noxious stimuli [*Chung* et al., 1979; *Craig and Kniffki,* 1982; *Fox* et al., 1980; *Giesler* et al., 1976; *Price* et al., 1978; *Willis* et al., 1974]. These are generally classed as low threshold, wide dynamic range (or multireceptive), and high threshold (or nociceptive-specific) cells, respectively. Another class of spinothalamic tract cell responds best or only to stimulation of receptors located in deep tissues, such as muscle. These neurons are classified as 'deep' spinothalamic tract cells.

The spinothalamic tract cells that are most likely to play an important role in signalling pain are the wide dynamic range (multireceptive) and the high threshold (nociceptive-specific) cells. An example of the types of responses of each of these that are evoked by graded intensities of mechanical stimulation of the skin are shown in figure 5/13. It is not clear whether the cells of these two classes have similar signalling roles, or if their signalling roles are distinct. For example, the tactile responses of the wide dynamic range cells might be a kind of biological noise that is ignored at higher levels of the sensory processing system. However, it is likely that the tactile responses provide meaningful signals since there are few low threshold spinothalamic tract cells, and yet the spinothalamic tract (or at least tracts in the anterolateral quadrant) can mediate touch [*Foerster and Gagel,* 1932; *Noordenbos and Wall,* 1976].

Many spinothalamic tract cells of both the wide dynamic range (multireceptive) and high threshold classes respond in a graded fashion to graded intensities of noxious heat applied to their receptive fields in the skin [*Kenshalo* et al., 1979]. The stimulus-response functions of these cells are indistinguishable, and they resemble the psychophysical functions reported for human subjects (see Chapter 3) [*LaMotte and Campbell,* 1978]. An example of the responses of a wide dynamic range and of a high threshold spinothalamic tract cell to a series of noxious heat pulses are shown in figure 5/14, columns I and II.

Spinothalamic tract cells can also sometimes be excited when noxious cold is applied to the skin [*Kenshalo* et al., 1982; *Willis,* 1983]. An example of the responses of one such neuron are shown in figure 5/14, column III.

Noxious chemical stimulation has been shown to activate spinothalamic tract cells. For example, *Levante* et al. [1975] found that intra-arterial injection of bradykinin into the hindlimb produced a powerful excitation of primate spinothalamic tract cells recorded in the lumbar enlargement.

Noxious stimuli applied to muscle can also excite certain spinothalamic tract cells. For example, injection of algesic chemicals, like bradykinin or serotonin into the arterial circulation of a muscle can excite spinothalamic tract cells that have receptive fields that include the muscle [*Foreman* et al., 1977, 1979b], as shown in figure 5/15A, B. Furthermore, injection of hypertonic sodium chloride solution into a muscle or a tendon will activate spinothalamic tract cells, as shown in figure 5/15C. Comparable intramus-

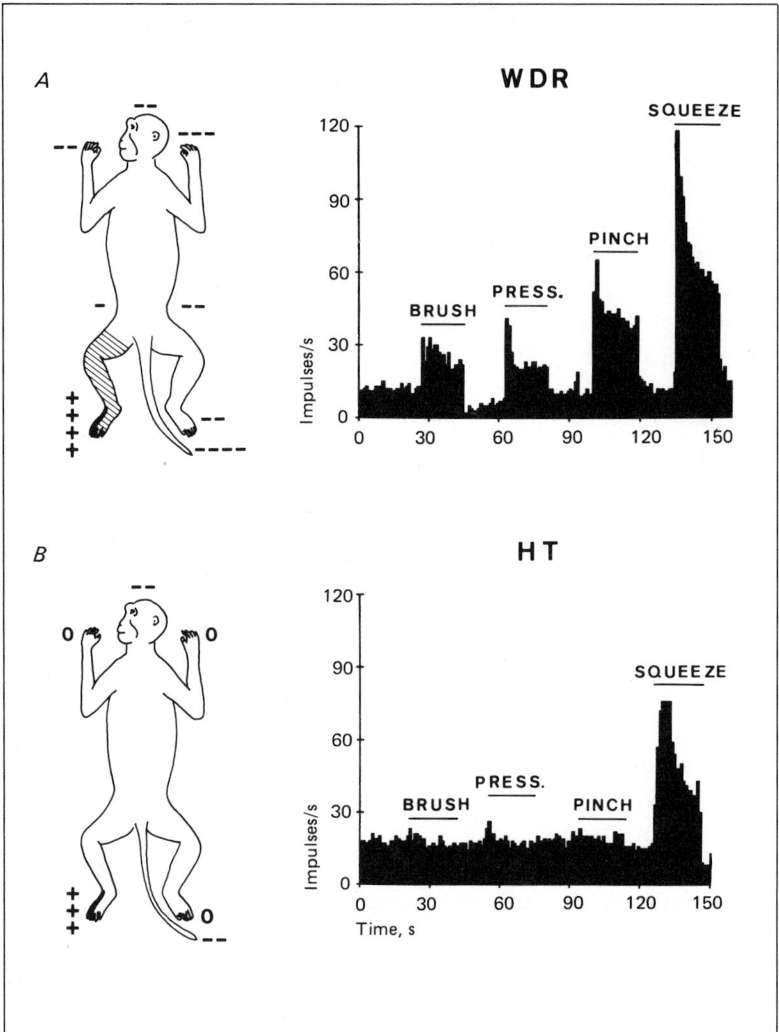

Fig. 5/13. The responses of a 'wide dynamic range' or multireceptive spinothalamic tract cell are shown in *A* and of a high threshold spinothalamic tract cell in *B*. The figurines indicate the excitatory (plus signs) and inhibitory (minus signs) receptive fields. The graphs show the responses to graded intensities of mechanical stimulation. Brush is with a camel's hair brush repeatedly stroked across the receptive field. Pressure (Press.) is applied by attachment of an arterial clip to the skin. This is a marginally painful stimulus to a human. Pinch is by attachment of a stiff arterial clip to the skin and is distinctly painful. Squeeze is by compressing a fold of skin with forceps and is damaging to the skin. [From *Willis*, 1981.]

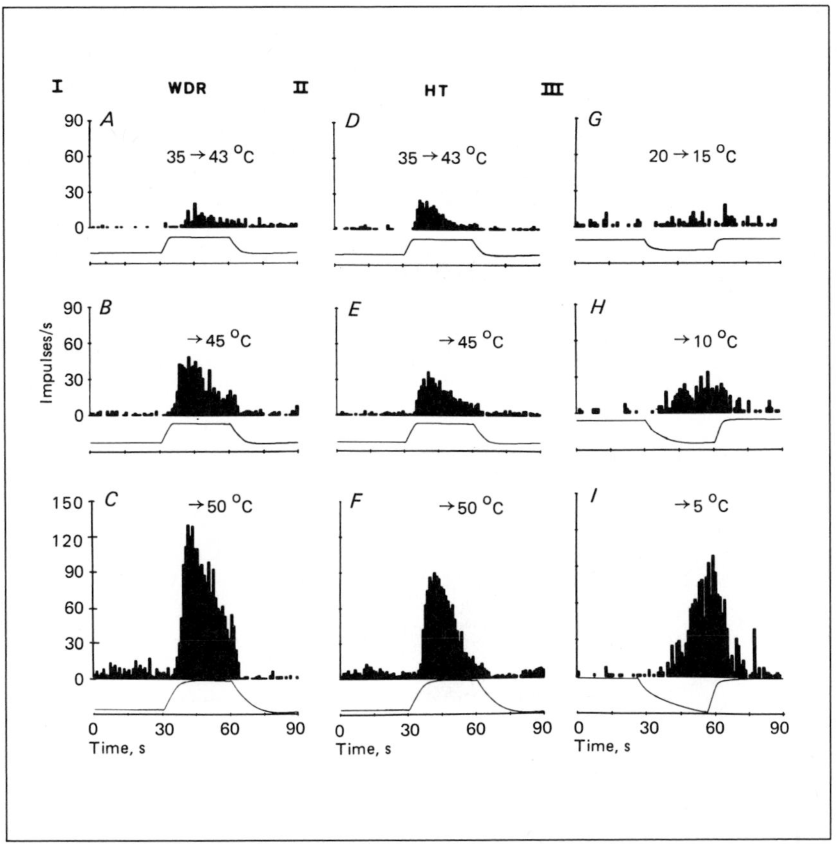

Fig. 5/14. Responses of spinothalamic tract cells to graded noxious thermal stimuli. Column I shows the responses of a 'wide dynamic range' or multireceptive cell to noxious heat pulses (to 43, 45 and 50 °C). Column II shows similar responses of a high threshold cell to the same stimuli. Column III shows responses of another cell to noxious cold (15, 10 and 5 °C). [Columns I and II from *Kenshalo* et al., 1979; column III from *Willis,* 1983.]

cular injections of hypertonic saline in human subjects produce an aching pain [*Lewis,* 1942].

Another form of noxious stimulation that has been shown to excite spinothalamic tract cells at the appropriate segmental level is visceral stimulation. For example, stimulation of cardiac sympathetic nerve fibers can excite spinothalamic tract cells at upper thoracic segmental levels [*Blair* et al., 1981; *Foreman and Weber,* 1980]. It is interesting that these cells often

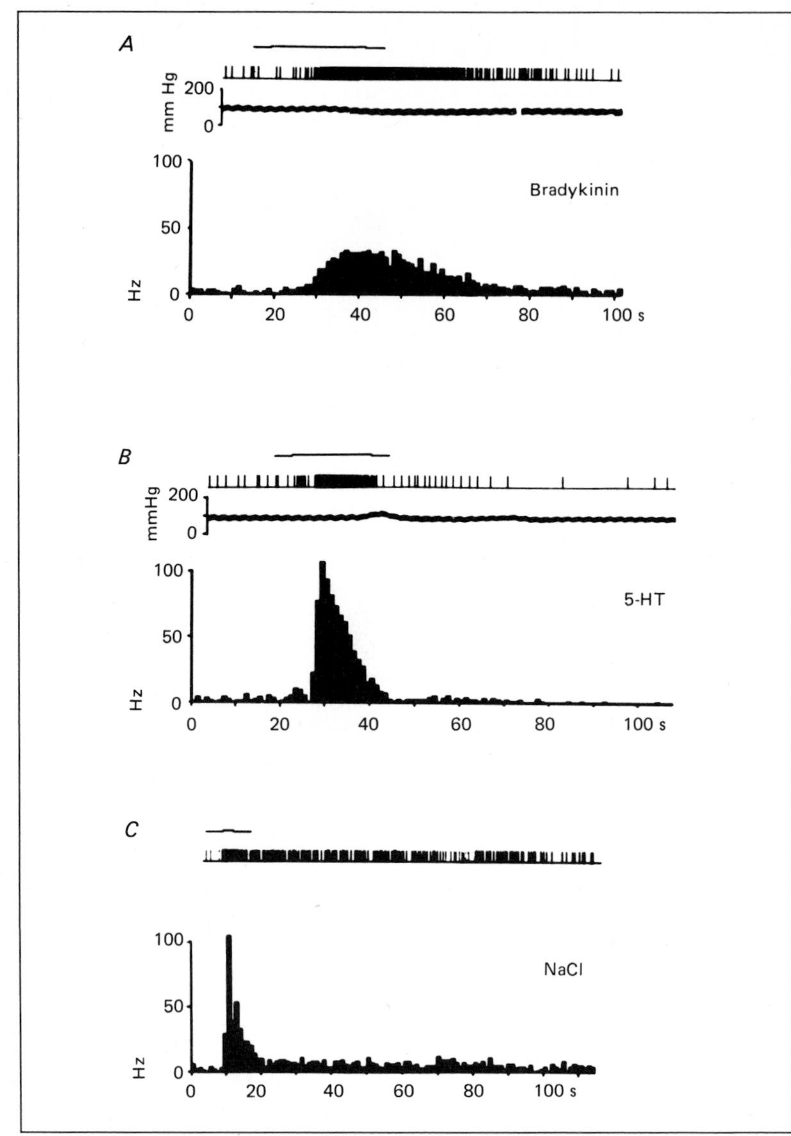

Fig. 5/15. The effect of algesic chemicals, bradykinin and serotonin, injected intra-arterially into the circulation of the triceps surae muscles on the discharges of a spinothalamic tract cell are shown in *A* and *B,* along with recording of the systemic arterial blood pressure. The responses are shown both by a pen recording of window discriminator pulse and by a histogram. *C* shows the excitation of the cell by injection of hypertonic NaCl into the triceps surae muscles. [From *Foreman* et al., 1977b.]

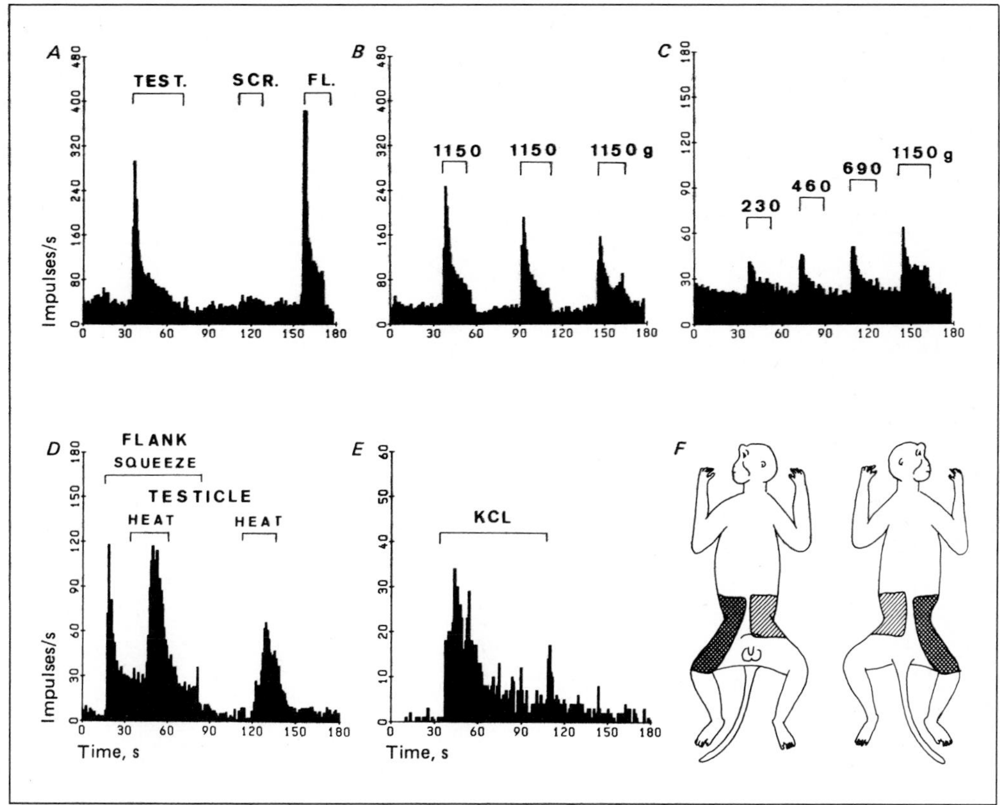

Fig. 5/16. Responses of wide dynamic range (multireceptive) spinothalamic tract cells to noxious visceral stimulation. The histogram in *A* shows the response of a cell to testicular compression and to squeezing the skin over the flank, but not to squeezing the skin of the scrotum. *B* shows reduction of the responses of same cell to repeated testicular compressions. *C* shows the graded responses of another cell to increasing forces of testicular compression. *D* shows the responses of another spinothalamic cell to squeezing the skin over the flank and heating the testicular surface to 51 °C or to heating alone. *E* shows the response of the same cell to application of isotonic KCl to the surface of the testicle. *F* shows the cutaneous receptive fields of two of the cells. [From *Milne* et al., 1981.]

have a cutaneous receptive field that extends along the inner aspect of the forearm, suggestive of the pattern of pain referral in angina pectoris. Spinothalamic tract cells located in upper lumbar segments of the spinal cord can be excited by noxious stimulation of the testicle such as crush, application of KCl or heat to the surface of the testicle (fig. 5/16) and by overdis-

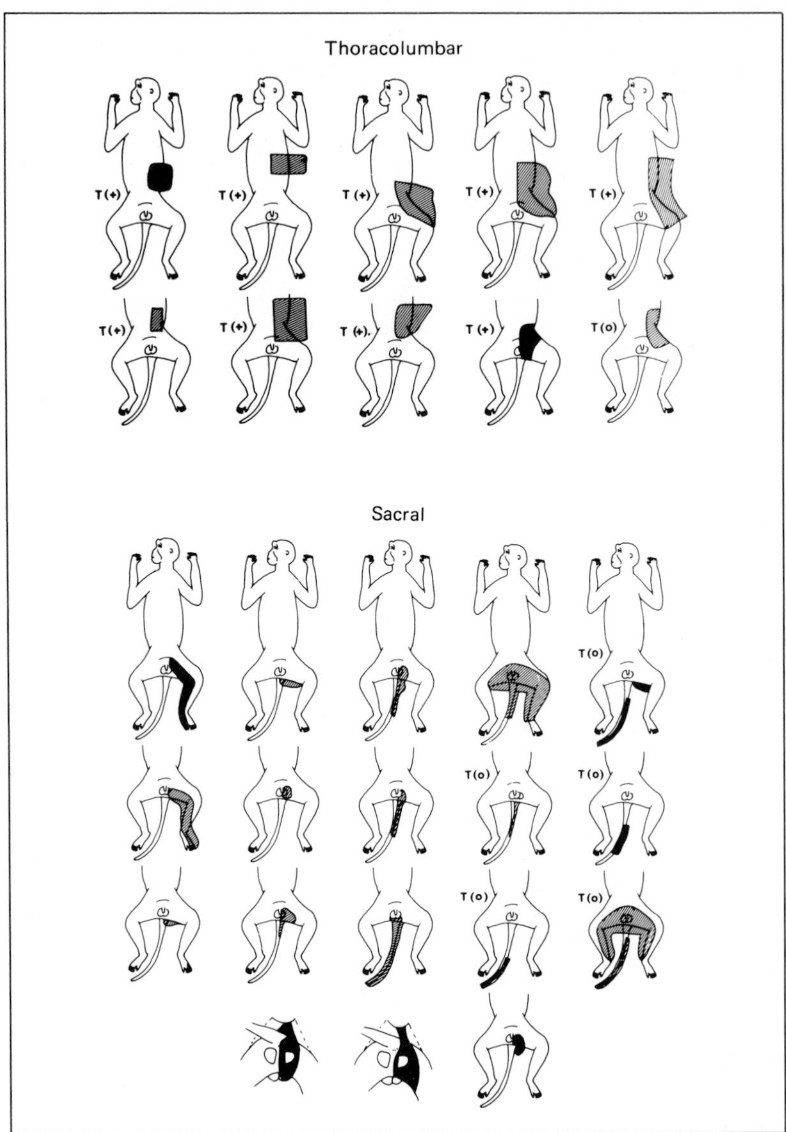

Fig. 5/17. Cutaneous excitatory receptive fields for spinothalamic tract cells in the lower thoracic and upper lumbar cord (thoracolumbar) or sacral cord. All of these cells had an excitatory input from the urinary bladder and/or ipsilateral testicle. Flaps indicate that the receptive field extended onto the dorsal surface. High threshold fields are shown by black. T(+) means testicular input; T(o) means absence of testicular input (others not tested). [From *Milne* et al., 1981.]

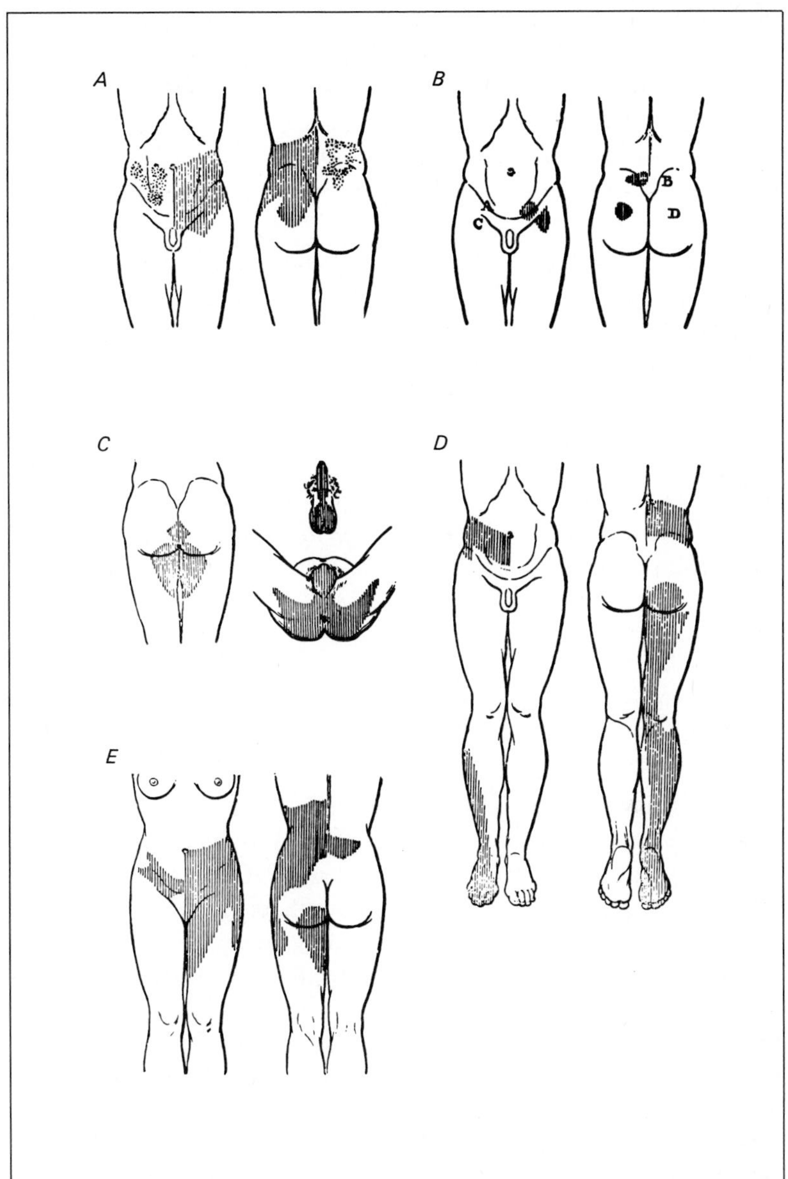

Fig. 5/18. Distribution of referred cutaneous tenderness in human subjects with different visceral disorders. *A* Severe renal colic. *B* Gonorrheal epididymitis. *C* Bladder pain. *D* Acute inflammation of right lobe of prostate. *E* Second stage of labor. [From *Head,* 1893.]

tention of the bladder [*Milne* et al., 1981]. Spinothalamic tract cells located in the sacral spinal segments can be excited by bladder distension, but not by noxious stimulation of the testicle. For cells in the lower thoracic and upper lumbar cord, the cutaneous receptive fields are over the flank, but cells in the sacral cord instead have cutaneous receptive fields in the peroneal region, on the medial aspect of the hindlimbs, or on the tail (fig. 5/17).

It seems evident that the patterns of input to spinothalamic tract cells from nociceptors in the skin and in viscera reflect the patterns of referral of pain and tenderness in man (fig. 5/18) [*Head,* 1893; *Lewis* 1942]. Although one mechanism that may contribute to an explanation of pain referral is the branching of individual nociceptors to supply receptive fields in different tissues [*Bahr* et al., 1981; *Perl,* 1984; *Pierau* et al., 1982; *Taylor and Pierau,* 1982], it seems likely that the patterns of convergent input onto somatosensory tract cells, such as the spinothalamic tract, are also likely to constitute another important mechanism [*Foreman and Weber,* 1980; *Foreman* et al., 1981; *Hancock* et al., 1975; *Milne* et al., 1981] (see Chapter 3).

Responses of Spinothalamic Tract Cells Projecting to the Medial Thalamus

Spinothalamic tract cells that project to the medial thalamus have been studied in the monkey, the cells being identified by antidromic activation from the region of the central lateral nucleus [*Giesler* et al., 1981b]. The cells tended to be located more ventrally than those that project to the lateral thalamus. Many of the medially projecting spinothalamic tract cells had very large receptive fields, often covering all of the surface of the body and face. The cells were often classified as high threshold neurons, although some were wide dynamic range (multireceptive) cells. Figure 5/19 shows the responses of one of these cells to the application of noxious heat to various sites on the skin.

The source of the input to these cells from much of the excitatory receptive field appears to depend upon a neural loop that involves the brain stem. Evidence for this statement is provided by experiments such as that illustrated in figure 5/20. The spinal cord was transected at an upper cervical level. When this was done, the receptive field diminished to include only the surface of the ipsilateral hindlimb. Evidently, input from other parts of

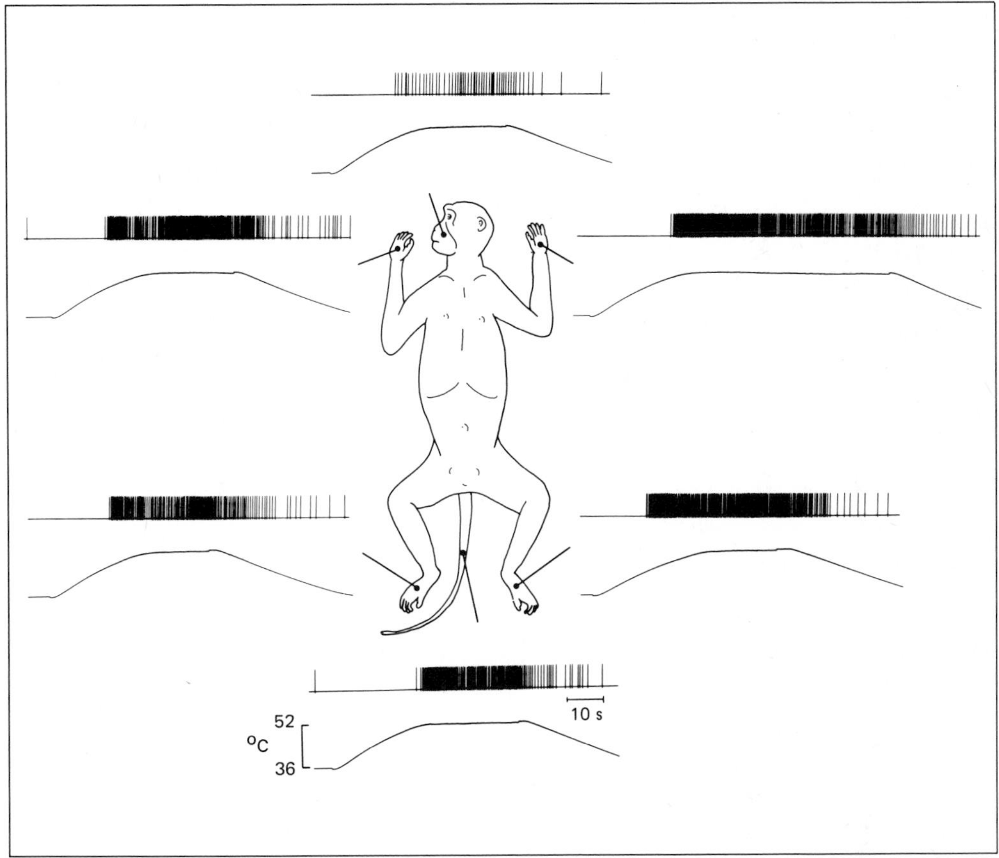

Fig. 5/19. Responses of a spinothalamic tract cell that projected to the medial thalamus to noxious heat stimuli applied to various sites on the surface of the body and face. The stimuli were from an adapting temperature of 36 to 52 °C. Records are pen recordings of window discriminator pulses triggered by the action potentials of the cell. [From *Giesler* et al., 1981b.]

the body and face caused activation of this neuron, which was located in the lumbosacral enlargement, by way of pathways descending from the brain. When medially projecting spinothalamic tract cells were examined while electrical stimuli were applied through an electrode inserted into the brain stem reticular formation, it was easy to demonstrate a strong excitatory drive [*Giesler* et al., 1981b].

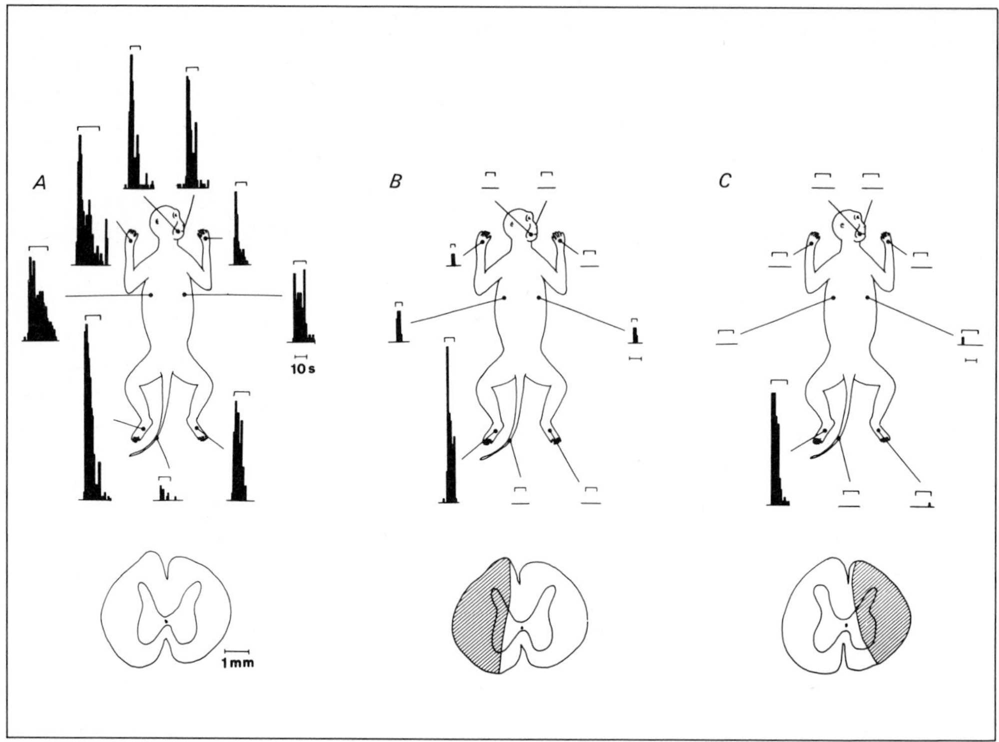

Fig. 5/20. The excitatory receptive field of spinothalamic tract cell projecting to the medial thalamus is constricted to just the ipsilateral hindlimb after transection of the spinal cord at an upper cervical level *(A)*. The transection was incomplete, and it was done in two stages *(B, C)*. [From *Giesler* et al., 1981b.]

Spinothalamic tract cells in laminae VI–VIII of the cat spinal cord often project to the medial thalamus, and they can be labeled retrogradely by horseradish peroxidase injected into the region of the nucleus parafascicularis [*Comans and Snow,* 1981]. The cells have been studied in detail by *Meyers and Snow* [1982a, b]. They could often be excited by noxious stimuli or by deep receptors. Most had large bilateral receptive fields. Several were labeled by intracellular injection of horseradish peroxidase.

Medially projecting spinothalamic tract cells of lamina I in the cat have been investigated by *Craig and Kniffki* [1982]. Almost all were nociceptive-specific (17 of 19 cells); the other 2 included a cell that responded to cooling the skin and another that was a wide dynamic range (multireceptive) cell. It was speculated that these neurons of lamina I might project to the nucleus submedius [*Craig and Burton,* 1981] (see Chapter 6).

Effects of Capsaicin

In recent experiments, it has been possible to demonstrate that application of capsaicin to a peripheral nerve in monkeys causes an initial activation of spinothalamic tract cells, followed by a block in conduction in many of the primary afferent fibers that are responsible for the effects of noxious heat in exciting these cells (fig. 5/21) [*Chung* et al., 1984c]. If C polymodal nociceptors are selectively blocked by capsaicin in the monkey, as they are in the rat [*Petsche* et al., 1983], this evidence would be consistent with a role for C polymodal nociceptors in conveying the effects of noxious heat stimuli on spinothalamic tract cells [cf. *Chung* et al., 1979]. However, the Aδ volley is partially blocked by capsaicin in the monkey, and so some Aδ fibers could also be involved [*Chung* et al., 1984c]. These observations are consistent with the reports of *LaMotte* et al. [1982] and *Meyer and Campbell* [1981] that C polymodal nociceptors are likely to be the main nociceptors involved in signalling noxious heat stimuli applied to normal skin.

Influence of Anesthesia

The possibility that anesthesia reduces the responsiveness of spinothalamic tract cells under experimental conditions has been tested by recording from dorsal horn neurons that could be activated antidromically from the upper cervical spinal cord in unanesthetized, decerebrate or spinalized animals [*Hori* et al., 1984]. The receptive fields in spinalized animals were indistinguishable from those of identified spinothalamic tract cells. The cells were likely to have been spinothalamic tract cells, since they had similar locations in the dorsal horn and comparable conduction velocities. Whether or not they were all spinothalamic tract cells, it was nevertheless possible to show that the injection of small doses of barbiturate did not reduce their responsiveness to stimulation of their receptive fields (fig. 5/22). In fact, the cells became more responsive to C fiber volleys [*Hori* et al., 1984].

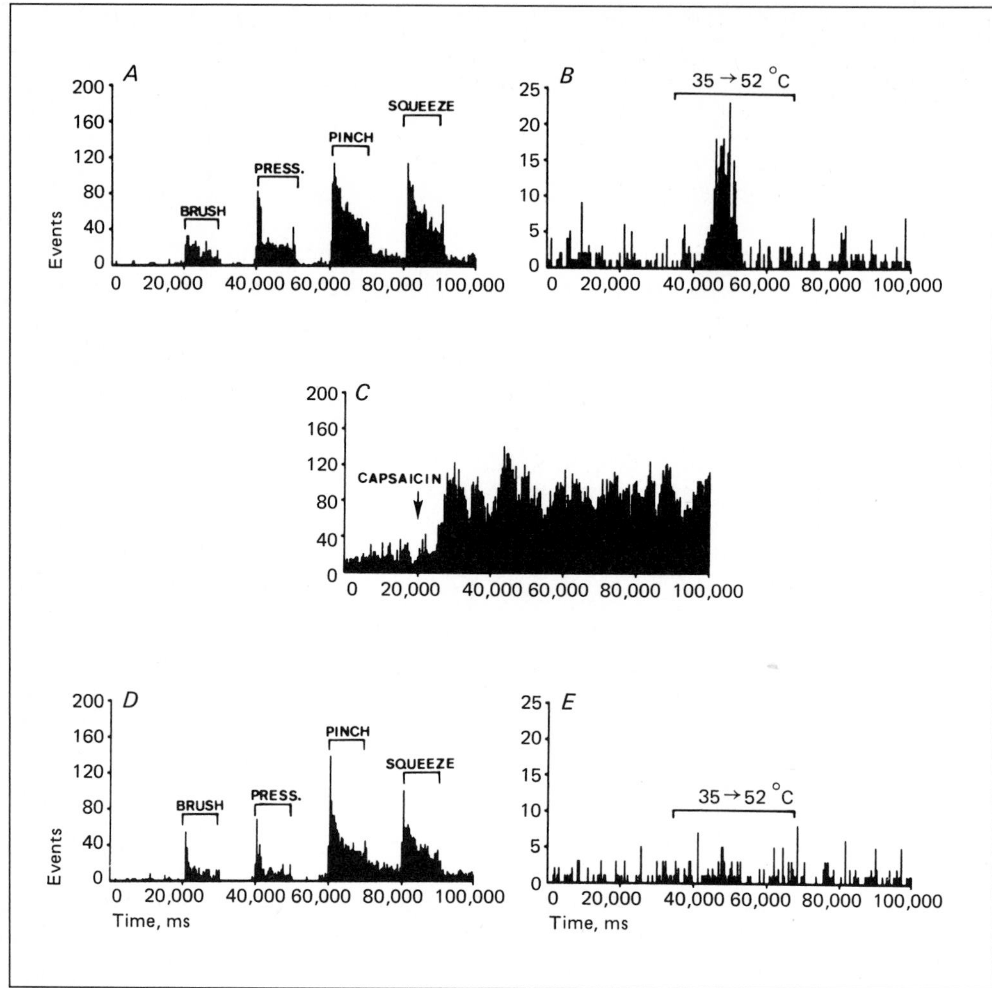

Fig. 5/21. Effects of capsaicin applied to a peripheral nerve on the activity of a spinothalamic tract cell. The histograms in *A* and *B* show the responses of a cell to graded intensities of mechanical stimulation and to noxious heat before capsaicin was applied to the sural nerve (tibial and common peroneal nerves cut). *C* shows that application of capsaicin causes a prolonged activation of the cell. *D* and *E* were taken after capsaicin-evoked discharges had quieted down. The graded mechanical stimuli were little changed, although the response to squeezing the skin may have been reduced somewhat. However, the response to noxious heat was nearly eliminated. Recordings of the compound action potential showed a marked reduction in the C fiber volley and some reduction in the Aδ volley. [From *Chung* et al., 1984c.]

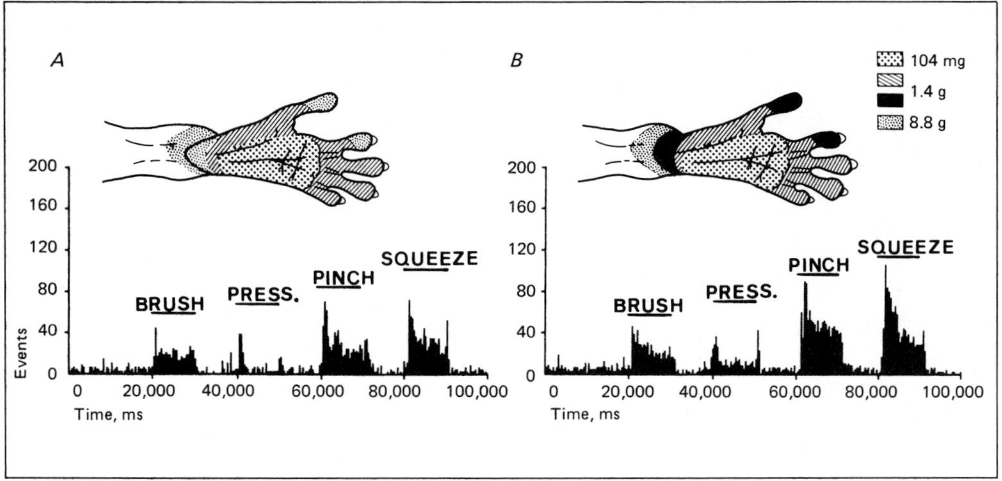

Fig. 5/22. Effect of barbiturate on 'spinothalamic tract-like' cell of dorsal horn. The cell was recorded in a decerebrate animal, and it was identified by antidromic activation from the contralateral upper cervical cord. The receptive field and responses to graded intensities of mechanical stimulation *(A)* were recorded before administration of 10 mg/kg of sodium pentobarbital. After the injection of the drug *(B)*, the receptive field became somewhat more sensitive (as shown by the spread of the area responding to 1.4 g von Frey filaments in black), and the responses to graded mechanical stimuli were, if anything, enhanced. [From *Hori* et al., 1984.]

Receptive Field Organization

Most spinothalamic tract cells that project to the hindlimb part of the ventral posterior lateral nucleus of the thalamus have restricted receptive fields on the contralateral hindlimb (fig. 5/23A) [*Giesler* et al., 1981b]. However, spinothalamic tract cells that project just to the medial thalamus, to the region of the central lateral nucleus of the intralaminar complex, often have very large receptive fields that may include all of the surface of the body and face (fig. 5/23B) [*Giesler* et al., 1981b]. Spinothalamic tract cells that branch and supply both the ventral posterior lateral nucleus and the central lateral nucleus have receptive fields like those of cells that project only to the ventral posterior lateral nucleus [*Giesler* et al., 1981b].

In addition to excitatory receptive fields, spinothalamic tract cells often have inhibitory receptive fields on other parts of the body surface. Wide dynamic range (multireceptive) cells usually have very large inhibitory

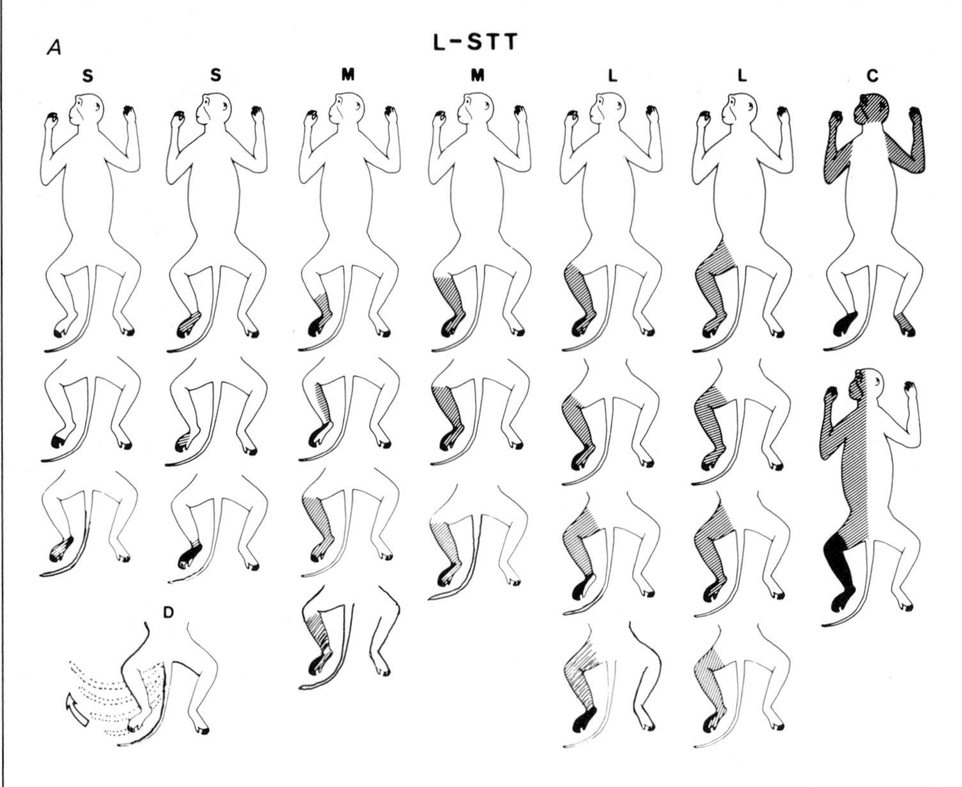

Fig. 5/23. A shows receptive fields of a series of spinothalamic tract cells that projected to the lateral thalamus (L-STT cells). *B* shows receptive fields of spinothalamic tract cells projecting to the medial thalamus (M-STT cells). [From *Giesler* et al., 1981b.]

receptive fields, extending over much of the surface of the body and face, apart from the excitatory field (fig. 5/24) [*Gerhart* et al., 1981]. The inhibitory receptive fields of high threshold (nociceptive-specific) spinothalamic tract cells are less easily demonstrated, but they can occupy a similarly extensive region. Uncommonly, there may be an inhibitory receptive field in the same region as the excitatory field and the inhibitory actions can be due to activation of sensitive mechanoreceptors [*Milne* et al., 1981]. However, usually, the inhibitory influences are restricted to nociceptive inputs [*Gerhart* et al., 1981]. These observations for spinothalamic tract cells are

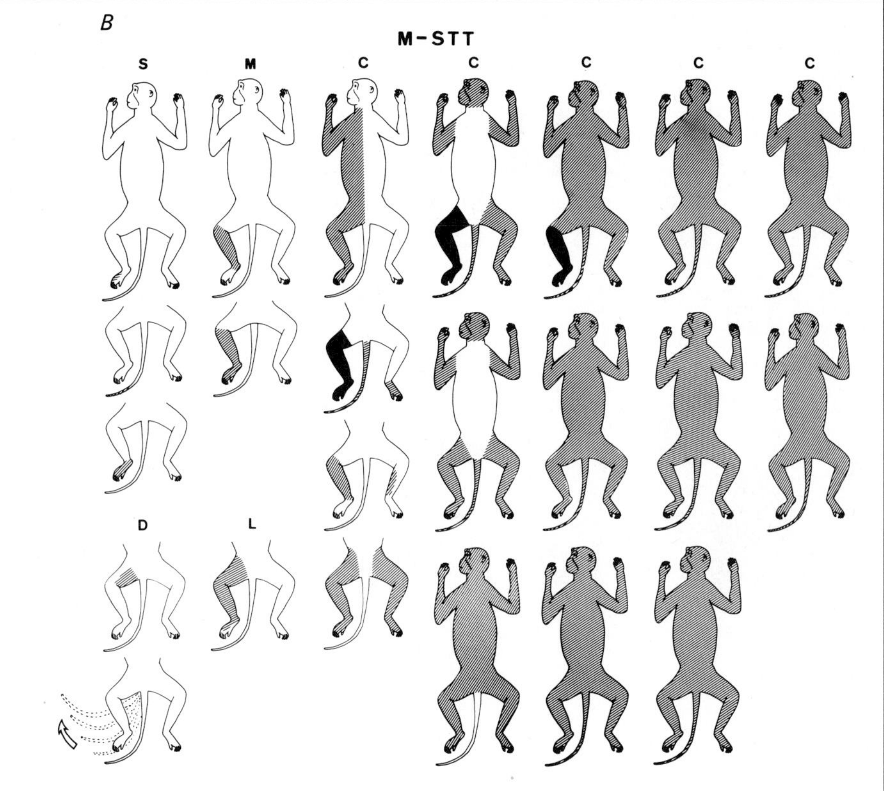

similar to those made for the 'diffuse noxious inhibitory controls' described by *Besson's* group for dorsal horn interneurons in the rat [*LeBars* et al., 1979a, b]. However, the high threshold inhibitory receptive fields in the monkey do not outlast the stimulus and are only partially reduced by transection of the spinal cord (fig. 5/25), whereas the 'DNIC' system in the rat causes inhibition that outlasts stimulation and depends upon a supraspinal loop. Recently, an inhibitory system within the rat spinal cord has been described in addition to the DNIC mechanism [*Cadden* et al., 1983; *Fitzgerald,* 1982].

The DNIC system has been proposed to be involved in the coding of nociceptive information [*LeBars* et al., 1979b]. The scheme suggested is that the population of wide dynamic range (multireceptive) ascending tract cells that are not excited by a noxious stimulus because the stimulus is out-

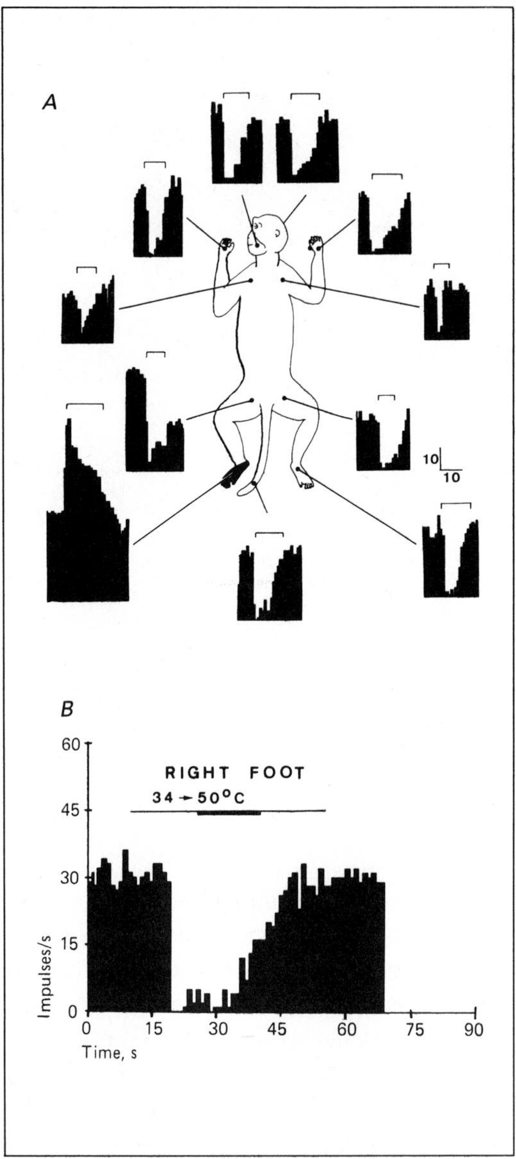

Fig. 5/24. A Inhibitory receptive field of a wide dynamic range (multireceptive) spinothalamic tract cell. The excitatory field was on the left foot, as indicated by the blackened area in the figurine. Pinching the skin anywhere else on the body or face caused inhibition of the background discharges of the cell. *B* Noxious heating of the right foot also caused inhibition of the cell. [From *Gerhart* et al., 1981.]

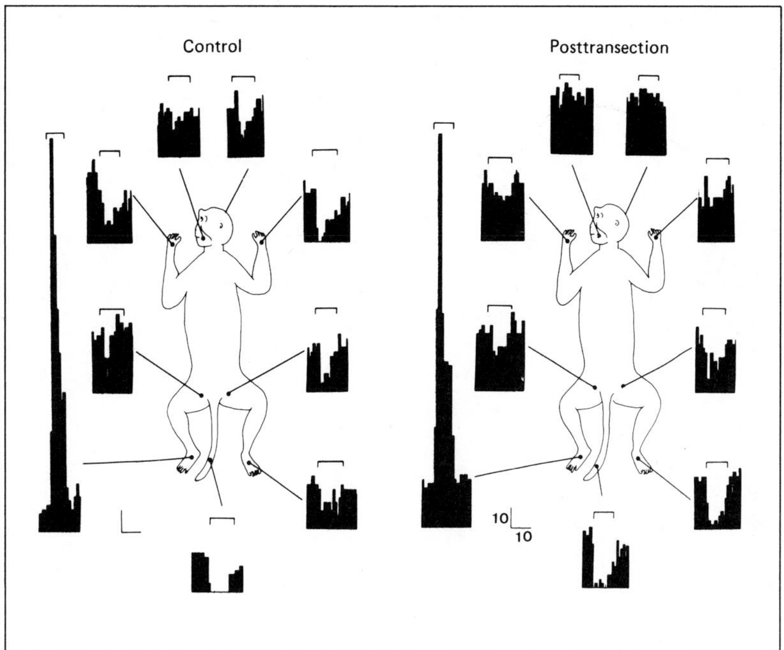

Fig. 5/25. Excitatory and inhibitory receptive fields of a spinothalamic tract cell before and after transection of the spinal cord at an upper cervical level. [From *Gerhart* et al., 1981.]

side their excitatory receptive field are instead inhibited. Thus, a noxious stimulus simultaneously excites one population of cells and inhibits another, larger population, producing contrast within the sensory system for nociception. Recent papers by *LeBars* and his colleagues provide additional arguments for this concept [*LeBars* et al., 1981; *LeBars and Chitour,* 1983].

Prolonged Inhibition following Peripheral Nerve Stimulation

When a peripheral nerve is stimulated repetitively for minutes, the responses of spinothalamic tract cells can be shown to be inhibited, also for minutes [*Chung* et al., 1984a, b]. For example, in figure 5/26, the common peroneal nerve was stimulated repetitively at 2 Hz for 15 min. The

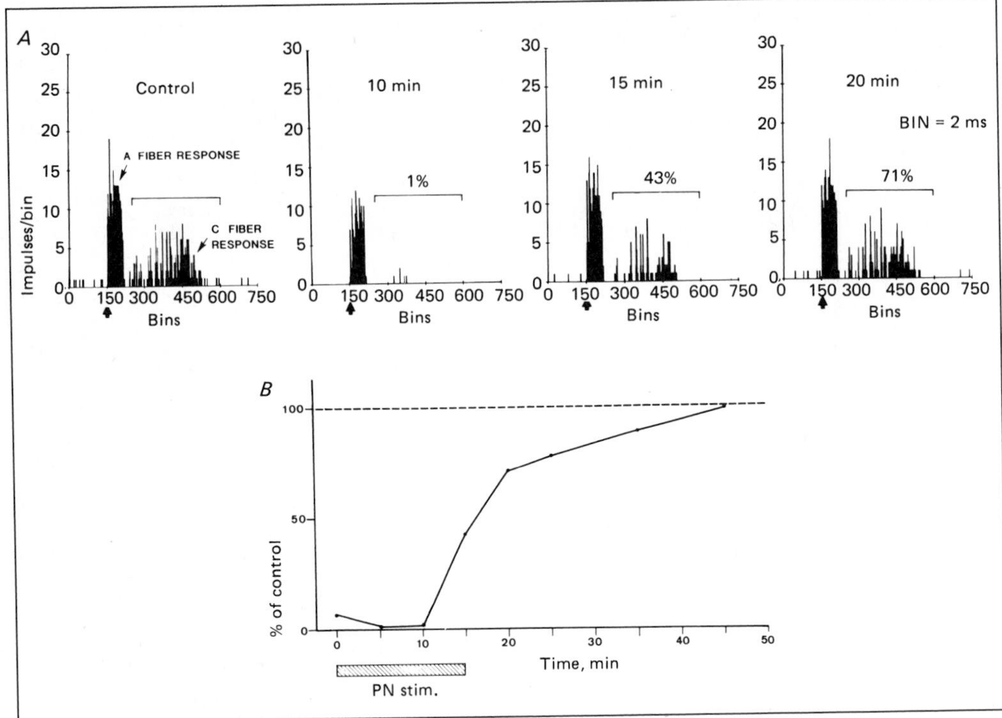

Fig. 5/26. The graphs in *A* show the responses of a spinothalamic tract cell to a volley in the A and C fibers of the sural nerve, before (Control), during (10 min) and after (15 and 20 min) conditioning stimulation of the common peroneal nerve at 2 Hz (C fiber strength) for 15 min. The graph in *B* plots the amount of inhibition of the C fiber response caused by the repetitive peripheral nerve (PN) stimulation. [From *Chung* et al., 1984a.]

responses of a spinothalamic tract cell to afferent volleys in A and C fibers of another peripheral nerve, the sural, were tested before, during and after conditioning stimulation. The responses to C fiber volleys were inhibited more than were the responses to A fiber volleys. There was a prolonged inhibition of the C fiber response that outlasted the period of conditioning stimulation by some 20–30 min.

By testing for the potency of various fiber groups and frequencies of stimulation, it could be shown that peripheral nerve stimulation-evoked inhibition was produced most effectively by Aδ fibers activated at rates of 20 Hz (higher frequencies were not tested for Aδ volleys). Aβ fibers were much less effective, although stimulation at a rate of 100 Hz was more

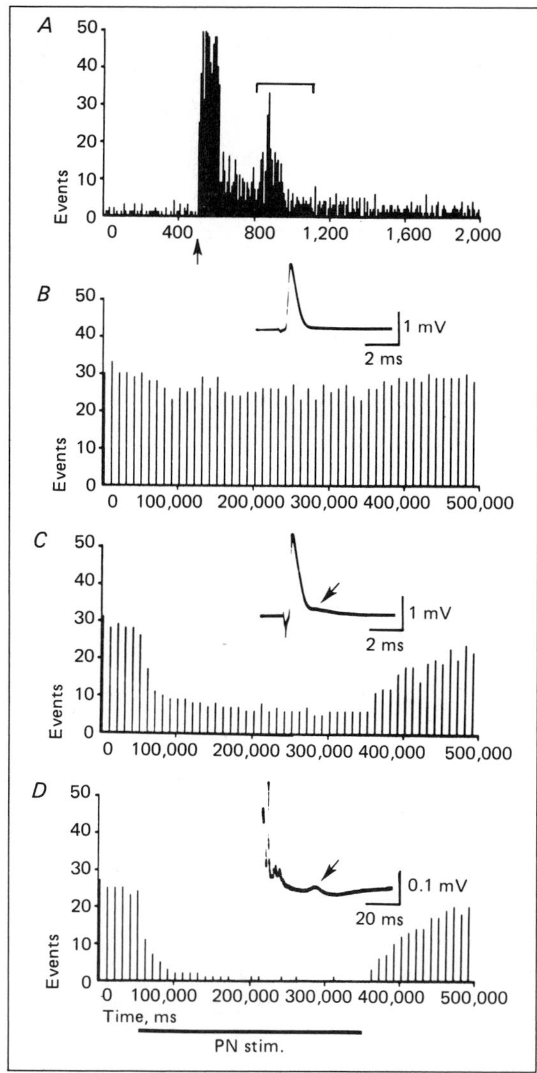

Fig. 5/27. A Inhibition of the response of a spinothalamic tract cell to C fiber volleys by conditioning stimulation of a peripheral nerve at different strengths. The C fiber responses were evoked by stimulation of the sural nerve. Conditioning stimulation was applied to the tibial nerve at 2 Hz for 5 min. The responses shown as vertical lines in *B–D* were gated counts of the C fiber responses. The insets show the afferent volleys recorded from a filament of the tibial nerve, indicating that the strength of stimulation activated just Aβ fibers in *B*, Aδ fibers in *C* and C fibers in *D*. Conditioning stimulation was on during the period indicated by the horizontal line at the bottom of the figure. [From *Chung* et al., 1984b.]

effective than stimulation at 2 Hz. C fiber volleys produced an increment in inhibition beyond that caused by Aδ fibers (fig. 5/27) [*Chung* et al., 1984b].

It was suggested that this type of peripheral nerve stimulation-produced inhibition resembles the effects of transcutaneous nerve stimulation and also acupuncture in producing analgesia [*Chung* et al., 1984b].

Pharmacology of Spinothalamic Tract Cells

The effects of iontophoretically released drugs have been tested upon spinothalamic tract cells [*Jordan* et al., 1978; *Willcockson* et al., 1984a, b]. The agents used have included several of the substances known to occur in dorsal root ganglion cells or in neurons of the dorsal horn. For example, glutamate excites spinothalamic tract cells, and substance P usually excites, but sometimes inhibits, these neurons (fig. 5/28) [*Willcockson* et al., 1984b]. Met- and leu-enkephalin inhibit spinothalamic tract cells, and this effect can be antagonized by naloxone (fig. 5/29) [*Willcockson* et al., 1984b]. The actions of somatostatin, neurotensin, ATP, and other substances intrinsic to the dorsal horn or contained in the terminals of primary afferents are not yet known. Some transmitters present in the endings of axons descending from the brain, such as serotonin, norepinephrine, and dopamine, are inhibitory [*Jordan* et al., 1978; *Willcockson* et al., 1984a], as is glycine, γ-aminobutyric acid and acetylcholine [*Willcockson* et al., 1984a].

Role of Spinothalamic Tract in Pain Transmission

The responses of spinothalamic tract cells to afferent volleys in Aδ and C fibers and to noxious stimulation of the skin, of muscle and of certain viscera seem to make this tract a strong candidate for mediating pain sensation in primates. Coupled with clinical observations, it seems reasonable to conclude that this pathway is likely to be of major significance in human pain. However, there is no reason to exclude important roles for other pathways as well.

Some of the pain-related phenomena for which the response properties of spinothalamic tract cells could serve as at least partial explanations include motivational-affective responses, hyperalgesia, referral, and central

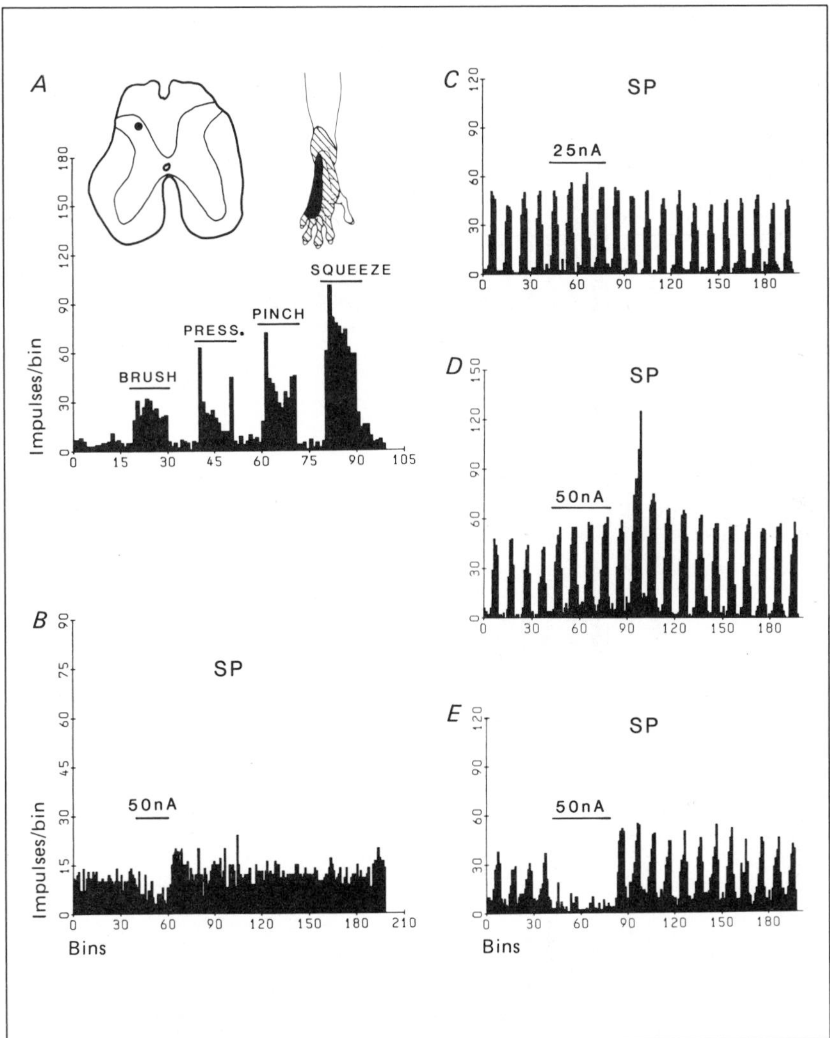

Fig. 5/28. Excitation of a wide dynamic range (multireceptive) spinothalamic tract cell by glutamate and by substance P applied iontophoretically. The responses of the cell to graded mechanical stimuli, the location of the recording point, and the receptive field are shown in *A*. The histogram in *C* shows the burst discharges produced by pulses of glutamate released from a multibarrel microelectrode array, and the enhancement of these discharges when substance P (SP) was concurrently released from another barrel. *D* shows the effect of increasing the ejection current for SP. The inhibitory effect of SP on the same cell (followed by an excitatory effect) is shown using glutamate pulses to evoke activity in *E* and pinch in *B*. [From *Willcockson* et al., 1984b.]

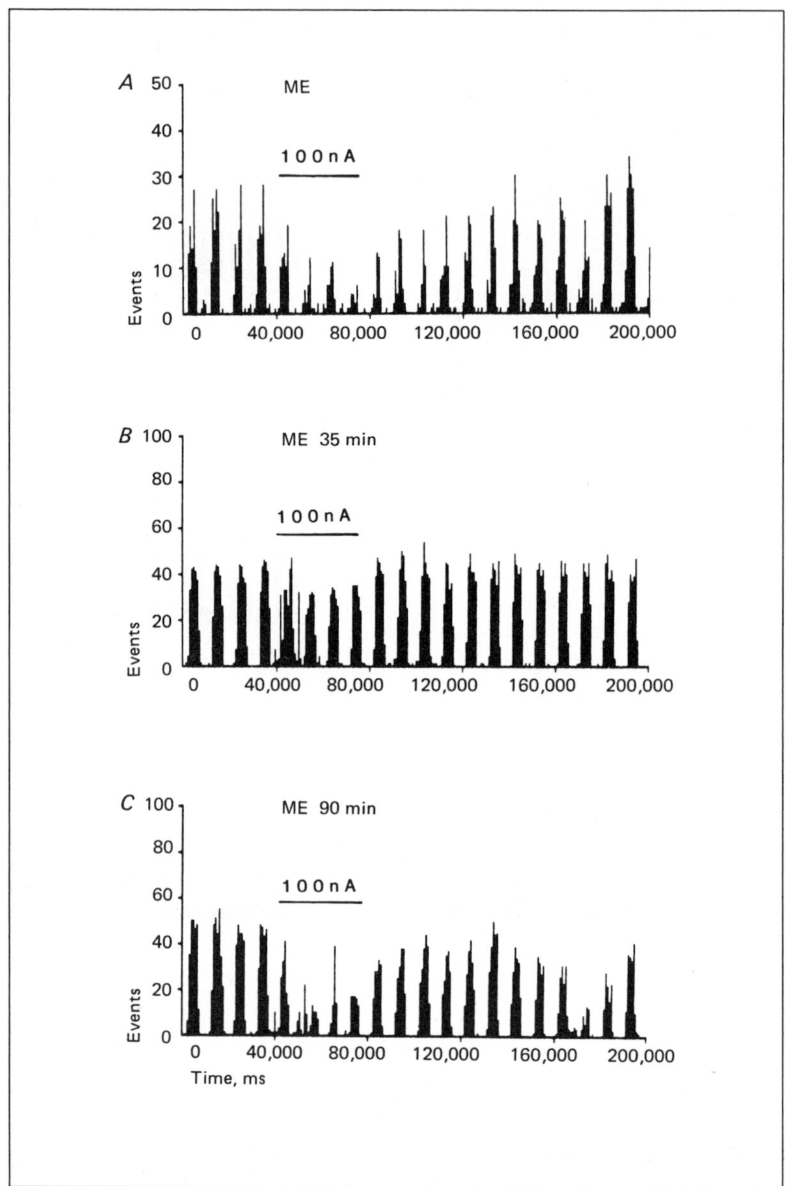

Fig. 5/29. Inhibitory effect of met-enkephalin (ME) on the glutamate-evoked activity of a spinothalamic tract cell is shown in *A*. Naloxone administered systemically (0.05 mg/kg, intravenously) antagonized the action of the met-enkephalin, as shown in *B*. The inhibitory effect returned, as seen in *C*. [From *Willcockson* et al., 1984b.]

pain. The projections of the spinothalamic tract to the reticular formation, by collaterals, and to several medial thalamic nuclei, including the central lateral nucleus, suggest a role for this pathway in arousal and in other motivational-affective reactions to noxious stimuli. The connections of the central lateral nucleus to motor systems [*Jones and Leavitt,* 1974] is consistent with motor responses to painful input, and the widespread connections of this nucleus to the cerebral cortex are in keeping with a possible role in arousal [*Jones and Leavitt,* 1974].

As discussed in Chapter 3, hyperalgesia can be subdivided into primary hyperalgesia and secondary hyperalgesia. A contribution is probably made to hyperalgesia by a central nervous system mechanism, since the details of the time course of development of hyperalgesia cannot be accounted for on the basis of the responses of peripheral nociceptors [*LaMotte* et al., 1983]. Spinothalamic tract cells show an increased responsiveness to tactile stimuli applied to an area of their receptive field outside of that which is damaged by noxious heat [*Kenshalo* et al., 1982], suggesting that an increase in sensitivity of tactile pathways has occurred, perhaps because of a 'wind-up' like phenomenon. Alternatively, a peripheral sensitization mechanism may have contributed [cf. *Fitzgerald,* 1979].

The relationship between the pain produced in human subjects by graded noxious heat pulses before and after thermal damage was discussed in Chapter 3 [*LaMotte* et al., 1983]. Figure 3/22 shows that hyperalgesia results from the thermal damage (50 °C for 100 s), at least for noxious heat pulses up to 49 °C. However, the pain caused by a 50 °C test pulse is essentially unchanged. Interestingly, spinothalamic tract cells show a similar phenomenon (fig. 5/30) [*Kenshalo* et al., 1979]. Two series of noxious heat stimuli were given, each stimulus lasting 120 s. The highest temperature reached was 50 °C, and so the last stimulus of the first series would have produced a comparable amount of thermal damage to that induced by *LaMotte* et al. [1983]. The responses of the spinothalamic tract cells were enhanced for stimuli of 43, 45 and 47 °C, but not 50 °C. The parallel with the human sensation curve is striking.

Pain referral may be explained in part on the basis of the responses of wide dynamic range (multireceptive) neurons of tracts like the spinothalamic tract. These cells receive a convergent input from nociceptors in the skin, muscle and viscera. Presumably, the brain would have difficulty in recognizing the source of a noxious input, and so one would attribute the input to whatever sources most often used the pathway during development, generally a source on the body surface.

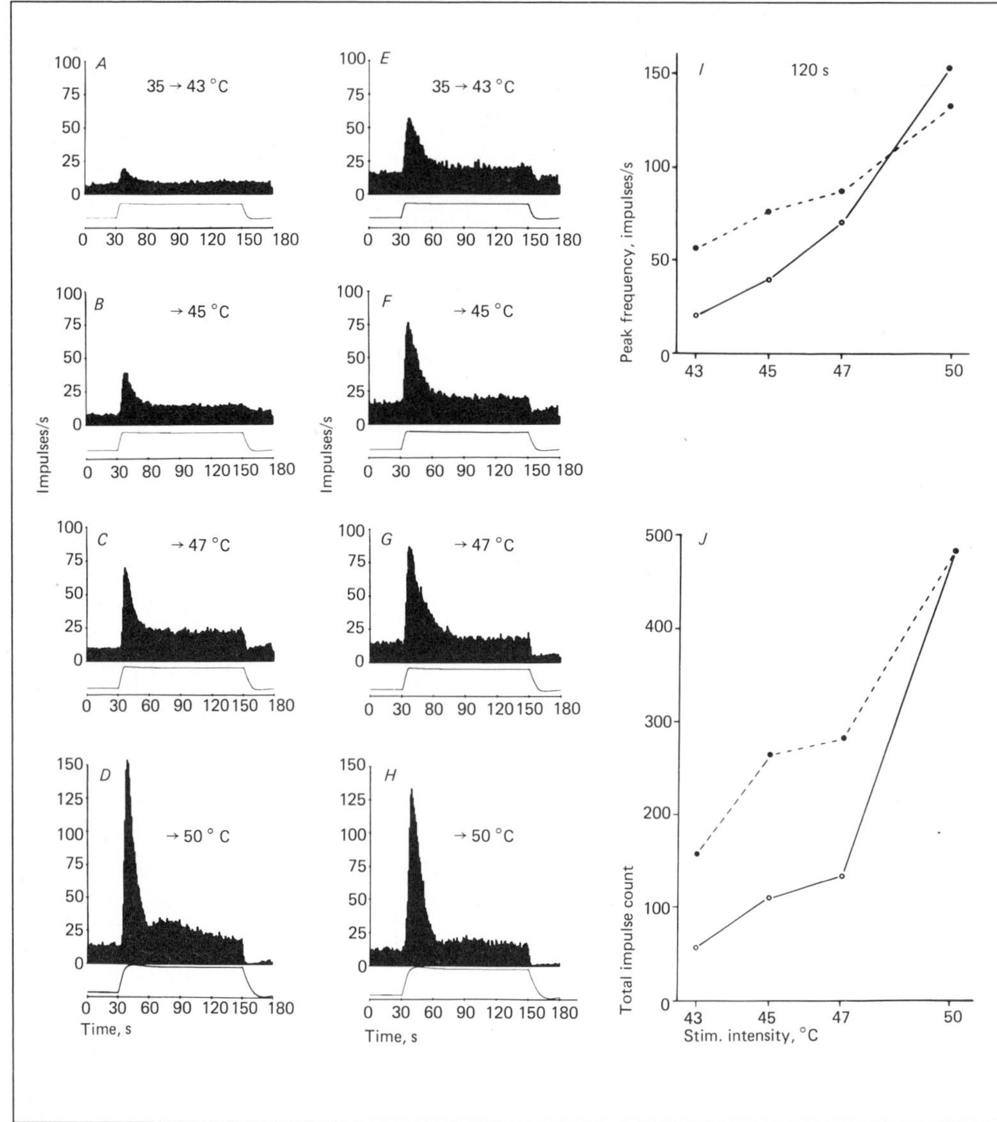

Fig. 5/30. The histograms are averaged responses of 11 spinothalamic tract cells to 120 s duration noxious heat stimuli. *A–D* are responses to the first series of stimuli, and *E–H* are those to the second series. The graphs in *I* and *J* show the peak frequencies and total impulses from onset of response to peak frequency plotted against temperature for the first (solid lines, open circles) and second (dashed lines, filled circles) series of stimuli. [From *Kenshalo* et al., 1979.]

Central pain might sometimes be due to activity in spinothalamic tract cells. Some pathways descending from the brain excite spinothalamic tract cells, while other pathways inhibit them. It would be easy to conceive of lesions that might alter the balance of excitatory or inhibitory inputs onto these cells in such a way as to enhance their background discharges and so to provide a constant signal to the brain that might be interpreted as pain originating from the periphery [*Giesler* et al., 1981b; *Haber* et al., 1980; *Yezierski* et al., 1983]. There is no direct evidence for this view, but presumably something of this sort could happen in cases of central or 'deafferentation' pain. The site of abnormal discharge might, of course, be anywhere along the pain transmission system. The possibility of an involvement of the spinothalamic tract is speculative.

Spinoreticular Tract

Cells of Origin

The locations of neurons that give rise to the spinoreticular tract have been mapped by antidromic activation in the cat [*Fields* et al., 1977; *Levante and Albe-Fessard,* 1972; *Maunz* et al., 1978] and monkey [*Haber* et al., 1982], and have also been mapped by retrograde labeling using horseradish peroxidase in the rat [*Andrezik* et al., 1981; *Chaouch* et al., 1983; *Kevetter and Willis,* 1982, 1983], cat [*Abols and Basbaum,* 1981] and monkey [*Kevetter* et al., 1982]. The distribution in these different species is generally comparable. In the monkey [*Kevetter* et al., 1982], more neurons were labeled in the cervical than in the lumbosacral enlargement. However, the greatest concentration of spinoreticular neurons was in the upper cervical spinal cord (fig. 5/31). Most spinoreticular neurons were in laminae VII and VIII, although some were in other laminae, including laminae I and V. About equal numbers of spinoreticular neurons in the cervical enlargement projected to the ipsilateral as to the contralateral side of the brain stem. However, more spinoreticular cells projected contralaterally than ipsilaterally in the lumbosacral enlargement [*Kevetter* et al., 1982; *Haber* et al., 1982; however, cf. *Kerr and Lippman,* 1974]. Some of the neurons had strikingly long dendrites oriented longitudinally with respect to the spinal cord [*Kevetter* et al., 1982].

Experiments employing a double labeling technique, using different retrograde transported substances, showed that some neurons in the spinal cord project to both the reticular formation and to the thalamus [*Kevetter*

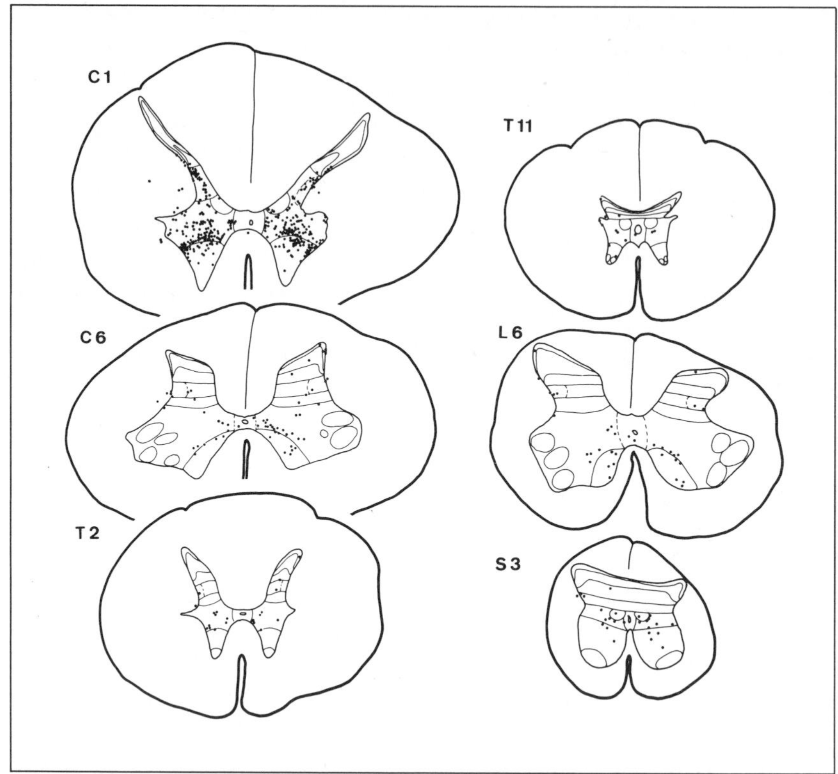

Fig. 5/31. Locations of cells of origin of the spinoreticular tract in the monkey. The distribution of cells labeled by a large injection of horseradish peroxidase into the pontomedullary reticular formation (bilaterally) is plotted on drawings of sections from different levels of the spinal cord. [From *Kevetter* et al., 1982.]

and Willis, 1982, 1983]. A similar observation has been made in electrophysiological experiments, using the antidromic activation technique. Some neurons can be activated antidromically from both the reticular formation and the thalamus or midbrain [*Fields* et al., 1977; *Giesler* et al., 1981b; *Haber* et al., 1982]. Thus, some neurons are both spinoreticular and spinothalamic or spinomesencephalic tract neurons. This observation is in keeping with earlier proposals of an input to the reticular formation from long ascending pathways [*Magoun,* 1963]. Some of the neurons that can be doubly labeled from the reticular formation and thalamus are in a special ventral horn nucleus, as investigated by *Carstens and Trevino* [1978b] in the upper cervical spinal cord.

A count of the spinoreticular neurons that were labeled by injections of horseradish peroxidase into the reticular formation (total of 1,481 cells) suggested that the number of these cells is of the same order of magnitude as the number of spinothalamic tract cells, at least in the monkey spinal cord [*Kevetter* et al., 1982; cf. *Willis* et al., 1979]. This is a surprising observation, since lesions of the anterolateral quadrant of the spinal cord produce substantially more degeneration in the reticular formation than in the thalamus [*Bowsher,* 1961, 1976; *Mehler,* 1969]. One explanation is that not all of the cells of origin of the spinoreticular tract were labeled, whereas a larger proportion of the cells giving rise to the spinothalamic tract may have been. Alternatively, the spinoreticular tract may arise from a similar number of cells as the spinothalamic tract, but each axon may provide more terminals to the reticular formation than are contributed by spinothalamic axons to the thalamus. *Chaouch* et al. [1983] also noticed that a relatively small number of spinoreticular neurons could be labeled in the rat spinal cord, even with large injections of horseradish peroxidase into the reticular formation.

Organization of the Spinoreticular Tract in the Spinal Cord and Brain Stem

The spinoreticular tract accompanies the spinothalamic tract in the ventrolateral quadrant of the spinal cord [*Mehler* et al., 1960]. As it enters the medulla, it occupies a zone just dorsolateral to the pyramid [*Mehler* et al., 1960]. Some of the fibers continue rostrally in a lateral position, many terminating in the lateral reticular nucleus, a cerebellar relay nucleus [*Kerr,* 1975b; *Mehler* et al., 1960]. Other spinoreticular axons ascend in a more medial position, and these terminate in the medial reticular formation.

Nuclei of Termination in the Reticular Formation

A number of investigators have described the terminations of the spinoreticular tract in the medial reticular formation of the monkey (fig. 5/32) [*Bowsher,* 1961; *Kerr,* 1975b; *Mehler,* 1969; *Mehler* et al., 1960]. Similar results have been obtained for the human [*Bowsher,* 1957; *Mehler,* 1962] and the opossum, rat, rabbit and cat [*Mehler,* 1969]. In the caudal medulla of the monkey, there are spinoreticular endings in the retroambiguous nucleus, the nucleus supraspinalis, and the dorsal and ventral parts of the nucleus medullae oblongatae centralis [*Mehler* et al., 1960]. More rostrally,

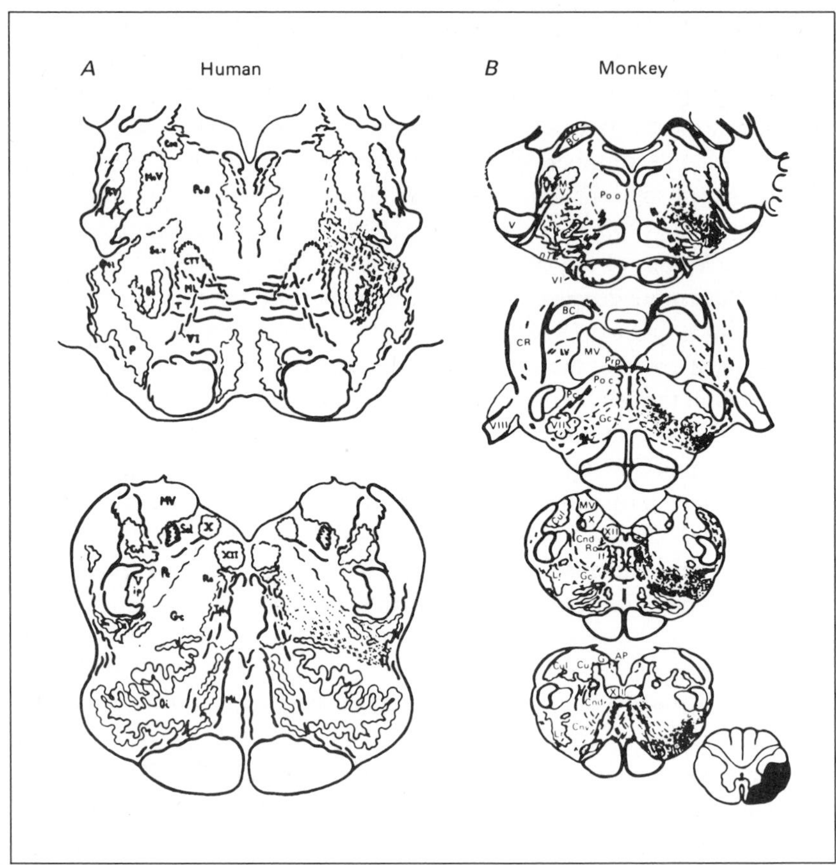

Fig. 5/32. Terminations of the spinoreticular tract in the human *(A)* and monkey *(B)*. [From *Mehler,* 1962; *Mehler* et al., 1960.]

there are terminations in the lateral reticular nucleus and in the nucleus gigantocellularis, the nucleus interfascicularis hypoglossi, and the nucleus of Roller. There are also spino-olivary projections, which can be considered a separate system. Rostral to the inferior olive, the spinoreticular tract projects to the nucleus paragigantocellularis lateralis and dorsalis and to the raphe. In the caudal pons, connections can be demonstrated to the nucleus pontis centralis caudalis and the nucleus subcoeruleus ventralis. More rostrally, there are endings in the nucleus pontis centralis oralis and in the subcoeruleus complex.

Identification of Spinoreticular Neurons in Physiological Experiments

As for the cells of origin of the spinothalamic tract, it is feasible to identify neurons of the spinoreticular tract by antidromic activation [*Fields* et al., 1975, 1977; *Haber* et al., 1982; *Levante and Albe-Fessard,* 1972; *Maunz* et al., 1978]. One or more stimulating electrodes can be placed in the reticular formation in order to activate axons projecting to the vicinity of the electrode. However, a problem with this approach is that axons passing by the electrode but activated by it may actually terminate some distance away. Thus, there is a possibility for misidentification of neurons as spinoreticular that actually have other destinations than the reticular formation.

Electrophysiological Response Properties of Spinoreticular Neurons and Role in Pain

Many spinoreticular neurons identified by antidromic activation are nociceptive [*Fields* et al., 1975, 1977; *Haber* et al., 1982; *Maunz* et al., 1978]. Some spinoreticular neurons have receptive fields that resemble those of spinothalamic tract cells; in fact, some spinoreticular neurons are also spinothalamic tract cells [*Haber* et al., 1982; *Giesler* et al., 1981b]. Thus, the same information that is transmitted to the thalamus by such neurons is also available within the reticular formation. However, it is not clear from experiments in which recordings have been made from reticular formation neurons if the sensory information content present in the discharges of spinothalamic-like spinoreticular neurons is preserved in the discharges of the reticular formation neurons. Clearly, many reticular neurons respond preferentially to noxious stimuli [*Casey,* 1969, 1971a; *Guilbaud* et al., 1973; *Wolstencroft,* 1964]. Furthermore, stimulation within the reticular formation produces aversive behavior [*Casey,* 1971b]. Nevertheless, it seems more likely that the reticular formation participates in the motivational-affective components of the response to a painful stimulus than in the sensory-discriminative component [*Melzack and Casey,* 1968; *Price and Dubner,* 1977; however, cf. *Bowsher,* 1976]. Ascending pathways from the reticular formation to the medial thalamus and to limbic structures are consistent with this hypothesis [*Bowsher,* 1975; *Bowsher* et al., 1968; *Nauta and Kuypers,* 1958; *Robertson* et al., 1973; *Scheibel and Scheibel,* 1958].

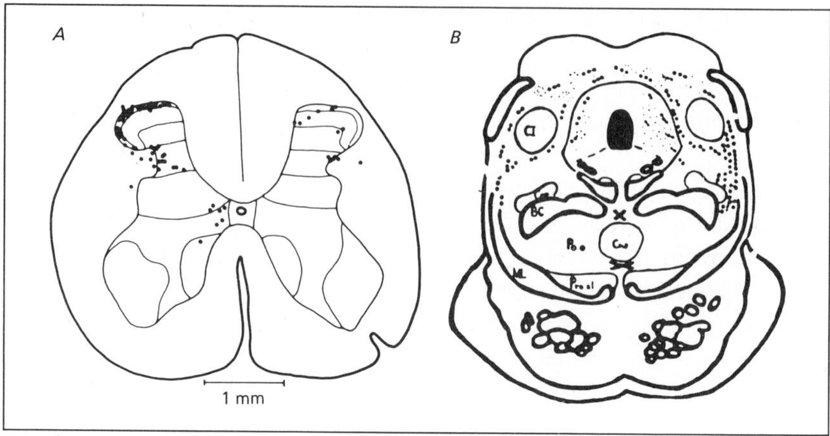

Fig. 5/33. A shows cells of origin of the primate spinomesencephalic tract labeled after an injection of horseradish peroxidase into the midbrain. [From *Willis* et al., 1979.] *B* shows the terminals of the spinomesencephalic tract in a monkey after anterolateral cordotomy. Terminal degeneration is shown by fine stippling and axons of passage by coarse stippling. [From *Mehler* et al., 1960.]

Spinomesencephalic Tract

Cells of Origin

The cells of origin of the spinomesencephalic tract have been mapped using the horseradish peroxidase technique in the monkey (fig. 5/33A) [*Trevino,* 1976; *Willis* et al., 1979], cat [*Wiberg and Blomquist,* 1981] and rat [*Liu,* 1983; *Menétrey* et al., 1982] and by the antidromic activation technique in the rat [*Menétrey* et al., 1980]. The cells have a distribution like that of the cells projecting to the ventrobasal complex (and not like that of cells projecting to the rhombencephalic reticular formation). Interestingly, more cells can be labeled in lamina I of the rat spinal cord (as well as in the lateral spinal nucleus or nucleus of the lateral funiculus) after midbrain than after lateral thalamic injections of horseradish peroxidase [*Menétrey* et al., 1982]. Large numbers of labeled cells are also seen in lamina I in the cat [*Wiberg and Blomquist,* 1981] and in the monkey [*Willis* et al., 1979].

The similarity in the locations of neurons projecting to the midbrain and to the ventral posterior lateral nucleus of the thalamus in horseradish peroxidase studies is consistent with the observation by *Price* et al. [1978] that the same neurons can often be activated antidromically following stimulation in both the thalamus and the region of the periaqueductal gray.

Organization of the Spinomesencephalic Tract in the Spinal Cord and Brain Stem

The spinomesencephalic tract ascends in the ventrolateral white matter of the spinal cord, in company with the spinothalamic and spinoreticular tracts [*Mehler* et al., 1960; *Mehler,* 1962, 1969]. However, there is evidence that there are spinomesencephalic projections ascending in the dorsolateral funiculus in the rat spinal cord [*Zemlan* et al., 1978; cf. *McMahon and Wall,* 1983]. The spinomesencephalic tract accompanies the spinothalamic tract to the level of the isthmus [*Mehler,* 1969]. At this level, the spinomesencephalic fibers course medially and dorsally to reach their terminal zones in the midbrain.

Nuclei of Termination in the Midbrain

The spinomesencephalic tract has comparable termination zones in the midbrains of the opossum, rat, rabbit, cat, monkey and human (fig. 5/33B) [*Boivie,* 1979; *Bowsher,* 1957; *Kerr,* 1975b; *Mehler,* 1969; *Mehler* et al., 1960]. There are projections to the deep layers of the superior colliculus, the nucleus intercollicularis, the lateral periaqueductal gray, the nucleus cuneiformis, the nucleus of Darkschewitz, and the Edinger-Westphal nucleus.

Identification of Spinomesencephalic Neurons in Physiological Experiments

It is feasible to identify spinomesencephalic tract cells by antidromic activation, provided that the stimulating electrode is placed sufficiently medially and that the stimulating current is kept low enough to avoid activation of the axons of the spinothalamic tract [*Price* et al., 1978; *Menétrey* et al., 1980].

Electrophysiological Response Properties of Spinomesencephalic Tract Cells and Role in Pain

There have been very few studies of the activity of spinal neurons that project to midbrain sites ventral to the tectum. Apart from the observations of *Price* et al. [1978], the main evidence is the study by *Menétrey* et al. [1980] in the rat. These neurons were largely nociceptive.

It seems highly likely that spinomesencephalic projections play an important role in pain mechanisms, as evidenced by the results of midbrain surgery in humans. Although pain can be partially relieved by interrupting the spinothalamic tract at a midbrain level [*Walker,* 1942], it has been found that pain relief is often followed by the development of central dysesthesias [*Nashold* et al., 1974]. However, surgical lesions placed medial to the location of the spinothalamic tract resulted in relief of suffering, suggesting an involvement of medial parts of the midbrain in the transmission system responsible for chronic pain [*Spiegel and Wycis,* 1948]. Lesions of the periaqueductal gray in animals also reduce nociceptive responses [*Melzack* et al., 1958; *Skultety,* 1958].

Other evidence for a role of midbrain nuclei in pain transmission comes from stimulation studies. Electrical stimulation near the periaqueductal gray in humans may cause a sensation of diffuse pain referred to central parts of the body or a feeling of fear [*Nashold* et al., 1969, 1974], and comparable stimulation in animals often elicits rage reactions and vocalization [*Skultety,* 1963; *Spiegel* et al., 1954].

In other studies, it has been found that stimulation in the region of the periaqueductal gray causes analgesia [*Mayer and Liebeskind,* 1974; *Reynolds,* 1969; see review by *Willis,* 1982].

Presumably, spinomesencephalic projections could participate in pain mechanisms by activating midbrain neurons that contribute to sensory processing, including relaying information to the somatosensory thalamus. There are ascending pathways from the midbrain to the ventrobasal thalamus that might serve this role [e.g. *Barbaresi* et al., 1982; *Hamilton,* 1973]. Other projections of the periaqueductal gray connect with the medial thalamus and the limbic system. These projections may serve a role in motivational-affective responses [*Chi,* 1970; *Hamilton,* 1973; *Hamilton and Skultety,* 1970]. Another possibility is that the nociceptive discharges of spinomesencephalic neurons might engage descending analgesia systems and thus limit nociceptive input, for example at a spinal cord level [see *Basbaum and Fields,* 1978].

Dorsally Situated Ascending Pathways That May Be Nociceptive

As mentioned in the historical section of this Chapter, the return of pain after an initially successful cordotomy in humans (and in monkeys) requires an explanation. One possibility is that the return of pain results from a plastic change involving ascending tracts in other parts of the spinal cord than the anterolateral quadrant contralateral to the source of the pain. For example, the pain could be conducted by way of an ipsilateral component of the spinothalamic tract. Alternatively, one or more pathways in the dorsal part of the cord may be involved. Consistent with this idea is evidence that the cat depends to a considerable extent on dorsal cord pathways for pain transmission [*Casey* et al., 1981; *Kennard,* 1954]. The most likely pathways in the dorsal part of the spinal cord to be involved in nociceptive transmission in the cat are the spinocervical tract and the postsynaptic dorsal column pathway. There is evidence in the rat that nociceptive transmission may depend as much upon propriospinal connections as upon long tracts [*Basbaum,* 1973]. Thus, propriospinal conduction must also be considered as a possible alternative to account for the return of pain after cordotomy.

Spinocervical Tract

Lateral Cervical Nucleus

It is not yet certain if there is a significant spinocervical tract in the human. The spinocervical tract is known to synapse in the lateral cervical nucleus in the cat [see *Willis and Coggeshall,* 1978]. The lateral cervical nucleus is an isolated cell group found just adjacent to the dorsal horn in the first two cervical segments. Similar nuclei are found in the dog, raccoon and monkey. Recently, a lateral cervical nucleus has also been described in the rat [*Giesler* et al., 1979b; see also *Lund and Webster,* 1967]. However, it is not entirely clear if there is a lateral cervical nucleus in the human. Some human spinal cords have a nucleus that appears to be comparable to the cat nucleus, but other human cords do not [*Truex* et al., 1965]. It seems reasonable to speculate that human spinal cords, like monkey spinal cord, do in fact have a spinocervicothalamic system, but that the relay nucleus in the upper cervical spinal cord is sometimes included within the dorsal horn. Evidence from a degeneration study in a human spinal cord is consistent with the presence of a spinocervicothalamic pathway [*Kircher and Ha,* 1968].

Cells of Origin

The locations of the cells of origin of the spinocervical tract have been mapped by means of the antidromic activation technique and by retrograde transport of horseradish peroxidase (fig. 5/34). In some of the antidromic

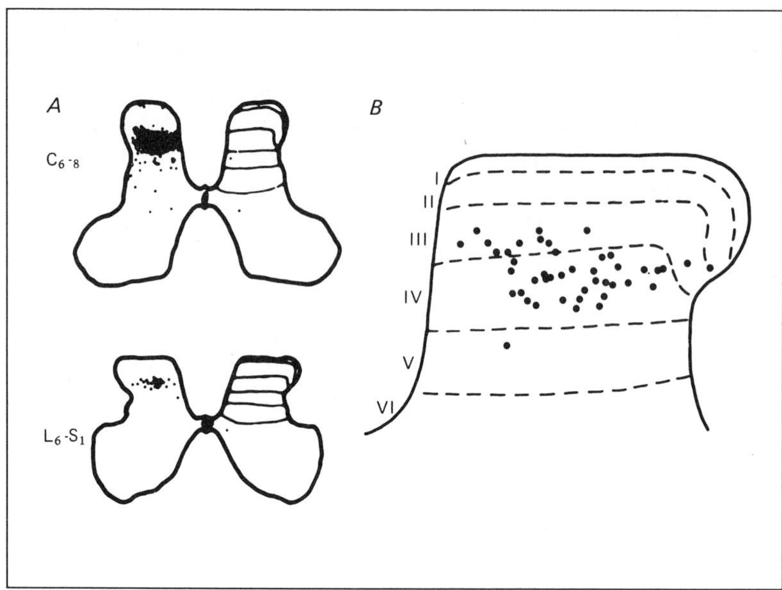

Fig. 5/34. Locations of the cells of origin of the spinocervical tract. The distribution of neurons labeled retrogradely by horseradish peroxidase after injection into the region of the lateral cervical nucleus are shown in *A*. The distribution of intracellularly injected spinocervical tract cells identified by antidromic activation are shown in *B*. [*A* is from *Craig,* 1976; *B* is from *Brown,* 1981.]

activation experiments, the recordings were made extracellularly, and so the locations of the spinocervical tract cells were not established precisely [*Bryan* et al., 1973, 1974]. In studies employing intracellular labeling, the exact locations of the spinocervical tract cells could be visualized (fig. 5/34B, 5/35) [*Brown* et al., 1977a]. One limitation of this type of study is the possibility of sampling error. Another approach is the retrograde labeling of spinocervical tract cells after injection of horseradish peroxidase into the region of the lateral cervical nucleus (fig. 5/34A) [*Brown* et al., 1980b; *Craig,* 1976, 1978]. Presumably all of the cell types belonging to the spinocervical tract would label, including any small neurons that might be hard to impale with a microelectrode. However, a different type of error that might occur is the inadvertent labeling of spinal neurons projecting to the dorsal horn at an upper cervical level, rather than to the lateral cervical nucleus, since the horseradish peroxidase is likely to spread into the adjacent dorsal horn [*Brown* et al., 1980b; *Craig,* 1976]. With these limitations

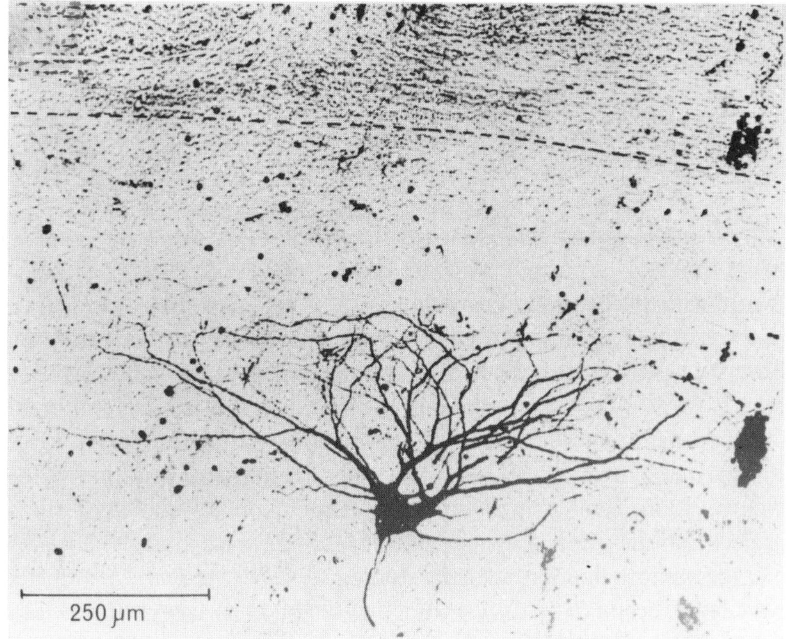

Fig. 5/35. Horseradish peroxidase stained spinocervical tract neuron in a sagittal section. The dorsal border of the dorsal horn is indicated by the dashed line. [From *Brown,* 1981.]

in mind, it can be said that the cells of origin of the spinocervical tract are found chiefly in laminae III, IV and V. *Brown* et al. [1980b] estimate that there are between 550 and 800 spinocervical tract cells on each side in the cat lumbar cord, and about 2,200 for each side of the entire cord.

The dorsally projecting dendrites of spinocervical tract cells ascend through lamina III, but they tend to turn horizontally near the border of laminae III and II [*Brown,* 1981]. This is seen well in sagittally sectioned material (fig. 5/35) [*Brown,* 1981].

Organization of Spinocervical Tract and Destination of Terminals

The spinocervical tract ascends in the dorsal lateral funiculus and ends in the lateral cervical nucleus at cervical segments 1 and 2. Neurons in the lateral cervical nucleus send their axonal projections to the contralateral ventral funiculus, where they turn rostrally to reach the brain stem, joining the medial lemniscus and ascending to the thalamus and to other targets

[*Berkley,* 1980; *Boivie,* 1970, 1980; *Ha,* 1971; *Ha and Liu,* 1966]. The projections of the lateral cervical nucleus are to the following structures: ventral posterior lateral nucleus, the medial part of the posterior complex and the midbrain [*Berkley,* 1980; *Blomqvist* et al., 1978; *Boivie,* 1970, 1980].

Identification of Spinocervical Tract Neurons

In neurophysiological experiments, it is now customary to identify spinocervical tract cells as those neurons that can be activated antidromically from the C3 segment but not from the C1 level or from the C1 level as well only if the latency increases enough to indicate that a collateral branch is likely to have been given off at the level of the lateral cervical nucleus [*Brown and Franz,* 1969]. There is a potential problem with this method of identification, since it is possible that some axons in the lateral funiculus project into the spinal gray matter at the C1–C2 segmental level. For example, the spinothalamic tract cells studied by *Carstens and Trevino* [1978b] have large receptive fields over much of the body surface. It is possible that some of this input reaches these cells by axons reaching the nucleus by a route through the lateral funiculus. If this speculation proves correct, then it might be very difficult to distinguish between spinocervical tract cells and neurons connecting with the spinothalamic tract cells of upper cervical segments.

Response Properties of Spinocervical Tract Cells

The original observations that led to the discovery of the spinocervical tract suggested that the cells have a tactile function [*Morin,* 1955]. However, some axons that projected in the vicinity of spinocervical tract were observed by *Lundberg and Oscarsson* [1961] to be excited by the flexion reflex afferents, as well as by low threshold mechanoreceptive afferents. Although many of the experiments on the spinocervical tract done in *Brown's* laboratory have emphasized the tactile properties of these cells, this body of work has also provided ample evidence that some spinocervical tract cells can be excited by C fiber volleys and by noxious intensities of stimulation [*Brown and Franz,* 1969; *Brown* et al., 1975]. Several studies in other laboratories have also demonstrated the ability of nociceptors, includ-

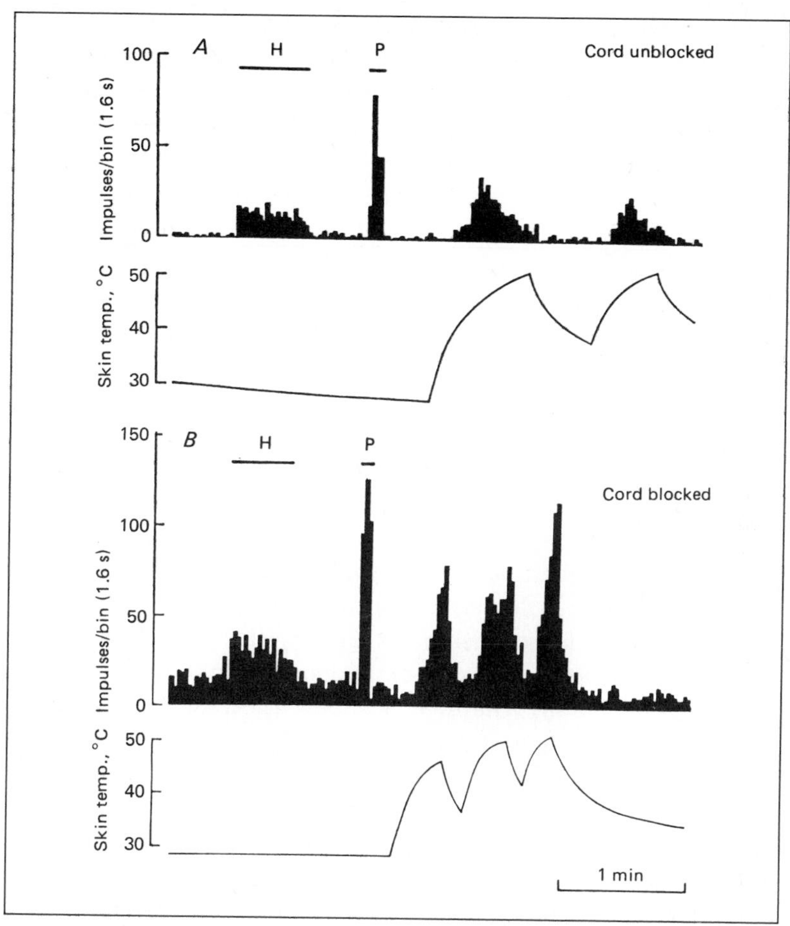

Fig. 5/36. Responses of a spinocervical tract cell to innocuous and noxious stimuli. In *A,* the spinal cord was intact in the anesthetized preparation. In *B,* the spinal cord was blocked by cold, reducing tonic descending inhibition. H shows the response to hair movement, P to pin prick; the application of noxious heat is shown by the temperature monitor. [From *Cervero* et al., 1977a.]

ing fine muscle afferents, to activate spinocervical tract cells (fig. 5/36) [*Cervero* et al., 1977a; *Hamann* et al., 1978; *Hong* et al., 1979; *Kniffki* et al., 1977; *Mendell,* 1966]. Visceral afferents supplying the urinary bladder do not appear to affect spinocervical tract cells in the lumbosacral cord [*Cervero and Iggo,* 1978].

A problem with the hypothesis that spinocervical tract cells mediate nociceptive information is the small proportion of definitive recordings of nociceptive responses in the lateral cervical nucleus [*Craig and Tapper,* 1978; *Horrobin,* 1966; *Kitai* et al., 1965; *Oswaldo-Cruz and Kidd,* 1964]. However, in the rat, many of the cells in the lateral cervical nucleus are nociceptive [*Giesler* et al., 1979b]. The receptive fields are large, and so the spinocervicothalamic system in the rat is unlikely to provide information about the location of a noxious stimulus.

The cervicothalamic projection is not only to the ventral posterior lateral nucleus [*Ha,* 1971], but also to the medial part of the posterior complex [*Berkley,* 1980; *Boivie,* 1970].

Postsynaptic Dorsal Column Pathway

Cells of Origin

The cells of origin of the postsynaptic dorsal column pathway have been demonstrated in the rat, cat and monkey in experiments using retrograde transport of horseradish peroxidase injected into the dorsal column nuclei or placed on severed axons of the dorsal column [*Bennett* et al., 1983; *Brown and Fyffe,* 1981; *Giesler* et al., 1984; *Lu* et al., 1984; *Rustioni and Kaufman,* 1977; *Rustioni* et al., 1979]. The cells are concentrated in laminae III and IV, although some are in adjacent laminae and along the medial border of the dorsal horn (fig. 5/37). A few of the cells are in other locations, including laminae I and II, as well as the ventral horn. Cells in lamina III may send dendrites dorsally as far as lamina I (fig. 5/38) [*Brown and Fyffe,* 1981]. A small proportion of the cells also project to the contralateral thalamus and so are both postsynaptic dorsal column and spinothalamic cells [*Hayes and Rustioni,* 1980].

Organization of the Postsynaptic Dorsal Column Pathway and Destination of Terminals

The axons of the postsynaptic dorsal column pathway can be traced to their destinations in the dorsal column nuclei by first transecting the dorsal roots chronically, allowing the debris of degeneration to be removed, and then transecting the dorsal column to interrupt the remaining axons that belong to the postsynaptic pathway [*Rustioni* et al., 1979; *Rustioni,* 1973, 1974]. Although most of the axons are in fact in the dorsal column, some

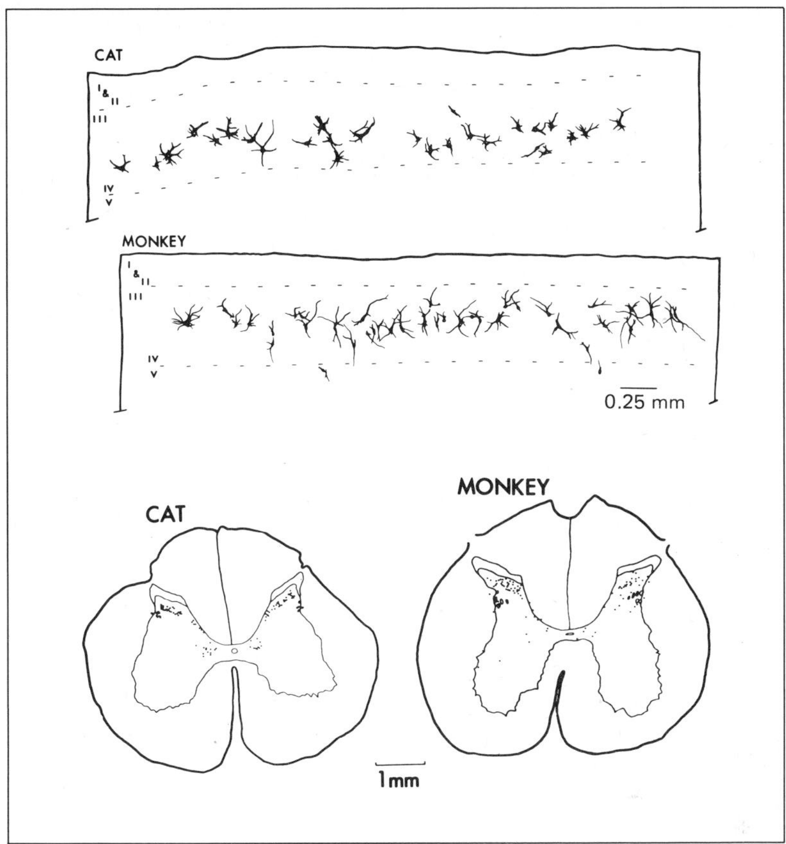

Fig. 5/37. Locations of the cells of origin of the postsynaptic dorsal column pathway in the cat and monkey. The sections above were cut sagittally, and those below transversely. [From *Bennett* et al., 1983.]

are in the dorsal lateral funiculus (see also *Giesler* et al. [1984] for differences in the pathway in the rat).

The axons of the postsynaptic dorsal column path seem to be relatively deep within the dorsal column, according to experiments in which recordings were made from both types of axons [*Uddenberg,* 1968a]. However, *Bennett* et al. [1983] did not notice any special distribution of labeled axons within the dorsal column.

The destination of the pathway is primarily to the dorsal column nuclei. The fibers are somatotopically organized, with those originating

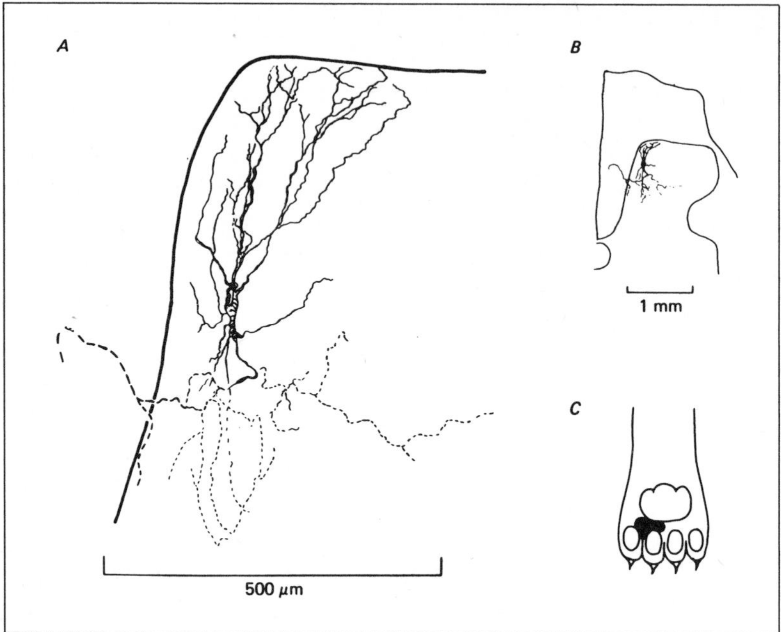

Fig. 5/38. A Reconstruction of horseradish peroxidase-filled postsynaptic dorsal column pathway neuron in the cat. The soma is in lamina III and the dorsally directed dendrites reach lamina I. *B* shows location in the dorsal horn and *C* the excitatory receptive field. [From *Brown and Fyffe,* 1981.]

from cells in the lumbar cord projecting to the nucleus gracilis and those from cells in the cervical cord projecting to the nucleus cuneatus [*Giesler* et al., 1984; *Rustioni,* 1973, 1974; *Rustioni* et al., 1979].

Evidence has recently been obtained that many postsynaptic dorsal column cells can be activated antidromically from the dorsal lateral funiculus and it is proposed that many of these neurons belong also to the spinocervical tract [*Lu* et al., 1984]. However, some of the cells are presumably distinct, since there are a number of differences in the morphology and receptive field organization of spinocervical and postsynaptic dorsal column cells [*Brown and Fyffe,* 1981].

A few of the cells that project to the dorsal column nuclei also project to the contralateral thalamus [*Hayes and Rustioni,* 1980]. Thus, some neurons are at once second-order dorsal column cells and spinothalamic tract cells.

Identification of Postsynaptic Dorsal Column Neurons

The recording of activity from neurons belonging to the second-order dorsal column pathway can be identified in electrophysiological experiments by activating the cells antidromically from the dorsal column nuclei. One technical problem with such experiments is that a stimulus applied in the dorsal column nuclei can activate the cells orthodromically, as well as antidromically, since the cells receive synaptic connections from collaterals of axons traveling in the dorsal columns [*Lu* et al., 1983]. Nevertheless, a number of studies have been done in which cells or axons belonging to this pathway have been identified [*Angaut-Petit,* 1975; *Bennett* et al., 1983; *Brown* et al., 1983; *Lu* et al., 1983; *Uddenberg,* 1968b].

Response Properties of Postsynaptic Dorsal Column Cells

Several different response types have been recognized in recordings from neurons belonging to this pathway. Some cells behave like the low threshold cells, while others resemble the wide dynamic range (fig. 5/39; multireceptive) neurons described in other tracts, including the spinothalamic tract [*Angaut-Petit,* 1975; *Brown* et al., 1983; *Lu* et al., 1983; *Uddenberg,* 1968b]. Thus, some of these cells are clearly nociceptive and could well contribute to pain mechanisms. However, only a few nociceptive-specific cells have been reported [*Angaut-Petit,* 1975], and none were seen by *Uddenberg* [1968b] or *Lu* et al. [1983]. Furthermore, none of the cells in the study by *Lu* et al. [1983] were activated by C fibers and none showed wind-up. However, this could have been a consequence of using barbiturate anesthesia.

Role of the Spinocervical Tract and of the Postsynaptic Dorsal Column Pathway in Pain

Since there is behavioral evidence for a dorsally placed nociceptive pathway in the cat [*Casey* et al., 1981; *Kennard,* 1954], it seems to be a reasonable hypothesis that the spinocervical tract and/or the second-order dorsal column pathway contribute to nociception in this animal. The observation that anterolateral cordotomy results in analgesia in primates and man suggests that the main pain transmission system in these forms is

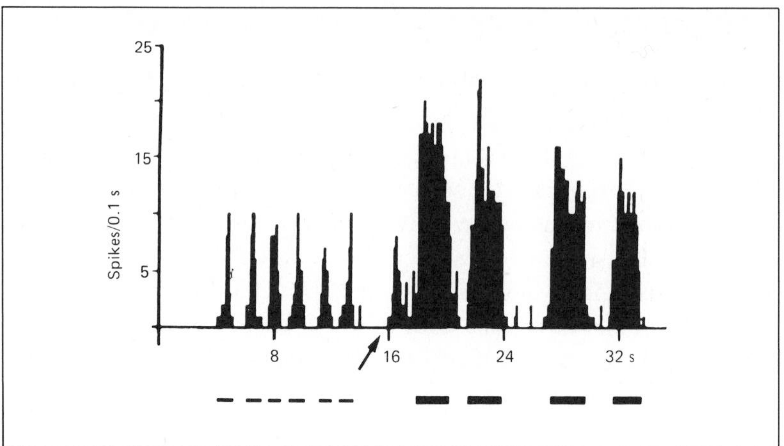

Fig. 5/39. Responses of neuron belonging to the postsynaptic dorsal column pathway. The cell was classified as a wide dynamic range (multireceptive) neuron because it responded to brushing the receptive field (thin horizontal lines under histogram), as well as to noxious pinching (thick horizontal lines). At the time of the arrow, a fold of skin was lifted by the forceps; the pinches were applied without changing the position of the skin fold. [From *Lu* et al., 1983.]

restricted to the ventral part of the cord. The recurrence of pain or nociceptive reactions some time after an initially successful cordotomy may be due to improved use of the dorsally situated nociceptive pathways. However, this is strictly speculative. The observation that this pathway may relay on cells in the dorsal column nuclei other than those that contribute to the medial lemniscus [*Angaut-Petit,* 1975; *Rustioni,* 1973, 1974] suggests that the functional role of the second-order dorsal column pathway may not be directly sensory. An alternative proposal, suggested by *Lu* et al. [1983], is that the dorsal column second-order pathway is involved in pain modulation. Dorsal column stimulation produces analgesia, and this could be in part due to volleys ascending in the second-order pathway.

Conclusions

1. The most important pathway signalling pain in man ascends in the anterolateral quadrant. Evidence for this derives from cordotomies and other lesions, from a patient with a nearly complete spinal cord transection

sparing only the anterolateral quadrant on one side and from stimulation of the anterolateral quadrant.

2. Candidate nociceptive pathways in the anterolateral quadrant are the spinothalamic tract, the spinoreticular tract, and the spinomesencephalic tract.

3. No information is available about the response properties of spinothalamic tract or other projecting neurons in humans. Therefore, inferences about their role depend upon animal experiments.

4. In animals, spinothalamic tract cells are often nociceptive. However, other sensory properties have also been described.

5. Many cells in the spinoreticular and spinomesencephalic pathways are also nociceptive.

6. Pain may return in humans after an initially successful cordotomy. Since there are nociceptive pathways in the dorsal part of the cord in animals like the cat, it is possible that dorsally situated pathways may provide an alternative route for pain transmission in humans after plastic rearrangements have occurred following cordotomy.

7. The cells of origin of the spinothalamic tract have been mapped using several different techniques. These cells are located in the marginal zone, deeper layers of the dorsal horn, intermediate region and ventral horn in the monkey and rat. In the cat, there are fewer cells in the deep part of the dorsal horn at the level of the lumbosacral enlargement, but the distribution in the cervical cord resembles that in the monkey. Spinothalamic tract cells projecting to the ventral posterior lateral nucleus are concentrated in laminae I and V, while cells projecting to the central lateral nucleus are more heavily concentrated in the ventral horn. Most spinothalamic tract cells are contralateral to the thalamic target. Special concentrations of ipsilaterally projecting spinothalamic tract cells are found in the upper cervical segments and in the sacral cord.

8. Spinothalamic tract axons give off collaterals to the pontomedullary reticular formation and to the region of the periaqueductal gray.

9. The total number of spinothalamic tract cells in the monkey spinal cord appears to be in the order of thousands.

10. Clinical evidence is divided on the level at which spinothalamic tract axons decussate. One opinion is that they decussate within the same segment, whereas another opinion is that the axons cross as many as five segments rostral to the cell body. Some spinothalamic axons in the monkey decussate immediately.

11. The spinothalamic tract has a somatotopic organization in the

anterolateral quadrant of the spinal cord. This is suggested in humans by the results of cordotomies and is shown in monkeys by the demonstration of the receptive fields of spinothalamic axons.

12. The more laterally placed spinothalamic axons in the rat spinal cord project to the ventrobasal complex, whereas the more medially placed spinothalamic axons project to the medial thalamus.

13. The spinothalamic tract ascends through the brain stem just dorsolateral to the inferior olivary nucleus and then adjacent to the lateral lemniscus.

14. The spinothalamic tract in the human and in the monkey terminates in the ventral posterior lateral nucleus of the thalamus. The same is true of the rat. However, the spinothalamic tract of the cat has few terminals in the VPL nucleus; instead, the projection to the ventral thalamus is to a 'shell' region adjacent to the VPL nucleus. Other spinothalamic endings in all of these species are found in the medial part of the posterior complex and the central lateral nucleus of the intralaminar complex. Other termination zones have been reported in the parafascicular, central medial, and submedius nuclei.

15. The spinothalamic tract projection to the ventral posterior lateral nucleus in the human and monkey is somatotopically organized. The projections to other thalamic nuclei are not.

16. Spinothalamic tract cells can be identified in electrophysiological experiments by antidromic activation of their axons from the thalamus. Spinothalamic tract cells would be difficult to study in unanesthetized animals. However, cells resembling them can be identified in decerebrate or spinalized preparations in unanesthetized preparations by antidromic activation from the contralateral spinal cord at an upper cervical level.

17. Spinothalamic tract cells can be activated by volleys in Aδ and C fibers. High threshold muscle and visceral afferents can also be shown to excite spinothalamic tract cells.

18. The responses of spinothalamic tract cells to graded intensities of mechanical stimulation of the skin allow these cells to be classified as low threshold, wide dynamic range (multireceptive) or high threshold (nociceptive-specific) cells. Other spinothalamic cells activated best or exclusively by receptors located in deep tissue, such as muscle, are classed as 'deep' cells.

19. Noxious stimuli other than mechanical can also excite many spinothalamic tract cells. Such noxious stimuli include noxious heating or cooling of the skin, injection of algesic chemicals into the arterial circulation

of muscle, injection of hypertonic saline into muscle, occlusion of the coronary artery, noxious mechanical, chemical or thermal stimulation of the testicle, and overdistention of the bladder. All can activate spinothalamic tract cells.

20. The pattern of cutaneous receptive fields of spinothalamic tract cells activated by visceral stimuli are similar to areas of pain referral in human subjects, suggesting a role of this convergence of visceral and somatic input in referred pain.

21. Spinothalamic tract cells that project to the medial thalamus are often located more ventrally than cells that project to the lateral thalamus. Furthermore, they often have very large receptive fields, over much or all of the surface of the body and face. The receptive fields, apart from that on the ipsilateral hindlimb, depend upon a supraspinal loop. The reticular formation may be the structure involved in this loop, since stimulation in the brain stem reticular formation produces a powerful excitation of these cells.

22. Application of capsaicin to a peripheral nerve causes a temporary excitation of primate spinothalamic tract cells. After the discharge slows, it can be shown that there is a substantial reduction in the C fiber volley (and some decrease in the Aδ volley). Although there is only a slight reduction in the responses to noxious mechanical stimuli, the response to noxious heat is substantially decreased.

23. The responses of cells resembling spinothalamic tract cells in location and axonal conduction velocity (backfired from the upper cervical cord contralaterally) are indistinguishable from those of identified spinothalamic tract cells. Injections of small doses of barbiturate do not decrease the responsiveness of the cells; in fact, in some cases the responses are increased.

24. In addition to excitatory receptive fields, spinothalamic tract cells often have inhibitory receptive fields. These are sometimes due to activation of sensitive mechanoreceptors in the same region as the excitatory field. However, more commonly, inhibition is produced only by noxious stimuli delivered to areas remote from the excitatory field. Such inhibitory fields are more readily identified for wide dynamic range (multireceptive) than for high threshold (nociceptive-specific) cells. The inhibitory fields may extend over most of the surface of the body and face. The fields differ from the 'Diffuse Noxious Inhibitory Controls' in that the inhibition does not outlast stimulation, and much of the inhibition remains after the cord is transected.

25. Prolonged stimulation of a peripheral nerve can result in a long-lasting inhibition of spinothalamic tract cells. The inhibition is most effectively produced by Aδ fibers, but Aβ fibers may have a slight effect and C fibers have an added action. The inhibition is increased as stimulus frequency is raised. The inhibition may contribute to the analgesia produced by transcutaneous nerve stimulation and acupuncture.

26. Spinothalamic tract cells can be excited by iontophoretic application of glutamate and substance P. They can be inhibited by substance P, met- and leu-enkephalin, serotonin, norepinephrine, dopamine, glycine, γ-aminobutyric acid, and acetylcholine.

27. The responses of spinothalamic tract cells seem appropriate for a contribution by this tract to pain sensation, the motivational-affective aspect of pain, hyperalgesia, pain referral, and possibly even central pain.

28. The spinoreticular tract originates from neurons that have a different distribution from those that project in that part of the spinothalamic tract that ends in the ventral posterior lateral nucleus of the thalamus. Most spinoreticular neurons are in laminae VII and VIII.

29. The spinoreticular projection from the cervical enlargement is both crossed and uncrossed. That from the lumbosacral enlargement is chiefly crossed.

30. Some spinoreticular cells are also spinothalamic neurons.

31. The number of spinoreticular neurons in the monkey appears to be similar to the number of spinothalamic neurons. The greater amount of degeneration produced in the reticular formation than in the thalamus following cordotomy may reflect the richness of axonal branching rather than the number of cells in the spinal cord.

32. The spinoreticular tract ascends with the spinothalamic tract in the anterolateral white matter of the cord.

33. The nuclei in which the spinoreticular tract ends (apart from the components that are related to such cerebellar relays as the lateral reticular and the inferior olivary nuclei) include the retroambiguous and supraspinalis nuclei; the dorsal and ventral parts of the nucleus medullae oblongatae centralis; the nuclei gigantocellularis, interfascicularis hypoglossi, and Roller; the nuclei paragigantocellularis lateralis and dorsalis; and the nuclei pontis centralis caudalis and oralis and the subcoeruleus complex.

34. Care must be taken in identifying spinoreticular neurons by antidromic invasion in electrophysiological experiments so as to avoid inadvertent stimulation of axons projecting more rostrally.

35. Some spinoreticular neurons have somatosensory receptive fields and these can often be activated by noxious stimuli.

36. It is unclear if reticular neurons could transmit sensory-discriminative information about pain to higher centers. However, it would seem likely that the reticular formation participates in the motivational-affective components of the pain response.

37. The cells of origin of the spinomesencephalic tract appear to be in similar locations to the spinothalamic neurons that project to the ventral posterior lateral nucleus. Spinomesencephalic neurons are concentrated in laminae I and V.

38. Some spinomesencephalic cells appear also to be spinothalamic tract cells, giving branches to both the midbrain and to the thalamus.

39. The spinomesencephalic tract accompanies the spinothalamic tract in its ascent through the spinal cord and brain stem, until it courses medially and dorsally as it reaches the isthmus.

40. Nuclei in which the spinomesencephalic tract terminates include deep layers of the superior colliculus; nucleus intercollicularis; lateral periaqueductal gray; nucleus cuneiformis; nucleus of Darkschewitz; Edinger-Westphal nucleus.

41. Many of the cells projecting in the spinomesencephalic tract in the rat have been found to respond to noxious stimuli.

42. The spinomesencephalic tract may play an important role in pain transmission. Neurosurgical procedures that interrupt the more medial parts of the midbrain are much more effective in reducing pain than are midbrain lesions restricted to the spinothalamic tract. Lesions of the periaqueductal gray in animals also cause hyperalgesia, and stimulation in this region may produce behavioral evidence of pain. However, stimulation at other locations in and around the periaqueductal gray causes analgesia. The spinomesencephalic tract could participate in both nociceptive signalling and in pain modulation.

43. Dorsal pathways may contribute to nociception, especially in animals like the cat. It is possible that dorsal pathways contribute to the return of pain after an initially successful cordotomy.

44. The spinocervical tract originates from neurons that are concentrated in laminae III, IV and V.

45. The spinocervical tract relays in the lateral cervical nucleus. Axons from this nucleus decussate and ascend through the brain stem in company with the medial lemniscus.

46. There may be some difficulty in the currently accepted technique

for identification of spinocervical tract cells. A problem is that axons might terminate in the gray matter at upper cervical segmental levels, rather than in the lateral cervical nucleus.

47. Although many spinocervical tract cells have tactile responses, some can also be activated by noxious stimuli. These wide dynamic range (multireceptive) neurons could play a role in pain mechanisms.

48. Additional work needs to be done on the possibility of nociceptive responses in the lateral cervical nucleus.

49. There is a postsynaptic dorsal column pathway that originates from neurons that are concentrated in laminae III and IV and that terminates mostly in the dorsal column nuclei. Some of the cells of origin are in deeper layers of the gray matter, and a few are in laminae I and II.

50. At least some postsynaptic dorsal column cells can be activated antidromically from both the dorsal column nuclei and the dorsal lateral funiculus. It is possible that many of these cells belong to both the postsynaptic dorsal column path and the spinocervical tract. However, there are anatomic and functional differences that suggest that at least some of the cells in these two pathways are distinct.

51. Identification of cells in the postsynaptic dorsal column pathway by antidromic activation can be difficult, since many of the cells are excited strongly by rapidly conducting fibers of the dorsal column.

52. Some neurons in the postsynaptic dorsal column pathway are low threshold cells and others are wide dynamic range (multireceptive) neurons.

53. It is possible that the postsynaptic dorsal column pathway contributes to nociception. However, if the pathway ends on cells in the dorsal column nuclei that do not project to the thalamus, then the role of the path may not be directly sensory. One possibility is that it contributes to pain modulation, since it is well known that dorsal column stimulation produces analgesia.

Chapter 6

Nociceptive Transmission to Thalamus and Cerebral Cortex

Historical Overview

According to *Sweet* [1981b], in his review of cerebral localization of pain, the notion that the cerebral cortex has at most a minimal role in the appreciation of pain can be attributed to the observations of *Head and Holmes* [1911], *Holmes* [1927], and *Head* [1920] on individuals with war injuries, as well as other conditions in which the cerebral cortex was damaged, and of *Penfield* and his colleagues [*Penfield and Boldrey,* 1937; *Penfield and Jasper,* 1954] on patients whose exposed cerebral hemispheres were stimulated electrically. Individuals with long-standing war wounds of the cerebral cortex had only incomplete loss of touch, pain or temperature sensibility. *Head and Holmes* [1911] and *Head* [1920] proposed that pain, thermal sense and gross touch are sensed in the thalamus, rather than in the cerebral cortex. Based on the small proportion of instances when individuals reported a sensation of pain following electrical stimulation of their exposed cerebral cortex during surgery to remove epileptic seizure foci (only 11 of more than 800 trials), *Penfield and Jasper* [1954] were led to the opinion that:

> 'It is the discriminative elements of sensation which find a representation in the cerebral cortex, while the appreciation of pain and temperature is subserved by thalamic circuits without essential detour to the cortex.'

However, the fact that some stimuli applied to the cortex in man did produce pain [*Penfield and Boldrey,* 1937] argues in favor of a role of the cerebral cortex in pain, as does the observation that lesions of the postcentral gyrus in man may cause at least a reduction in pain and temperature sensations [*Dejerine and Mouzin,* 1915; *Echols and Colclough,* 1947; *Erick-*

son et al., 1952; *Horrax,* 1946; *Lende* et al., 1971; *Lewin and Phillips,* 1952; *Marshall,* 1951; *Russell,* 1945; *Stone,* 1950; *Talairach* et al., 1960; *White and Sweet,* 1969; *Woolsey* et al., 1979]. The problem is more likely to be that pain has multiple representations in the cerebral cortex than that it has none.

Another line of evidence that favors the view that the cerebral cortex is involved in pain was emphasized by *Sweet* [1981b]. This is the occurrence of pain as an aura in occasional cases of epilepsy [*Garcin,* 1937; *Lewin and Phillips,* 1952]. Still another argument for cortical involvement in pain is the observation that a syndrome resembling 'thalamic' pain [*Dejerine and Roussy,* 1906] can be produced by lesions of the cerebral cortex [*Biemond,* 1956].

Pain is a potent means of producing arousal [*Magoun,* 1963], and so it produces widespread electroencephalographic changes. Thus, it should be no surprise that pain can increase brain metabolism over wide areas of the cerebral cortex, especially the frontal cortex [*Lassen* et al., 1978; *Tsubokawa* et al., 1981]. If much of the cerebrum reacts to a painful stimulus, how can one sort out which neural systems are primarily involved in the processing of information that leads to the perception of pain from those systems that are concerned with aspects of the response to a painful stimulus other than sensation?

One lead is provided by clinical studies in which lesions or stimulation of the thalamus can be shown to alter some aspect of pain. For example, pain relief can result from a lesion placed just ventral to the ventrobasal complex [*Hassler,* 1960; *Hassler and Riechert,* 1959], perhaps by interrupting the input to the ventrobasal complex. There have also been reports of pain relief following lesions of the ventrobasal complex itself [*Bettag and Yoshida,* 1960], with the best results occurring in cases of phantom limb pain and the thalamic syndrome. However, lesions in the ventral posterior lateral nucleus often give poor results in terms of pain relief, and there is the undesirable side effect of a substantial mechanoreceptive sensory deficit [*Mark* et al., 1960, 1963]. Better relief is seen for facial pain with a lesion in the ventral posterior medial nucleus, perhaps because such a lesion may also interrupt projections to the medial thalamus [*Mark* et al., 1960, 1963]. The lesions made by *Hassler's* group ventral to the ventrobasal complex are done in combination with medial lesions for this same reason [*Hassler,* 1960, 1970; *Hassler and Riechert,* 1959].

Pain relief has also been reported following lesions of the intralaminar complex of the thalamus [*Spiegel* et al., 1966; *Sano* et al., 1966; *Tsubokawa,*

1967; *Urabe and Tsubokawa,* 1965]. However, according to *Spiegel* et al. [1966], pain may recur despite initially successful relief. A lesion placed in the ventromedial thalamus in animals prevents the changes in blood flow in the frontal cortex produced by noxious stimulation, but lesions in the ventrolateral or dorsal parts of the thalamus do not [*Tsubokawa* et al., 1981]. Thus, it would seem that the widespread cortical activation produced by noxious stimuli is mediated by pathways linked to the cortex by way of the medial thalamus. Pain relief from lesions of the intralaminar complex may be a consequence of interference with the arousal mechanism.

Lesions in the anterior and medial dorsal nuclei of the thalamus may produce pain relief, presumably by a reduction in the affective component of pain, since pain sensation per se may remain intact [*Mark* et al., 1963]. Similarly, prefrontal lobotomies can make chronic pain tolerable, even though pain sensation appears to be unaltered [*Freeman and Watts,* 1950; *King* et al., 1950]. More limited lesions, such as cingulotomies or subcaudate tractotomies, have been reported to be helpful in the treatment of pain [*Corkin* et al., 1979; *Hurt and Ballantine,* 1974; *Sweet,* 1981b; *Foltz and White,* 1962].

Stimulation in the ventrobasal complex or between it and the medial geniculate body in humans may produce sensations of warmth, coolness, tingling, burning or pain [*Halliday and Logue,* 1972; *Hassler,* 1970; *Tasker* et al., 1976]. In the normal individual, stimulation in the medial thalamus appears to have little sensory effect. However, stimulation in the midbrain or medial thalamus in patients with deafferentation pain can trigger the pain [*Tasker* et al., 1983; see also *Sano* et al., 1966].

Most experimental work on thalamocortical mechanisms of pain has been directed at an understanding of the processing of nociceptive information and is likely to contribute primarily to our understanding of 'nociceptive pain'. Some attempts have begun to examine thalamocortical circuits in models of 'deafferentation' or 'chronic' pain, but it is still too early to know how valid the models are, much less the mechanisms at work.

The approach that will be taken in this Chapter is to review what is known about the anatomy and electrophysiology of those thalamocortical circuits that are most likely to be affected by activity in the pathways ascending in the anterolateral white matter of the spinal cord, especially the spinothalamic tract. However, it should be stated at the outset that we do not have a detailed knowledge of which pathways are in fact responsible for nociceptive responses recorded at the level of the thalamus or cerebral cortex.

Thalamic Nuclei Receiving Nociceptive Input from the Spinothalamic Tract: Ventral Posterior Lateral Nucleus

In the subhuman primate and in the rat, the spinothalamic tract projects to the ventral posterior lateral (VPL) nucleus (fig. 6/1, 6/2; see also fig. 5/4, 5/5) [*Berkley,* 1980; *Boivie,* 1979; *Bowsher,* 1961; *Chang and Ruch,* 1947; *Clark,* 1936; *Kerr,* 1975b; *Kerr and Lippman,* 1974; *Lund and Webster,* 1967; *Mehler* et al., 1960; *Pearson and Haines,* 1980; *Peschanski* et al., 1983; *Ralston,* 1984; *Walker,* 1938; *Zemlan* et al., 1978]. The same observation applies also to the human [*Bowsher,* 1957; *Mehler,* 1962; *Walker,* 1940].

There is a controversy as to whether or not the spinothalamic tract has a substantial projection to the VPL nucleus in the cat [*Rinvik,* 1968]. *Boivie* [1971], using silver techniques for demonstrating degenerating fibers of the spinothalamic tract, found that there were few, if any, endings in the cat VPL nucleus (fig. 6/3). He attributed previous positive reports to inadvertent interruption of the cervicothalamic tract by lesions placed in the uppermost cervical cord. The cervicothalamic tract sends a major projection to the VPL nucleus [*Berkley,* 1980; *Boivie,* 1970]. Much of the spinal cord projection was to the part of the ventral lateral nucleus that is adjacent to VPL [*Boivie,* 1971].

Jones and Burton [1974] obtained similar results to those of *Boivie* [1971]. The spinal projections were to part of the VL nucleus and to the PO_m nucleus (as well as to the central lateral nucleus), a region that appeared to form a shell around the VPL nucleus [*Jones and Burton,* 1974]. However, an important point was that the part of the VL nucleus that received a spinal cord projection had connections with the SI cortex, not the

Fig. 6/1. Terminations of the spinothalamic tract in the monkey diencephalon. The sections were cut in the stereotaxic plane. Terminal degeneration is shown by dots and degenerating fibers of passage by wavy lines. The lesion of the spinal cord was at a midcervical level. [From *Boivie,* 1979.]

Fig. 6/2. Terminals of the rat spinothalamic tract labeled by anterograde transport of wheat germ agglutinin-conjugated horseradish peroxidase injected either into the cervical *(A)* or the lumbar *(B)* enlargement. The drawings in *A* show the distribution of terminals following a cervical injection on the left and a lumbar injection on the right. [From *Peschanski* et al., 1983.]

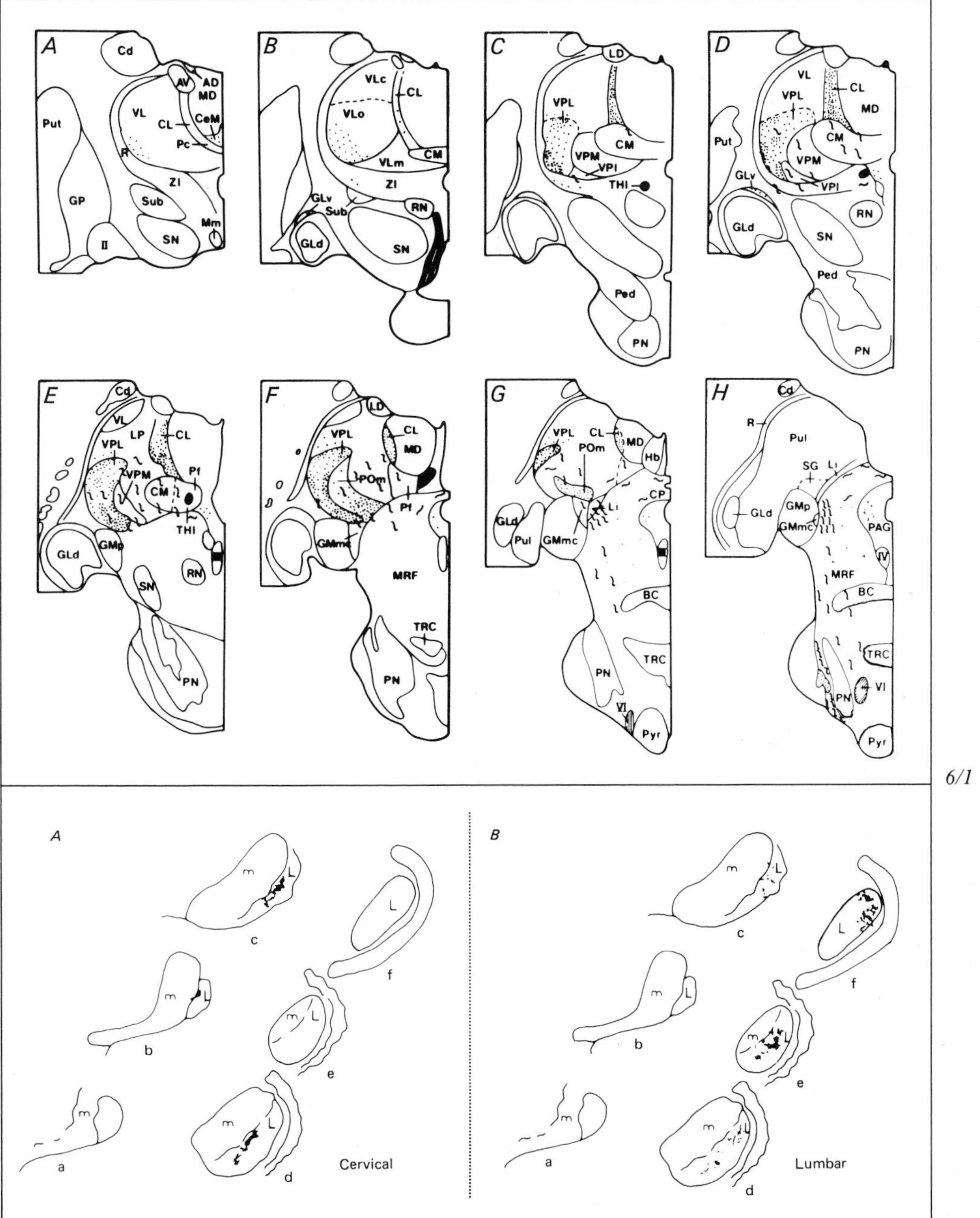
A
Cd
AD
AV
MD
VL
Put
CL
CeM
Pc
R
ZI
GP
Sub
Mm
SN
II
B
VLc
CL
VLo
VLm
CM
ZI
GLv
Sub
RN
GLd
SN
C
LD
VPL
CM
VPM
VPI
THI
Ped
PN
D
VL
CL
VPL
MD
Put
CM
VPM
GLv
VPI
RN
GLd
SN
Ped
PN
E
Cd
VL
LP
CL
VPL
VPM
Pf
CM
THI
GLd
GMp
RN
SN
PN
F
LD
VPL
CL
MD
POm
Pf
GMmc
MRF
TRC
PN
G
VPL
CL
MD
POm
Hb
CP
Li
GLd
Pul
GMmc
BC
TRC
PN
VI
Pyr
H
Cd
R
Pul
SG
Li
GLd
GMp
GMmc
PAG
IV
MRF
BC
TRC
VI
PN
Pyr
6/1
A
m
L
c
f
m
L
b
m
L
e
m
a
m
L
d
Cervical
B
m
L
c
L
f
m
L
b
m
L
e
m
a
m
L
d
Lumbar
6/2

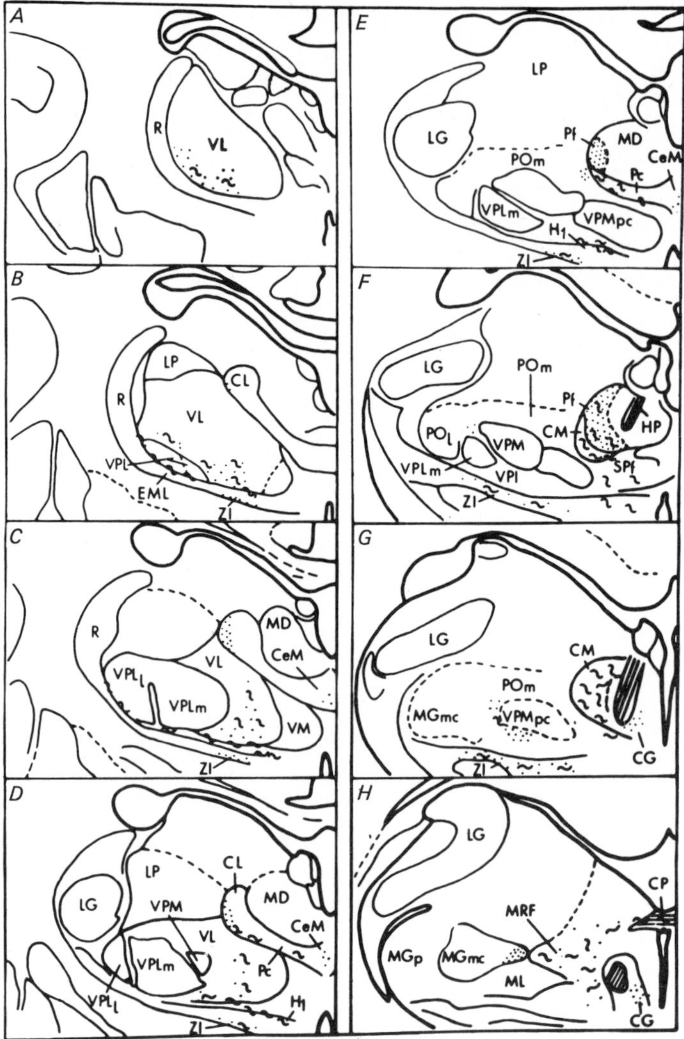

Fig. 6/3. Terminations of the spinothalamic tract in the diencephalon of the cat. The dots indicate degenerating terminals and the wavy lines fibers of passage. The lesion was placed at a midcervical level. Silver degeneration methods were used. [From *Boivie,* 1971.]

motor cortex [see also *Jones and Leavitt,* 1973]. Not being able to devise a better term, *Jones and Burton* [1974] referred to this region as the 'spinal' part of the ventrolateral complex.

Berkley [1980] also describes spinothalamic projections to a 'shell region' surrounding the VPL nucleus, including a region between the VPL and ventral lateral nucleus, as well as the posterior complex and the ventral posterior inferior nucleus.

Recently, there have been reports of a direct projection of the cat spinothalamic tract to the VPL nucleus in experiments using anterograde tracing with horseradish peroxidase (fig. 6/4) [*Burton and Craig,* 1983; *Mantyh,* 1983]. The question remains as to the significance of the projection to VPL in the cat, and there is also a problem in the definition of the boundary between VPL proper and a surrounding 'shell region'.

The spinothalamic projection to the VPL nucleus is somatotopically organized, with fibers from the lumbar cord distributing laterally to the endings of fibers from the cervical cord [*Boivie,* 1979; *Chang and Ruch,* 1947; *Clark,* 1936; *Lund and Webster,* 1967; *Mehler* et al., 1960; *Walker,* 1940]. The somatotopic organization is similar to that reported for the terminals of the medial lemniscus [*Boivie,* 1978; *Clark,* 1936; *Walker,* 1938].

The spinothalamic tract cells projecting to the VPL nucleus in the monkey are concentrated in laminae I and V on the contralateral side of the spinal cord [*Willis* et al., 1979]. The rat has a similar arrangement, although there are fewer spinothalamic tract cells in lamina I and there are many spinothalamic cells in the ventromedial part of the dorsal horn [*Giesler* et al., 1979a]. In the cat lumbar spinal cord, spinothalamic tract cells are mainly in laminae I, VII and VIII [*Carstens and Trevino,* 1978a; *Trevino and Carstens,* 1975; *Trevino* et al., 1972]. The lack of spinothalamic tract cells in lamina V may correlate with the sparse projection to the VPL nucleus in this species. However, there are spinothalamic tract cells in lamina V of the cervical enlargement of the cat spinal cord [*Carstens and Trevino,* 1978a].

The terminals of the spinothalamic tract in the monkey and human VPL nucleus are distributed in patches or clusters throughout the nucleus (fig. 5/4, 5/5) [*Boivie,* 1979; *Bowsher,* 1961; *Kerr,* 1975b; *Mehler,* 1962; *Mehler* et al., 1960; *Ralston,* 1984]. *Mehler* refers to the pattern of these endings as 'archipelagos' or 'pleides'. If the patchy termination zones are followed in serial sections, it becomes evident that they are actually rod-like zones of degeneration [*Boivie,* 1979; cf. *Jones,* 1983].

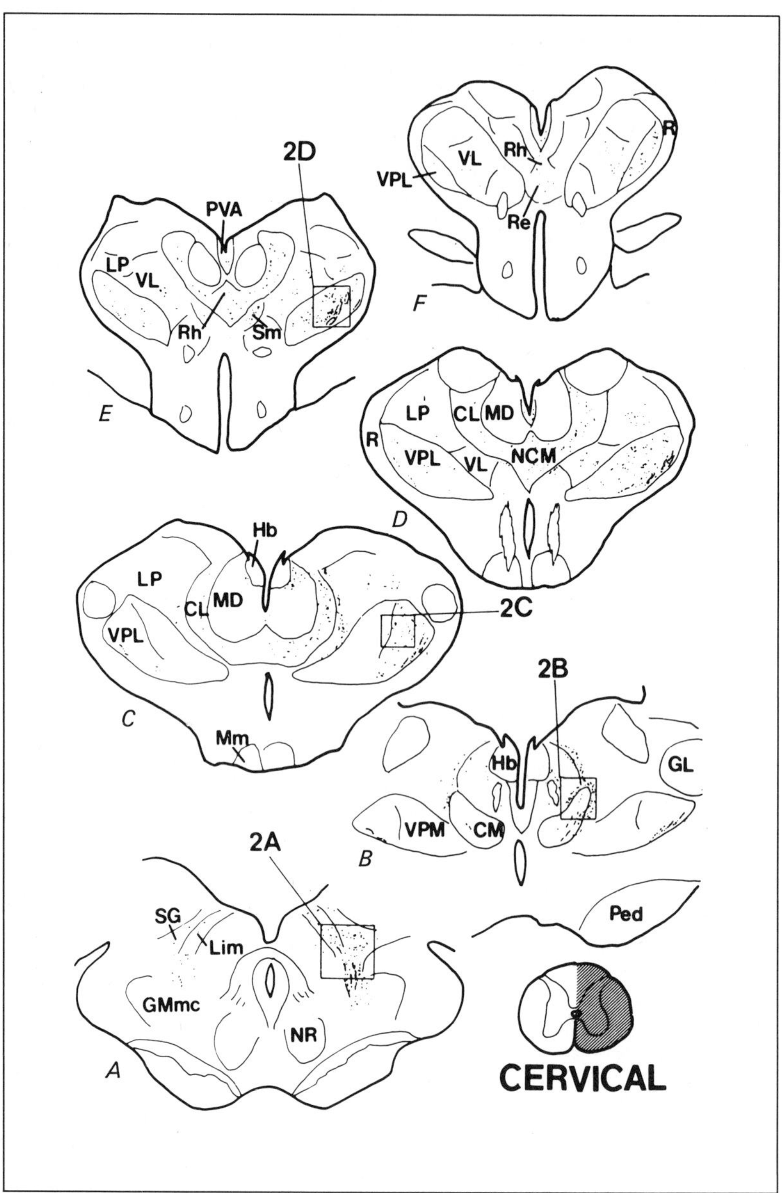

Fig. 6/4. Terminations of the spinothalamic tract in the cat as demonstrated by anterograde transport of wheat germ agglutinin-conjugated horseradish peroxidase. The injection site is shown at the lower right (hemisection with application of marker to cut axons). [From *Mantyh,* 1983.]

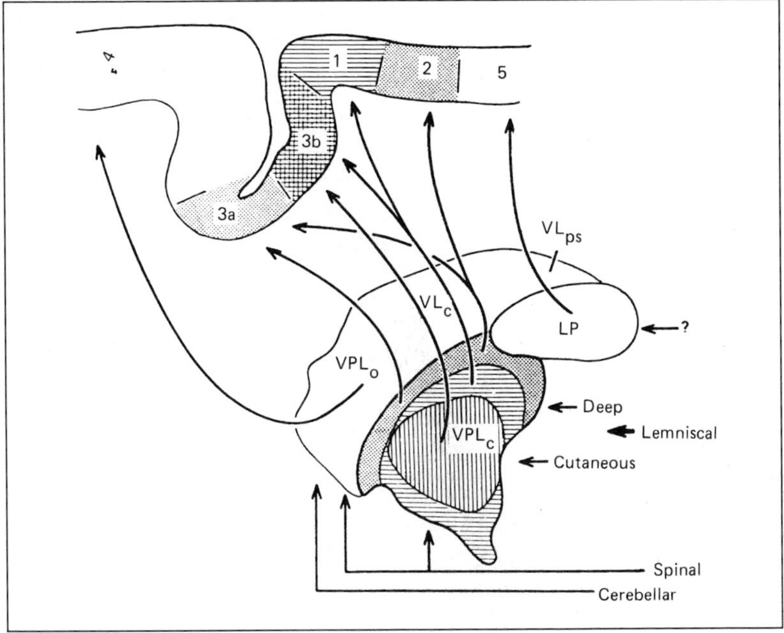

Fig. 6/5. Schematic diagram of the thalamic relay nuclei receiving projections from the medial lemniscus, spinal cord and cerebellum, and their cortical connections. [From *Jones and Friedman,* 1982.]

The VPL nucleus projects to both the SI and SII regions [*Marshall* et al., 1941; *Woolsey and Fairman,* 1946; *Woolsey* et al., 1942] of the somatosensory cerebral cortex and receives corticothalamic projections from these areas [*Burton and Jones,* 1976; *Donoghue* et al., 1979; *Friedman and Jones,* 1980; *Friedman* et al., 1980; *Hand and Morrison,* 1970; *Jones* et al., 1979, 1982; *Jones and Burton,* 1976; *Jones and Friedman,* 1982; *Jones and Leavitt,* 1973; *Jones and Powell,* 1969, 1970; *Landry and Deschenes,* 1981; *Macchi* et al., 1959; *Manson,* 1969; *Morrison* et al., 1970; *Rose and Woolsey,* 1943; *Saporta and Kruger;* 1977; *Whitsel* et al., 1978]. An individual thalamocortical relay cell may project to both areas [*Manson,* 1969; *Jones,* 1975b; *Spreafico* et al., 1981]. Within the SI cortex, there are separate termination zones in areas 3a, 3b, 1 and 2 [*Jones* et al., 1979].

Lin et al. [1979] suggest that area 2 receives its projection from a separate nucleus just dorsal to what they define as VPL and called by them ventralis intermedius. They propose that this nucleus may be equivalent to part of what others would call lateral posterior or even VPL$_o$. *Friedman and*

Jones [1981] and *Jones and Friedman* [1982], on the other hand, show that cells in the VPL nucleus proper (defined by the area of projection of the dorsal column nuclei) project to area 2 and also 3a, as well as to areas 3b and 1. The cells in question that connect to areas 3a and 2 are along the dorsal and rostral margin of VPL, and so no doubt included the ones retrogradely labeled by *Lin* et al. [1979] from area 2. Figure 6/5 shows the projection scheme of *Jones and Friedman* [1982]. The spinothalamic tract ends both in the 'cutaneous core' region that projects to areas 3b and 1 and in the VPL_0 nucleus, which is connected with the motor cortex. It is emphasized by *Jones* [1983] that the cerebellar projections to VPL_0 are onto separate neurons from those receiving spinothalamic projections.

If small injections of horseradish peroxidase are made into the SI cortex, cylindrical arrays of neurons are labeled in the VPL nucleus (fig. 6/6) [*Jones* et al., 1979; *Jones,* 1983; *Lin* et al., 1979]. It has been proposed that these arrays of VPL neurons project upon functional 'columns' of cortical neurons [*Jones* et al., 1979, 1982]. The columnar organization of the SI cortex was first described by *Mountcastle* [1957], who found that the receptive field properties of cortical neurons within vertical zones were similar; these vertical zones were suggested to be 'column-like' configurations of neurons receiving an input from the thalamocortical projections which enter the cortex in radially oriented bundles. The arrays of VPL neurons that label with horseradish peroxidase injected into the SI cortex would presumably correspond to 'thalamic columns' that project onto cortical 'columns'. If it is true that a given thalamocortical relay cell can supply terminals to more than one cytoarchitectural area of the SI cortex [*Landry and Deschenes,* 1981; *Lin* et al., 1979; however, see reservations of *Jones* et al., 1979, and *Jones and Friedman,* 1982], then a single thalamic rod should project to different areas (fig. 6/7) [*Lin* et al., 1979]. This hypothesis should be testable, using a multiple retrograde labeling technique.

Fig. 6/6. Rod-like 'thalamic column' demonstrated by horseradish peroxidase injection into SI cortex in monkey. The site of the injection is shown at the top left and the distribution of the HRP within the cortex at the top right. Units in the injected region had receptive fields on the little toe, as indicated at the bottom left. Cells were labeled in the VPL_c nucleus in the rod-like region shown in the series of frontal sections at the bottom right. [From *Jones,* 1983.]

Fig. 6/7. Schematic of possible connections of a rod-like group of neurons in the VPL nucleus with somatotopically corresponding cortical neurons in the hand representation of areas 3b and 1. [From *Lin* et al., 1979.]

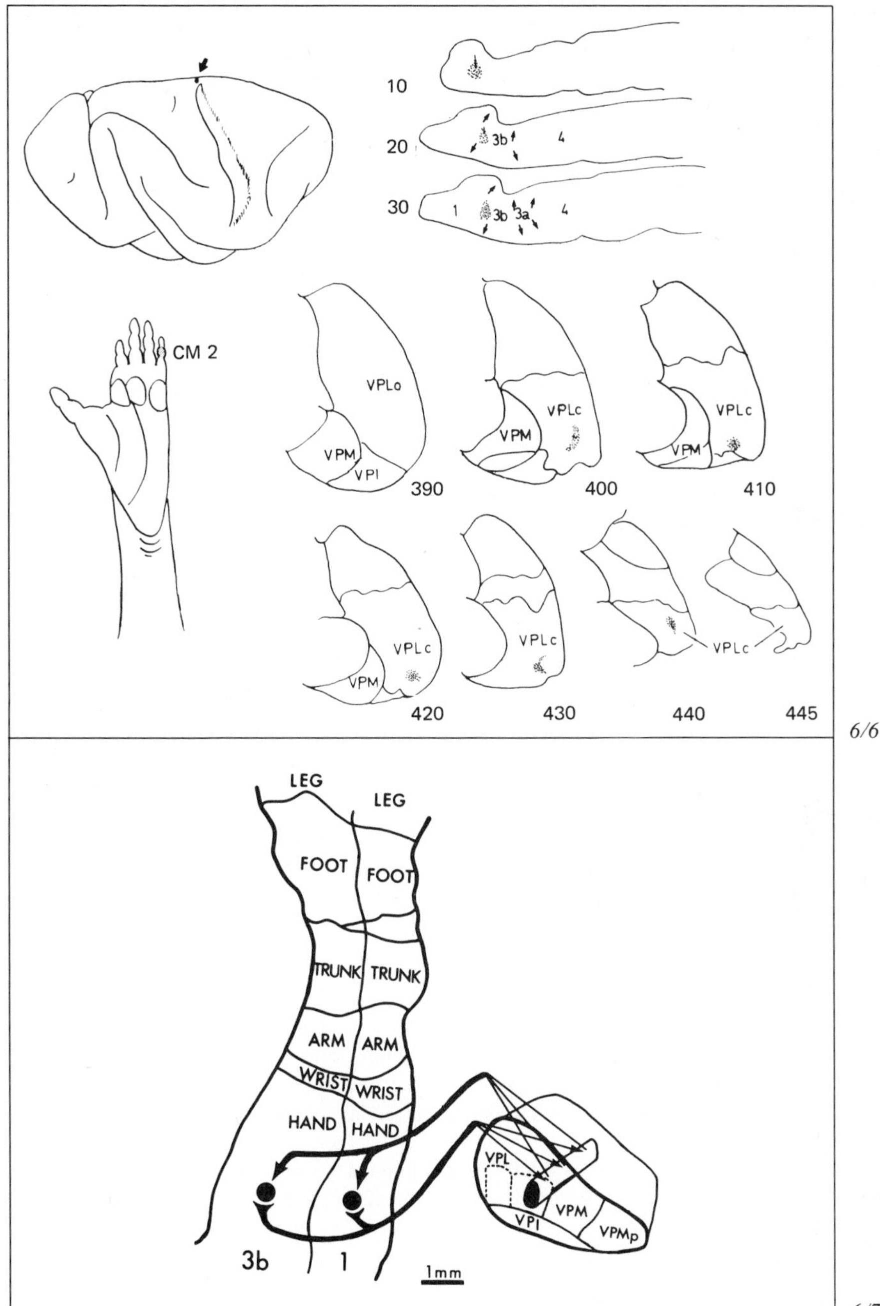
10
20
30
3b
4
1
3a
CM 2
VPLo
VPM
VPI
390
VPLc
400
410
420
430
440
445
LEG
FOOT
TRUNK
ARM
WRIST
HAND
VPL
VPMp
3b
1
1mm

6/6

6/7

It seems reasonable to suppose that the rod-like termination zones of spinothalamic tract fibers are in relation to 'thalamic columns'. If so, these might project to cortical columns that have a special relationship to spinothalamic tract input.

Landry and Deschenes [1981] and *Penny* et al. [1982] have presented evidence for separate populations of thalamic neurons that project to different cortical laminae in SI. Small VPL cells project all the way to layer I [*Penny* et al., 1982], whereas large VPL cells terminate in layers IIIb and IV. *Penny* et al. [1982] suggest the possibility that large and small VPL neurons may have separate functions based on inputs from large and small neurons in the dorsal column nuclei. In the case of a spinothalamic projection to VPL, the possibility of different connections linked to separate ascending tracts would also need to be considered.

Evidence that Nociceptive Spinothalamic Tract Axons End in the VPL Nucleus

Applebaum et al. [1979] followed the projections of individual spinothalamic tract cells into the thalamus by systematically mapping the sites from which the axon of a cell could be activated antidromically. It was assumed that the most rostral low threshold point for microstimulation was near the termination of the axon. Figure 6/8 shows several low threshold points (less than 50 μA) for the axon of a high threshold spinothalamic tract cell as it entered the thalamus (fig. 6/8A), apparently bifurcated (fig. 6/8B) and then ascended to its termination zones in both the VPL_c nucleus and the CL nucleus (fig. 6/8C, D). Figure 6/9I shows the most rostral low

---

Fig. 6/8. A Low threshold points for antidromic activation of a high threshold spinothalamic tract cell. The axon appeared to branch at the level shown in *B* and to terminate both in the VPL_c nucleus, *C*, and in the central lateral nucleus, *D*. [From *Applebaum* et al., 1979.]

Fig. 6/9. The most rostral low threshold points at which the axons of spinothalamic tract cells could be activated antidromically. (Small symbols, 15 μA or less; medium ones, 16–30 μA; large symbols, 31–50 μA.) I shows the approximate termination zones for high threshold cells; those in lamina I region are shown by open circles, those in laminae IV–VI by filled circles. II shows approximate termination zones for wide dynamic range (multireceptive) cells. Open and filled circles as for I. [From *Applebaum* et al., 1979.]

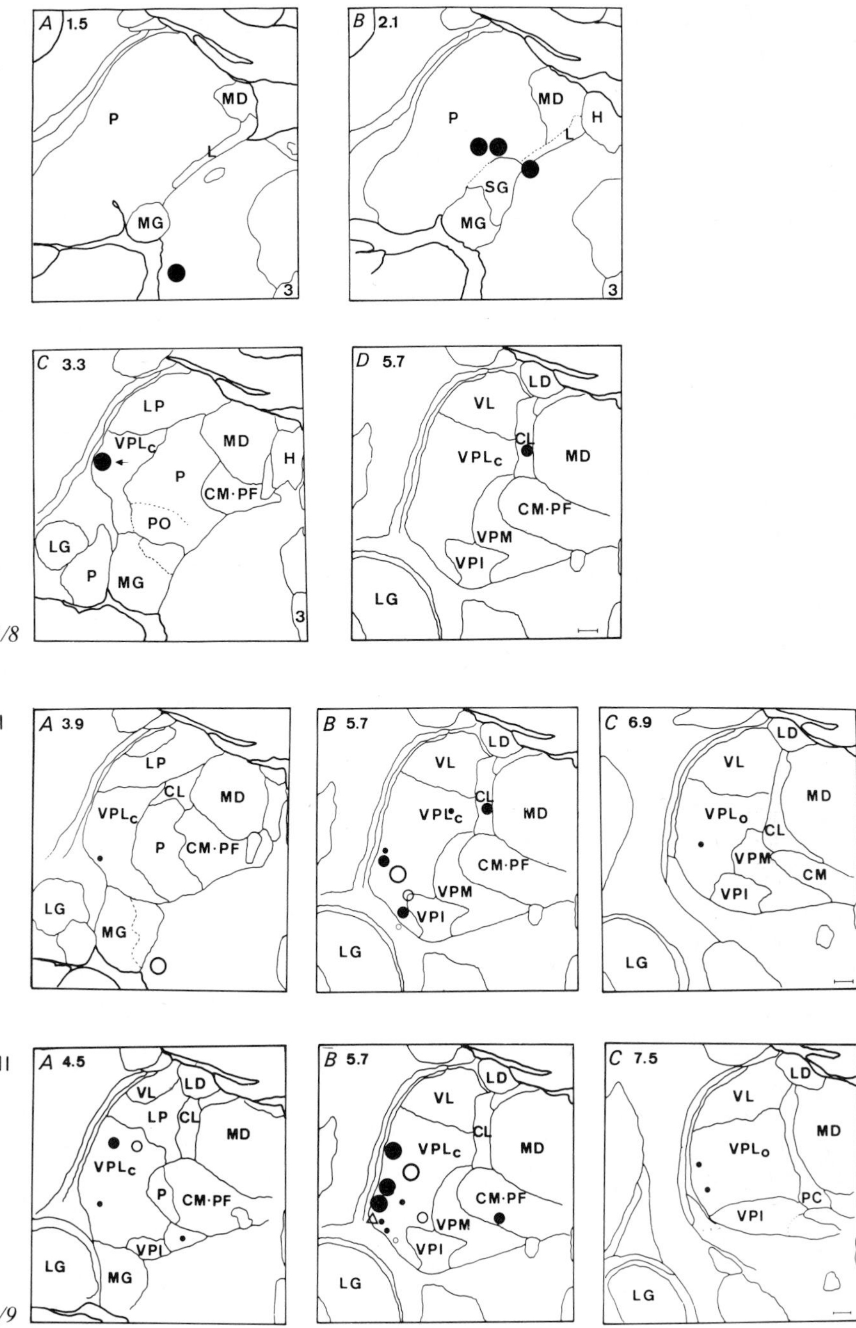
A 1.5
MD
P
L
MG
3
B 2.1
MD
H
P
L
SG
MG
3
C 3.3
LP
VPLc
MD
H
P
CM·PF
PO
LG
P
MG
3
6/8
D 5.7
LD
VL
CL
VPLc
MD
CM·PF
VPM
VPI
LG
I
A 3.9
LP
CL
MD
VPLc
P
CM·PF
LG
MG
B 5.7
LD
VL
CL
VPLc
MD
CM·PF
VPM
VPI
LG
C 6.9
LD
VL
MD
VPLo
CL
VPM
CM
VPI
LG
II
A 4.5
VL
LD
LP
CL
MD
VPLc
P
CM·PF
VPI
LG
MG
6/9
B 5.7
LD
VL
CL
VPLc
MD
CM·PF
VPM
VPI
LG
C 7.5
LD
VL
MD
VPLo
PC
VPI
LG

threshold points for 11 high threshold spinothalamic tract cells. These included neurons whose cell bodies were located either in the vicinity of lamina I (open circles) or in laminae IV–VI (filled circles). It should be noted that most of the apparent terminal zones were in the lateral part of the VPL_c nucleus. Similarly, wide dynamic range (multireceptive) neurons could be shown to end in the VPL_c nucleus (fig. 6/9II). Some appeared to end further rostrally in VPL_o.

Nociceptive Responses of Neurons in the VPL Nucleus

Many investigations of the response properties of neurons within the mammalian ventral posterior lateral nucleus have emphasized the responses of these cells to mechanoreceptive inputs, either from tactile receptors or from proprioceptors or other deep receptors [*Angel and Clarke,* 1975; *Baker,* 1971; *Berkley,* 1973; *Guilbaud* et al., 1977b; *Loe* et al., 1977; *Millar,* 1973b; *Mountcastle and Henneman,* 1952; *Mountcastle* et al., 1963; *Poggio and Mountcastle,* 1960, 1963; *Rose and Mountcastle,* 1952, 1954; *Tsumoto,* 1974; *Tsumoto and Nakamura,* 1974]. However, it is not evident from many of these studies that there are nociceptive cells within the same region. This may be attributed in many cases to inappropriate testing, since these investigators were not seeking nociceptive responses, and in the case of experiments on unanesthetized animals, it would have been unwarranted to attempt noxious stimulation. There have been sporadic reports of nociceptive responses or responses that can be interpreted as nociceptive of neurons in the region of the VPL nucleus of monkeys, cat and rats [*Angel and Clarke,* 1975; *Gordon and Manson,* 1967; *Harris,* 1978a, b; *Krauthamer* et al., 1977]. Recently, such observations have become more frequent and systematic. Some of the studies will be reviewed here.

Monkey

Gaze and Gordon [1954] were the first to report recording nociceptive responses in the VPL nucleus of the monkey. Their observations were limited, however, to just a single cell.

Perl and Whitlock [1961] recorded responses from neurons of the monkey VPL nucleus in animals that had an interruption of most of the spinal cord, except for the ventrolateral quadrant ipsilateral to the side of the thalamus being explored (fig. 6/10, upper left). In their study, the activity of 115 cells

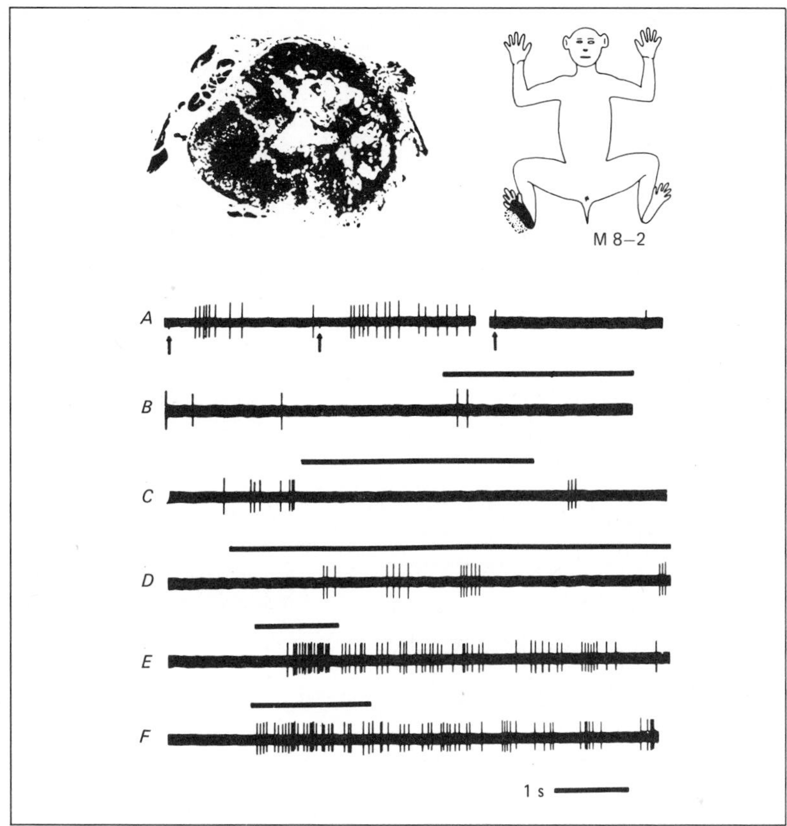

Fig. 6/10. The lesion of the monkey spinal cord is illustrated at the top left. All of the cord was transected except for one ventral quadrant at a high cervical level. The receptive field of a nociceptive specific neuron in the VPL nucleus is shown at the top right The records below are the responses to *A,* electrical stimulation of the right and left plantar nerves; *B, C,* tactile stimulation of the right sole; *D,* pressure applied to right sole; *E, F,* pinching the right sole. [From *Whitlock and Perl,* 1961; *Perl and Whitlock,* 1961.]

in various thalamic nuclei was investigated. Several important features were noted for neurons located within the VPL nucleus. The receptive fields bore a somatotopic relationship to the location of the cells within VPL. Most of the receptive fields were restricted in size and located on the contralateral side of the body. Many of the cells were activated by tactile stimuli, such as hair movement or brushing the skin. Some of the cells were excited best by a brisk stimulus such as a tap. Other neurons were activated

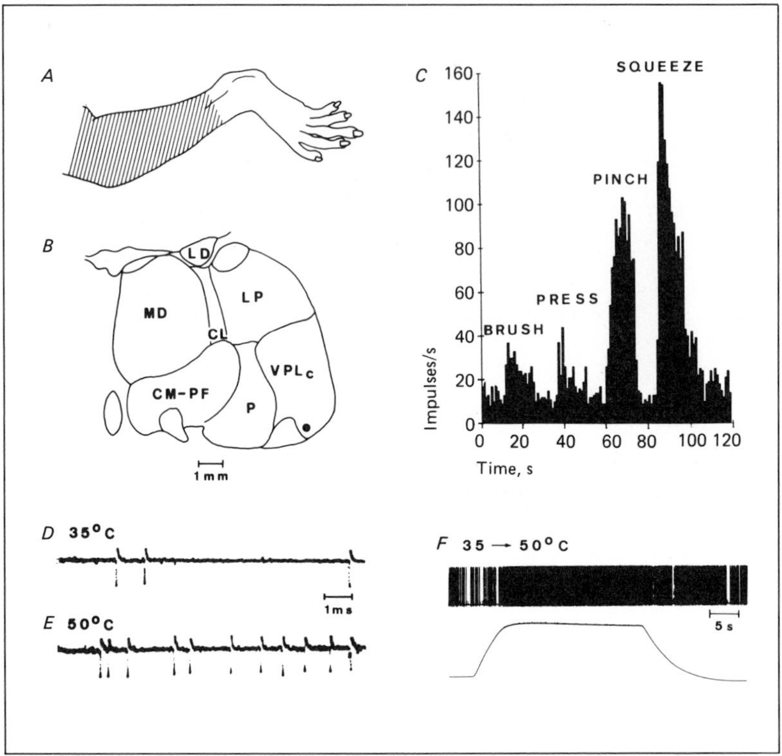

Fig. 6/11. Responses of a 'wide dynamic range' (multireceptive) VPL neuron. *A* shows the receptive field and *B* the location of the recording site. The histogram in *C* shows the responses to graded intensities of mechanical stimulation. *D* is the background discharge and *E* the activity during a noxious heat stimulus of 50 °C. F shows the discharges as pen recordings of window discriminator pulses during a noxious heat stimulus. [From *Kenshalo* et al., 1980.]

by pressure applied either to the skin or to subcutaneous tissues. Two units responded to muscle stretch or movement of fascia, and 4 were excited by joint movements. Two VPL neurons were excited exclusively by noxious stimuli. The cell whose responses are illustrated in figure 6/10 was activated both by noxious mechanical stimuli and by C fiber volleys in the plantar nerve. *Perl and Whitlock* [1961] concluded that the VPL nucleus receives input over the spinothalamic tract from hair and touch receptors and proprioceptors, as well as from nociceptors. Furthermore, the spinothalamic tract provides information about stimulus location.

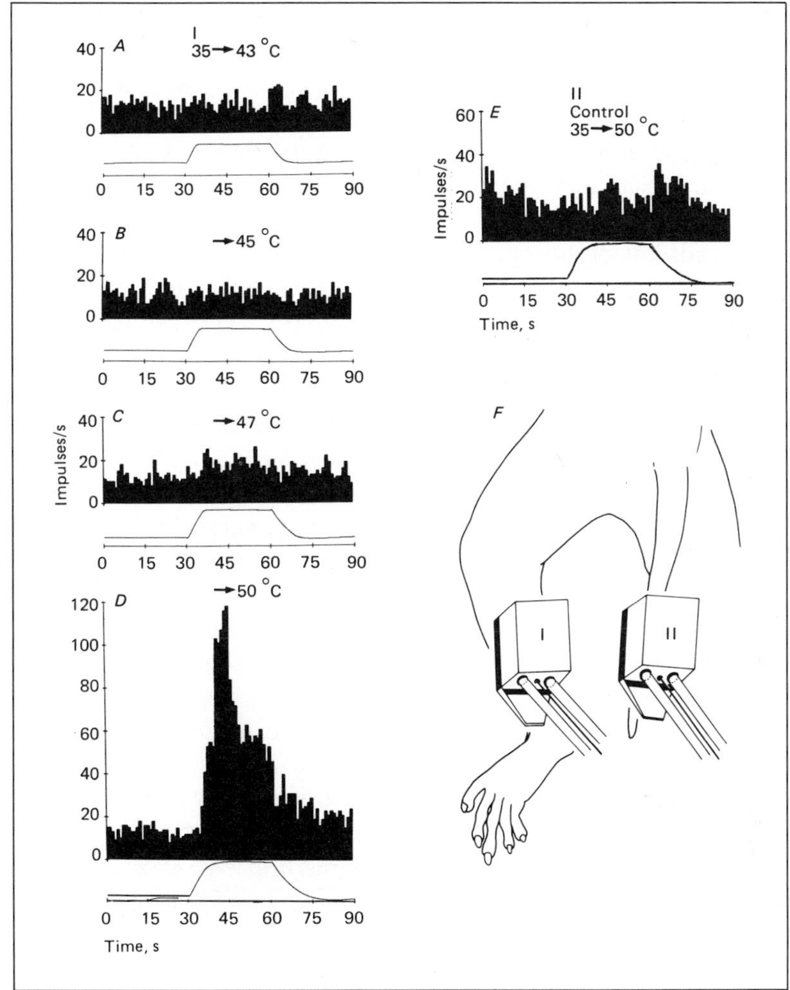

Fig. 6/12. Responses of the same neuron as that illustrated in figure 6/11 to graded noxious heat stimuli applied to the receptive field in I and to a 50 °C stimulus applied to the tail in II. [From *Kenshalo* et al., 1980.]

Pollin and Albe-Fessard [1979] recorded the responses of thalamic neurons in awake behaving monkeys in normal animals and after interruption of the dorsal columns and spinocervical tract at a thoracic level. In the normal animals, cells in the VPL nucleus usually had contralateral receptive fields and responded to stimuli of a single type, such as hair movement, light pressure or joint movement. Some units responsive to pin-prick and usually also light pressure were found. The locations of the units were somatotopically organized. During vertical tracks, cells at different depths within VPL responded in sequence to the following types of stimuli: movement; pressure and pin prick; light touch. A few cells in the posterior part of the VPL nucleus had ipsilateral or bilateral receptive fields. In animals in which the dorsal columns and one spinocervical tract had been interrupted, a few cells could still be found in the VPL nucleus that had peripheral receptive fields. These were nociceptive and responded to pressure, pricking or pinching, or noxious heat. None of the cells responded just to a tactile stimulus. Many of the cells in an area that would ordinarily have a hindlimb input could be activated from the forelimb.

Kenshalo et al. [1980] demonstrated that nociceptive neurons could be recognized in recordings from the monkey VPL nucleus in anesthetized preparations by using noxious heat stimuli in addition to noxious mechanical ones. The nociceptive neurons were greatly outnumbered by cells activated by innocuous stimuli. The responses of the nociceptive cells could be subdivided into 'wide dynamic range' (multireceptive) and high threshold categories, as for spinothalamic tract cells. Of the 54 neurons that could be classified, 48 were wide dynamic range and 6 were high threshold cells. The responses of one of the wide dynamic range VPL neurons are illustrated in figures 6/11 and 6/12. The neuron discharged much more vigorously to noxious mechanical stimuli than to brushing the hairy skin (fig. 6/10), and it was excited when noxious heat stimuli were applied to the receptive field on the leg but not to skin away from the receptive field, such as the tail (fig. 6/12). The cell could be activated by noxious heat even after interruption of the dorsal column and dorsal lateral funiculus on the side ipsilateral to the

---

Fig. 6/13. Effect of spinal cord lesions on the response of the same VPL neuron as illustrated in figures 6/11 and 6/12 to noxious heat stimuli. In *A*, the spinal cord was intact and the noxious heat stimulus produced a substantial discharge. In *B*, the response was not greatly changed by a lesion of the dorsal quadrant of the spinal cord on the side ipsilateral to the receptive field (although the background activity was elevated). The response was diminished by the lesion in *C* and eliminated by the lesion in *D*. [From *Kenshalo* et al., 1980.]

Fig. 6/14. Sites within or just beneath the SI cerebral cortex from which nociceptive thalamocortical neurons in the VPL nucleus could be activated antidromically. All thresholds were less than 50 μA. In *A*, the stimulation points were in the hindlimb area and in *B* the forelimb area. [From *Kenshalo* et al., 1980.]

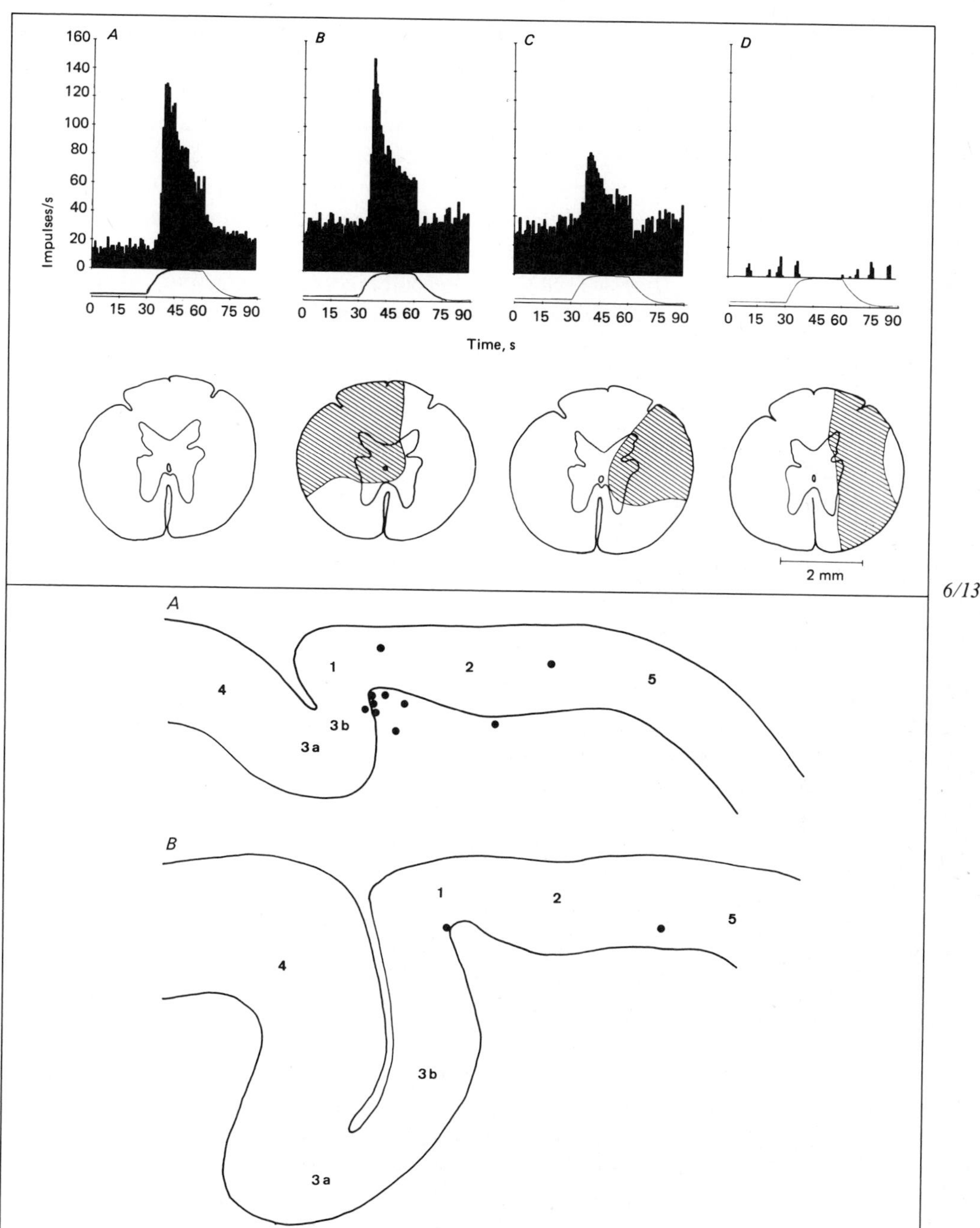
A
B
C
D
Impulses/s
0
20
40
60
80
100
120
140
160
0 15 30 45 60 75 90
0 15 30 45 60 75 90
0 15 30 45 60 75 90
0 15 30 45 60 75 90
Time, s
2 mm
A
1
2
5
4
3b
3a
B
1
2
5
4
3b
3a

6/13

6/14

receptive field (contralateral to the thalamic recording site), but the response was diminished or abolished by lesions of the lateral funiculus on the side contralateral to the receptive field (fig. 6/13). The nociceptive neurons recorded within the VPL nucleus were located in somatotopically appropriate regions of the nucleus, and the cells projected to the SI region of the cerebral cortex, as shown by antidromic activation. The sites from which most nociceptive VPL cells could be activated antidromically were near the junction of areas 3b and 1 (fig. 6/14).

Casey and Morrow [1983] have recently recorded the activity of neurons of the VPL nucleus in awake behaving monkeys. They found that 9 of 76 units that could be activated by cutaneous stimuli were maximally responsive to noxious stimuli. However, all of the nociceptive cells could also be excited by innocuous stimuli and so could be classified as wide dynamic range (multireceptive) cells. The receptive fields were small and located contralateral to the thalamic recording site.

Cat

Several studies have described the occurrence of nociceptive neurons in the vicinity of the VPL nucleus in the cat. The first such study was done by *Gaze and Gordon* [1954]. They found 6 neurons in the VPL region that were activated by Aδ fibers of the saphenous nerve and by strong stimuli, such as squeezing, pinching, tapping or pricking with a pin. Recently, *Gordon* [1983] has reexamined histologically the locations of 5 of these cells (discounting 1 that responded to tap). All 5 nociceptive specific cells were recorded at the borders of the VPL nucleus, either superficial, deep or rostral to the main body of the nucleus. Thus, the results of the 1954 study are consistent with those of more recent experiments by *Honda* et al. [1983] and *Kniffki and Mizumura* [1983]. *Gordon and Manson* [1967] found a number of neurons in the cat VPL nucleus that responded to noxious heat. All but one of the same cells were also responsive to the movement of hair within the receptive field. These cells would thus now be classified as 'wide dynamic range' or multireceptive cells.

Perl and Whitlock [1961] and *Whitlock and Perl* [1959] did an investigation in the cat similar to the one already described for the monkey. The spinal cord was transected at an upper cervical level except for the ventrolateral quadrant. Recordings were then made from neurons in various thalamic nuclei, including VPL. As in the monkey, the cat VPL region contained neurons that could be excited by innocuous mechanical stimuli, joint movement or noxious stimuli.

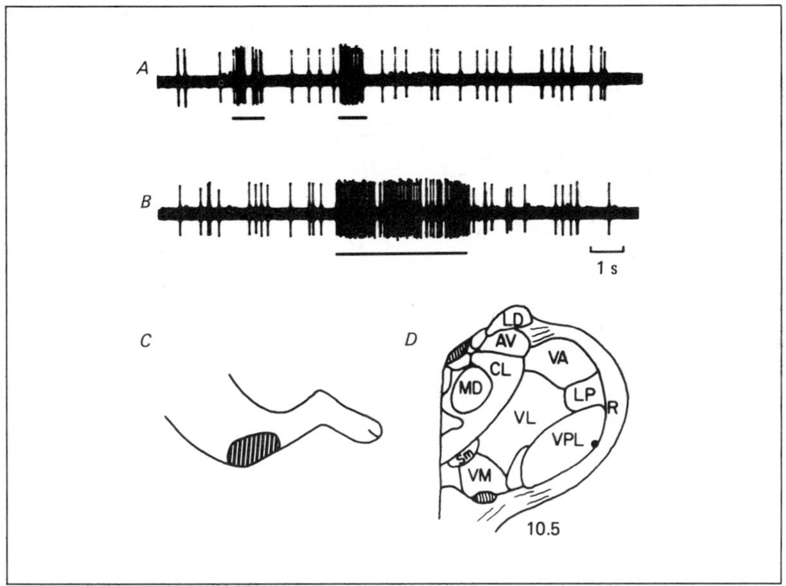

Fig. 6/15. Responses of a high threshold (nociceptive-specific) unit in the region of the cat VPL nucleus. The unit responded to Aδ fiber volleys in the sciatic nerve and was not activated by innocuous mechanical stimuli or by noxious thermal stimuli. The responses shown in *A* were evoked by pin prick and in *B* by pinching with forceps. The receptive field is drawn in *C*, and the location of the recording site in *D*. [From *Honda* et al., 1983.]

Recently, there have been two reports of recordings from nociceptive neurons located in the 'shell' region around the VPL nucleus in the cat. *Honda* et al. [1983] found that most of the units in their sample from the VPL region of the cat responded to mechanical stimulation of the skin or deep tissue or to joint movement. However, stronger stimuli were needed for some units. One group was activated by tap, whereas another population was nociceptive. The activity of one of 17 nociceptive-specific VPL units is illustrated in figure 6/15. No receptive field was detected for 5 units; they were identified solely by their responses to electrical stimulation of small afferent fibers in a peripheral nerve. The other 12 units had restricted receptive fields on the contralateral hindlimb and responded to noxious pinching, pin prick, sometimes noxious heating of the skin or noxious pressure applied to subcutaneous tissue. Marks at the recording sites indicated that the units were located near the dorsal or ventral boundaries of the VPL nucleus (fig. 6/16).

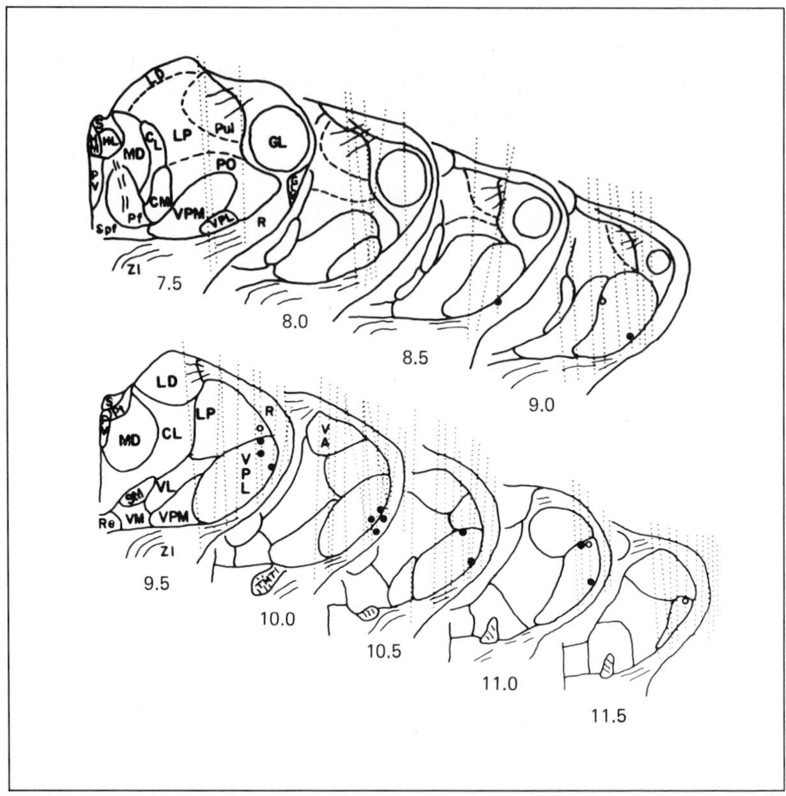

Fig. 6/16. Locations of sites from which recordings were made from 17 units judged to be high threshold (nociceptive-specific) cells in the region of the VPL nucleus in the cat. [From *Honda* et al., 1983.]

Similar findings were obtained by *Kniffki and Mizumura* [1983]. They searched for units that would respond to injection of algesic chemicals into the gastrocnemius-soleus muscles or Achilles tendon. 25 units of this kind were examined. Of these, 14 had an additional restricted cutaneous receptive field, and 4 had large or complex cutaneous fields; 7 had only deep receptive fields. Figure 6/17 shows the responses of one of the units with a convergent input from deep receptors and skin. The unit was excited by pinching or stroking the skin and by pinching the Achilles tendon. Injection of potassium ions into the arterial circulation of the gastrocnemius-soleus muscles had a strongly excitatory action. Bradykinin injection produced a small response. The unit was located on the ventral border of the VPL

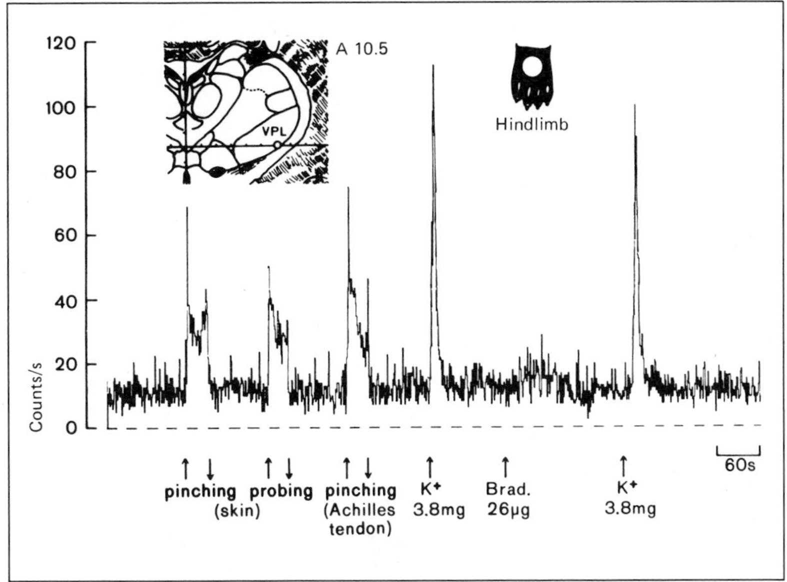

Fig. 6/17. Responses of a unit in the region of the cat VPL nucleus. The cell had a receptive field on the hindpaw, as shown in the inset. It responded to mechanical stimulation of the skin and tendon and to intra-arterial injections of potassium and bradykinin into the arterial circulation of the gastrocnemius-soleus muscles. [From *Kniffki and Mizumura,* 1983.]

nucleus. Most of the units investigated that had nociceptive responses were located on the dorsal, ventral or lateral boundaries of VPL, although 2 units were within VPL.

Comparable results have been reported for the trigeminal system by *Yokota and Matsumoto* [1983a, b], who recorded nociceptive-specific and wide dynamic range (multireceptive) responses from neurons arranged around the periphery of the ventral posterior medial (VPM) nucleus of the cat. Interestingly, the nociceptive-specific and wide dynamic range neurons were spatially segregated into different zones within the VPM nucleus.

Rat

Nociceptive responses have also been recorded in the VPL nucleus of the rat thalamus. *Mitchell and Hellon* [1977] reported recordings from 15 nociceptive neurons in the rat VPL nucleus. They compared the responses of these cells to those of nociceptive neurons of the dorsal horn and SI

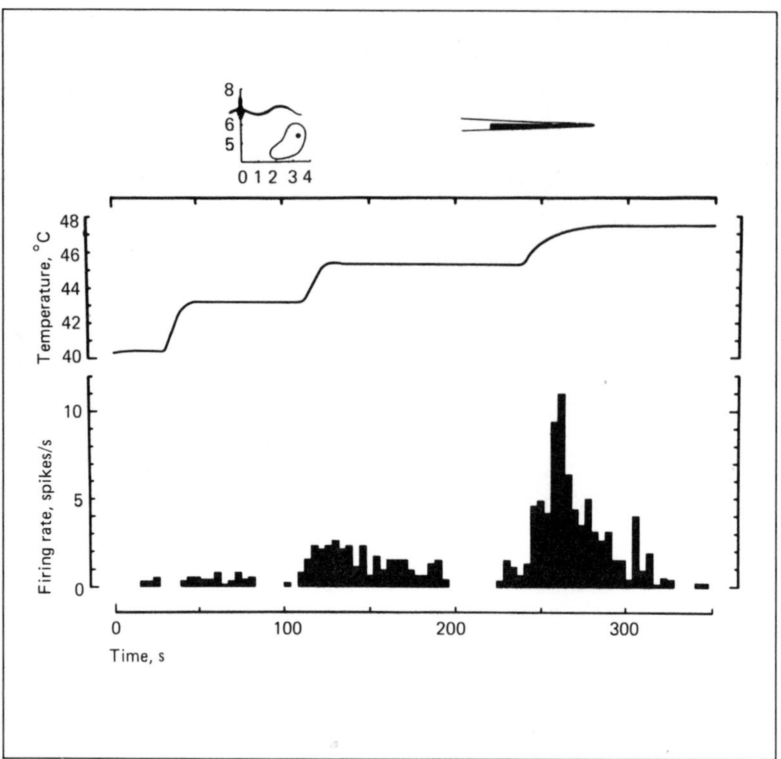

Fig. 6/18. Responses of a unit in the rat VPL nucleus to increases in noxious heat applied to the receptive field on the tail. The responses were not maintained in this unit, although static responses were observed for other units. [From *Mitchell and Hellon,* 1977.]

cerebral cortex. The thalamic units had receptive fields on the tail. They were excited by noxious heat (fig. 6/18) and also by noxious mechanical stimulation of tail skin. Innocuous stimuli, including tail position changes, were ineffective. The receptive fields covered various extents of the surface of the tail, but were often bilateral. The stimulus-response curves for graded intensities of noxious heat showed an increasing discharge for temperatures between 40 and 50 °C. Higher temperatures appeared to inactivate the receptors.

Guilbaud et al. [1980] and *Peschanski* et al. [1980b] have also investigated the responses of nociceptive neurons in the VPL nucleus of the rat. In the study of 163 units by *Guilbaud* et al. [1980], 93 could not be excited by

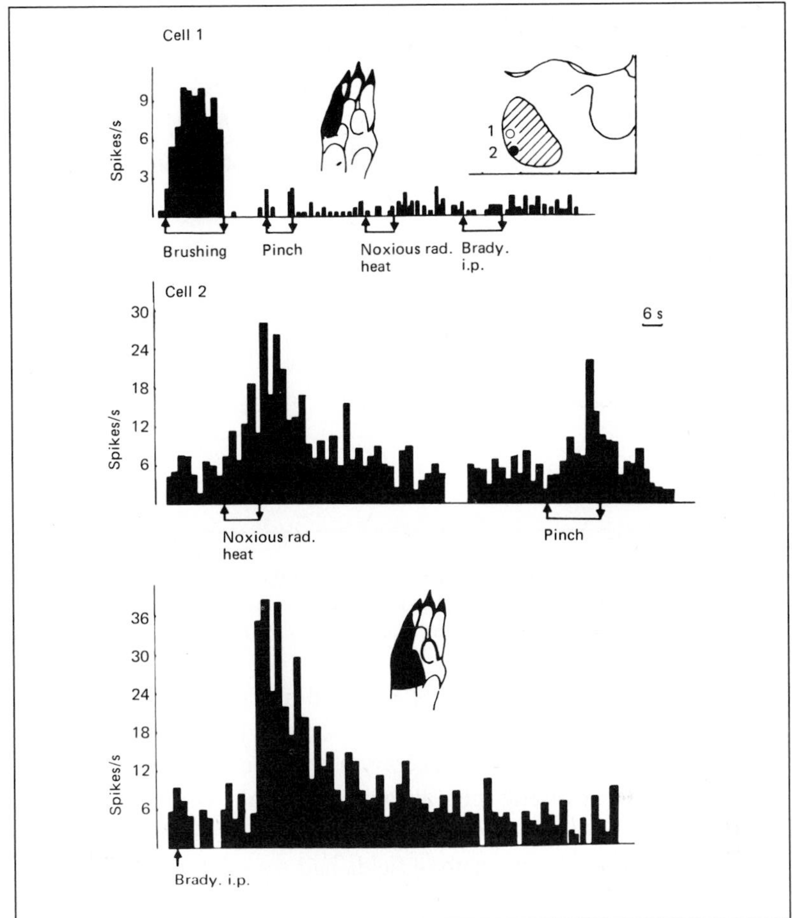

Fig. 6/19. Responses of 2 neurons of the rat VPL nucleus recorded in the same track. Cell 1 responded only to innocuous stimuli, whereas cell 2 was activated by noxious stimuli, including noxious heat, pinch, and intraperitoneal injection of bradykinin. [From *Guilbaud* et al., 1980.]

innocuous mechanical stimuli. These cells could be activated only by noxious intensities of stimulation, such as pinching or pin prick (fig. 6/19). Some of these cells were not excited by nerve volleys restricted to myelinated fibers, but they did show a response to C fiber volleys; others responded to volleys in Aδ fibers. Many of the cells were activated by noxious heat or by intraperitoneal injection of bradykinin (fig. 6/19). Nocicep-

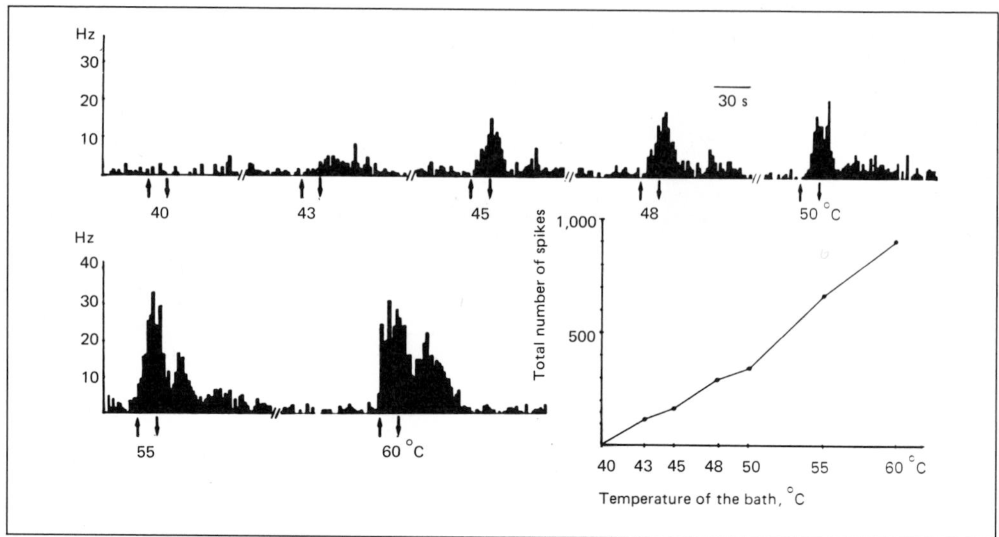

Fig. 6/20. Responses of a nociceptive neuron in the rat VPL nucleus to graded noxious heat stimuli. The area of skin exposed and stimulus duration were constant. The graph plots total number of spikes evoked by each stimulus. [From *Peschanski* et al., 1980b.]

tive-specific cells often had large or medium-sized receptive fields, and these were frequently bilateral. 19 of the cells in the sample were classified as wide dynamic range (multireceptive) cells, and 51 were low threshold cells. The low threshold cells always had small, contralateral receptive fields. Nociceptive cells were found throughout the VPL nucleus, not just the rostral part, and the receptive fields had a somatotopic relationship to different parts of the VPL nucleus (hindlimb and tail represented rostrally and forelimb caudally in the nucleus).

Peschanski et al. [1980b] examined the stimulus-response relationships for 24 nociceptive neurons in the rat VPL nucleus, using graded intensities of noxious heat. The stimuli were applied to the tail by immersing the tail in heated water. Thresholds varied between 40 and 50 °C, with a mean of 45.2 °C. Most of the units showed an increased discharge throughout the range of stimuli used, up to 60 °C (fig. 6/20). However, the discharge rates of some units reached a plateau before the maximum stimulus intensity was reached. Some cells showed an increased discharge when more of the tail was exposed, and some showed a more prolonged discharge as stimulus duration was increased. When stimuli exceeding 50 °C were given, the receptive field showed sensitization.

Thalamic Nuclei Other than VPL Receiving Input from the Spinothalamic Tract

The spinothalamic tract terminates in a number of nuclei in addition to the VPL nucleus. These include the medial part of the posterior complex, the central lateral nucleus of the intralaminar complex, and the nucleus submedius [*Berkley,* 1980; *Boivie,* 1971, 1979; *Craig and Burton,* 1981; *Jones and Burton,* 1974; *Kerr,* 1975b; *Mehler,* 1962; *Mehler* et al., 1960; *Pearson and Haines,* 1980]. There have also been reports of terminals in other thalamic nuclei, and there is a projection to the zona incerta in the subthalamus [*Berkley,* 1980; *Boivie,* 1971, 1979; *Burton and Craig,* 1983].

Medial Part of the Posterior Complex (PO_m)

The spinothalamic tract projects in part to the PO_m nucleus [*Berkley,* 1980; *Boivie,* 1971, 1979; *Jones and Burton,* 1974; *Pearson and Haines,* 1980], as does the medial lemniscus [*Berkley,* 1980; *Boivie,* 1978]. The posterior complex is a group of nuclei located in a region adjacent to the VPL nucleus [*Burton and Jones,* 1976; *Poggio and Mountcastle,* 1960; *Rioch,* 1929; *Rose,* 1942; *Rose and Woolsey,* 1943]. It contains cells often ascribed to the magnocellular part of the medial geniculate nucleus which receive a spinal projection [cf. *Burton and Jones,* 1976; *Pearson and Haines,* 1980] that is not somatotopically organized, according to most investigators.

There has been no attempt to determine the laminar distribution of spinothalamic tract cells projecting to the PO_m nucleus of the monkey. In the cat, the cells projecting to PO_m in one animal appeared to be concentrated in laminae I and IV–VII of the spinal cord [*Carstens and Trevino,* 1978a].

The cortical projection of the PO_m nucleus is to the retroinsular cortex in the monkey and to a comparable region of the cat [*Burton and Jones,* 1976; cf. *Graybiel,* 1973; *Heath and Jones,* 1971]. This cortical zone is adjacent to the SII cortex and forms an additional somatosensory receiving area independent of SI and SII.

Poggio and Mountcastle [1960] recorded from neurons of the posterior complex of anesthetized cats and found that a high percentage (71 of 123) required noxious intensities of stimulation of skin or deep tissue for activation. Some cells could be excited by innocuous stimuli, although not by joint movement. Only a few had restricted contralateral receptive fields.

Most had large or bilateral receptive fields. Some of the cells had a convergent input from the auditory system. *Poggio and Mountcastle* [1960] proposed that these cells played a role in pain, although the cells clearly lacked place-specific information. Localization of painful stimuli could be due to concomitant activation of mechanoreceptive pathways or of other nociceptive neurons projecting to the VPL nucleus.

Calma [1965] also observed that the posterior complex of the anesthetized cat received a bilateral input. However, he recorded evoked potentials, rather than unitary activity. He demonstrated that stimulation of a dissected dorsal funiculus could produce an evoked response in the posterior complex, in addition to that produced by axons in the lateral funiculus.

Casey [1966] recorded from several units in the posterior thalamus in unanesthetized behaving monkeys. Two neurons that appeared to be nociceptive-specific when studied at a time that the monkey was asleep became more like wide dynamic range (multireceptive) neurons when the animal was in quiet wakefulness.

Curry [1972] and *Curry and Gordon* [1972] reexamined the responses of neurons in the posterior complex in anesthetized cats. Most of the cells could be activated by stimulation of the dorsal column. The ventral quadrants were less effective. The cells usually responded to innocuous stimuli. Only 3% were nociceptive. Receptive fields were large and often had an ipsilateral component. *Berkley* [1973] found the cells in cat PO_m often have wide receptive fields, involving 2 or more paws. *Nyquist and Greenhoot* [1974] report that the cells in the posterior nucleus can often be excited from all 4 limbs. Only a few of their sample were activated by noxious stimuli.

Guilbaud et al. [1977b] did a similar study in cats, but used unanesthetized animals in which pain was suppressed by hyperventilation. Many of the cells in the posterior complex in such preparations were nociceptive. Of 76 cells, 41 were nociceptive-specific and 35 were wide dynamic range (multireceptive) cells. Receptive field sizes varied, but were not particularly large. Many fields were contralateral, although 36% were bilateral. Intra-arterial injections of bradykinin excited most of the cells examined (37/39). *Guilbaud* et al. [1977b] agreed with *Poggio and Mountcastle* [1960] that the posterior complex of the cat might be involved in pain mechanisms.

Brinkhus et al. [1979] found that some units in the posterior thalamic nucleus of the cat were responsive to tactile stimulation of small contralateral receptive fields. However, many of the cells had large receptive fields, and often they could be excited in a graded fashion by graded noxious heat stimuli. Heat-responsive neurons could also be activated by noxious mechanical stimuli. Some of the nociceptive cells were nociceptive-specific, others were the wide dynamic range (multireceptive) type of cell.

Intralaminar Nuclei

There are abundant terminations of the spinothalamic tract in the central lateral nucleus of the intralaminar complex [*Berkley,* 1980; *Boivie,* 1971, 1979; *Jones and Burton,* 1974; *Kerr,* 1975b; *Mehler,* 1962; *Mehler* et al., 1960; *Pearson and Haines,* 1980]. There are also terminals in the primate on cell groups in the adjacent medial dorsal nucleus that *Olszewski* [1952] termed the subnuclei multiformis and densocellularis [*Kerr,* 1975b; *Mehler,* 1962; *Mehler* et al., 1960; *Pearson and Haines,* 1980]. Contemporary opinion is that these components of the medial dorsal nucleus should be grouped with the central lateral nucleus, since they are more related in connectivity to this nucleus than to a nucleus projecting to the agranular frontal cortex [e.g. see *Pearson and Haines,* 1980]. The spinothalamic projection to the central lateral nucleus does not appear to have a somatotopic organization.

Spinothalamic tract cells that project to the medial part of the thalamus, presumably including the central lateral nucleus, appear to be concentrated in deeper layers of the spinal cord gray matter than are spinothalamic tract cells projecting to the VPL nucleus [*Carstens and Trevino,* 1978a; *Giesler* et al., 1979a; *Willis* et al., 1979].

The central lateral nucleus receives an input from the cerebellum [*Mehler and Nauta,* 1974] and the reticular formation [*Robertson* et al., 1973]. The central lateral nucleus has abundant connections to the striatum, but in addition it projects to wide areas of the cerebral cortex, including the sensorimotor cortex (fig. 6/21) [*Bentivoglio* et al., 1981; *Jones and Leavitt,* 1974]. Individual neurons in the central lateral nucleus do not appear to project to both the basal ganglia and the cerebral cortex [*Steriade and Glenn,* 1982].

There are, at best, only sparse endings of the spinothalamic tract in the centre median and parafascicular nuclei [*Boivie,* 1971, 1979; *Burton and Craig,* 1983; *Jones and Burton,* 1974; *Kerr,* 1975b; *Mehler,* 1962; *Mehler* et al., 1960]. However, noxious inputs could still gain access to the centre median-parafascicular complex by way of the reticular formation [*Bowsher* et al., 1968; *Nauta and Kuypers,* 1958; *Robertson* et al., 1973; *Scheibel and Scheibel,* 1958]. A major projection from the centre median-parafascicular

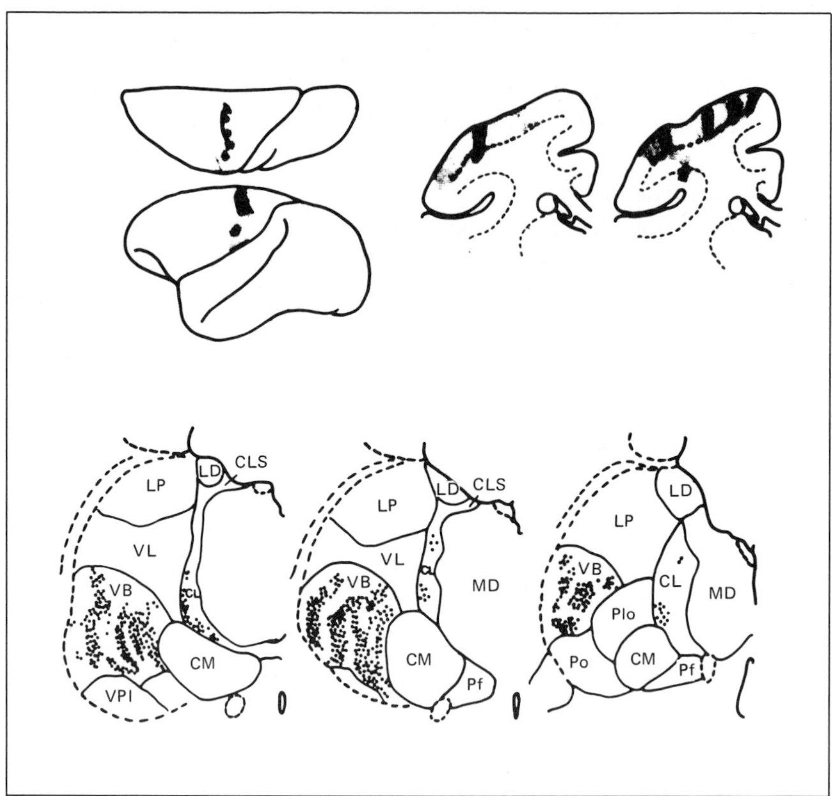

Fig. 6/21. Projection of the central lateral nucleus to the SI cortex of the monkey. Horseradish peroxidase injected into the SI cortex labeled large number of cells in the ventrobasal complex (each dot represents 3–4 cells) and some cells in the central lateral nucleus (each dot is for a single cell). [From *Jones and Leavitt,* 1974.]

complex is to the basal ganglia [*Jones and Leavitt,* 1974], and so the role of this thalamic complex in pain is more likely to be related to motor reactions than to sensory experience.

Perl and Whitlock [1961] recorded from 6 neurons in the intralaminar region in monkeys that had had all of the spinal cord transected except for one ventral quadrant. The neurons had an excitatory input from several limbs, and they showed an after-discharge following noxious stimulation.

Albe-Fessard and Kruger [1962] examined the responses of neurons in the intralaminar region of chloralose-anesthetized cats. The cells were not activated by innocuous stimuli, but they could be discharged when noxious

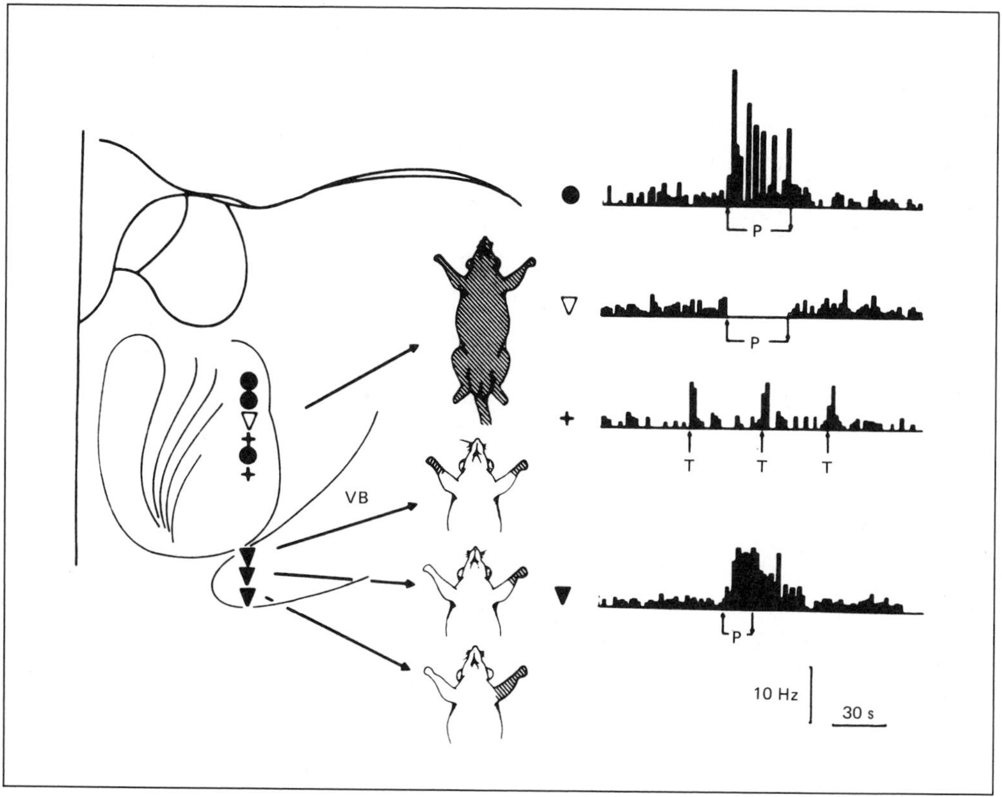

Fig. 6/22. Responses of neurons in the posterior intralaminar region and in the VPL nucleus of the rat thalamus. The recording sites along an electrode track are shown. Typical receptive fields are indicated on the figurines. The histograms show some response types, coded by symbols corresponding to the units from which the records were made. Response types include, from above down, excitation by noxious pinch; inhibition by noxious pinch; excitation by tapping; excitation by noxious pinch. [From *Peschanski* et al., 1981.]

stimuli were employed. Many of the cells had inputs from anywhere on the body surface. The responses remained despite interruption of the dorsal columns, and so it was proposed that the input was conveyed by the spinothalamic or spinoreticular tracts. Similar responses were reported by *Urabe* et al. [1966] in cats that were unanesthetized and immobilized.

Casey [1966] observed three cells in the intralaminar region of unanesthetized behaving monkeys that responded like the ones already described in the posterior complex. They were nociceptive-specific when the animal was asleep, but the activity was more like that of wide dynamic range neurons when the animal was in quiet wakefulness.

Other investigators in a variety of animal preparations have confirmed that neurons in the intralaminar region are often nociceptive, have very large receptive fields that may include part of the ipsilateral body surface, and receive input over pathways ascending in the ventral part of the spinal cord [*Chang,* 1973; *Dong* et al., 1978; *Nyquist and Greenhoot,* 1974; *Urabe* et al., 1966]. Comparable neurons have been described in man [*Ishijima* et al., 1975].

Recently, *Peschanski* et al. [1981] have described the responses of neurons in the intralaminar region of the anesthetized rat. Of 126 cells, 98 were affected by noxious stimuli (fig. 6/22). Some cells were exclusively excited by noxious input, while others responded to tapping as well. Some neurons were inhibited rather than excited by noxious stimuli (or by noxious stimuli and tapping). There was often a convergent input from receptors in the skin and in subcutaneous tissue. Noxious heat could often activate the cells. However, the threshold was often high, and the stimulus-response curves indicated that the cells did not encode graded noxious heat stimuli well. Furthermore, repeated stimuli often did not evoke comparable responses. It was thought unlikely that these cells could contribute sensory-discriminative information about noxious stimuli. However, they might play a role in the detection of novel stimuli, especially potentially harmful ones.

The large receptive fields of many of the spinothalamic tract cells that project to the medial thalamus [*Giesler* et al., 1981b] are consistent with the responses of neurons recorded in this region.

Nucleus Submedius

There is a dense projection from the spinothalamic tract to the nucleus submedius [*Craig and Burton,* 1981; *Ralston,* 1984]. The location of the nucleus submedius is shown in fig. 5/6. Spinothalamic tract cells projecting to the nucleus submedius in the monkey and the cat appear to be concentrated in lamina I [*Craig and Burton,* 1981]. The nucleus submedius projects to the orbitofrontal cortex [*Craig* et al., 1982]. There is a somatotopic organization in the nucleus submedius. The rostral part of the nucleus receives projections from the lumbar cord, whereas the middle part of the nucleus has projections from the cervical cord [*Craig and Burton,* 1981]. Preliminary recordings suggest that neurons within the nucleus submedius may be nociceptive [*Craig and Burton,* 1981].

Thalamic Reticular Nucleus

There appear not to be any direct projections of the spinothalamic tract to the reticular nucleus of the thalamus [*Boivie,* 1971; cf. *Bowsher,* 1961], but there would be indirect projections, since axons passing through the reticular nucleus en route from the VPL nucleus to the cortex or from the cortex to the thalamus give off collaterals to the reticular nucleus [*Jones,* 1975a; *Scheibel and Scheibel,* 1966]. The projections are topographic and a given region within the reticular nucleus receives projections from interconnected parts of thalamus and cortex. Thus, any stimuli that activated cells of the VPL nucleus, including spinothalamic inputs, could presumably have an effect on neurons of the reticular nucleus.

The reticular nucleus does not appear to project to the cerebral cortex. Instead, it projects back into the dorsal thalamus, and a particular part connects to the same regions of the thalamus from which it receives connections [*Jones,* 1975a; *Minderhoud,* 1971]. It is generally believed that the reticular nucleus consists of inhibitory interneurons.

A number of electrophysiological studies have demonstrated that neurons in the reticular nucleus can respond to noxious stimuli [*Angel,* 1964; *Peschanski* et al., 1980a; *Sugitani,* 1979]. Interestingly, in the sample of 62 such cells of *Peschanski* et al. [1980a], the neurons were uniformly inhibited by noxious stimuli of the skin. If these cells are inhibitory interneurons, it is more likely that the nociceptive responses are concerned with modulation of pain transmission than with signalling pain.

Thalamic Nuclei in Which Alternative Nociceptive Tracts End

Nociceptive pathways other than the spinothalamic tract include the spinoreticular and spinomesencephalic tracts and possibly the spinocervical and postsynaptic dorsal column pathways (see Chapter 5). The rhombencephalic reticular formation projects to the medial thalamus, with terminals in many of the same nuclei that receive a direct spinothalamic projection. In addition, there is a heavy projection to the centre median-parafascicular complex [*Bowsher* et al., 1968; *Nauta and Kuypers,* 1958; *Scheibel and Scheibel,* 1958]. The mesencephalic reticular formation shares some of the same projections [*Chi,* 1970; *Hamilton,* 1973; *Hamilton and Skultety,* 1970; *Nauta and Kuypers,* 1958], and in addition there are projections from the periaqueductal gray to the ventrobasal complex [*Barbaresi* et al., 1982; *Hamilton,* 1973]. The spinocervicothalamic pathway ends not only in the

VPL nucleus but also in the PO_m nucleus [*Berkley,* 1980; *Boivie,* 1970, 1980]. It is not clear if the postsynaptic dorsal column pathway synapses upon cells in the dorsal column nuclei that project to the thalamus or that have other targets. If the pathway does in fact have direct access to the thalamus, the most likely nuclei are those receiving a medial lemniscus input, which include both the VPL and PO_m nuclei [*Berkley,* 1980; *Boivie,* 1978].

Nociceptive Responses of Neurons in the Somatosensory Cerebral Cortex

The neurons in the ventral posterior lateral nucleus of the thalamus (apart from interneurons) project to the SI and SII regions of the cerebral cortex. Thus, it would be anticipated that nociceptive cells could be found in the SI and/or SII regions of the cortex. However, given the scattered nature of the spinothalamic tract terminals in the VPL nucleus, and the preponderance of VPL neurons that respond only to innocuous stimuli (at least in the monkey), one might expect most somatosensory cortical neurons to respond just to innocuous stimuli and only occasional cells to be nociceptive. If the speculation that the clustered terminals of spinothalamic axons are upon neurons forming thalamic 'columns' is correct, then nociceptive neurons in the somatosensory cortex might occur in scattered nociceptive cortical columns.

As in the case of the VPL nucleus, most of the studies of the somatosensory cortex have emphasized the responses of neurons to innocuous mechanical stimulation [*Amassian,* 1953; *Armstrong-James,* 1975; *R.E. Bennett* et al., 1980; *Dreyer* et al., 1974; *Ferrington and Rowe,* 1980; *Gardner and Costanzo,* 1980a, b; *Hyvärinen and Poranen,* 1978; *Mountcastle,* 1957; *Mountcastle* et al., 1957; *Mountcastle and Powell,* 1959; *Paul* et al., 1972; *Powell and Mountcastle,* 1959; *Sur* et al., 1982; *Towe and Amassian,* 1958; *Welker,* 1971; *Whitsel* et al., 1978]. Responses to noxious stimuli were not observed in most studies (or were described as high threshold responses without further specification). In addition to the sampling problem alluded to above, anesthesia could have interfered with the occurrence of nociceptive responses in many of the experiments. In the case of investigations using unanesthetized animals, noxious stimuli were either not applied or were kept at a minimal intensity. However, several investigations have demonstrated the presence of nociceptive neurons in the SI cortex.

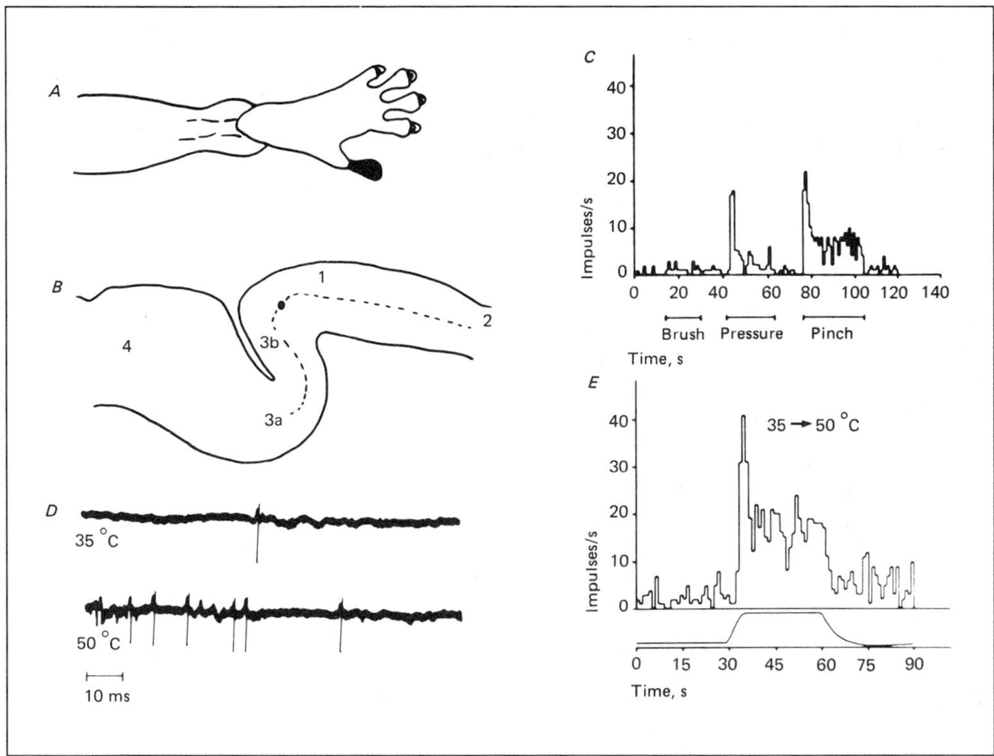

Fig. 6/23. Responses of a neuron in the SI cortex of the monkey to noxious stimuli. The cell was classified as a high-threshold neuron. The receptive field is shown in *A*, and the location of the recording site in *B*. The histogram in *C* shows the absence of an effect of an innocuous brush stimulus, and an excitatory action of pressure and pinch stimuli. In *D* are action potential sequences before and during application of a noxious heat stimulus. The histogram in *E* shows the excitation of the cell by noxious heat. [From *Kenshalo and Isensee,* 1983.]

Monkey SI Cortex

Mountcastle and Powell [1959] reported finding some neurons in the SI cortex in monkeys that responded to noxious stimuli. These cells often had large receptive fields that sometimes included parts of the ipsilateral body surface. The cells were thus unlikely to signal stimulus location.

Kenshalo and Isensee [1983] were able to demonstrate nociceptive responses in 68 neurons of the SI cortex of the monkey. Many adjacent neurons were excited in a slowly adapting fashion by pressure applied to the

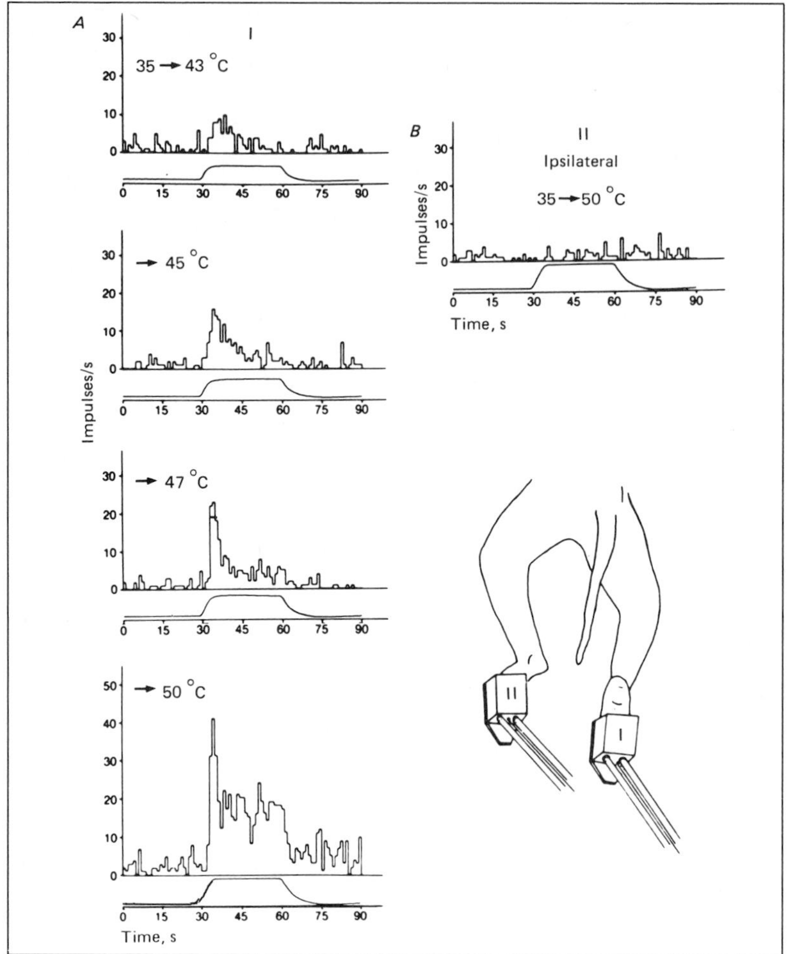

Fig. 6/24. I. Responses of the cell illustrated in figure 6/23 to graded intensities of noxious heat stimuli. No response was evoked when noxious heat was applied to the opposite foot (II). [From *Kenshalo and Isensee,* 1983.]

skin. The nociceptive cells could be classified as either wide dynamic range (multireceptive) cells or as high threshold cells. Of those that were so classified, 37 were wide dynamic range and 31 high threshold cells. 34 of these nociceptive neurons had restricted contralateral receptive fields, whereas 12 had very large receptive fields that could include the entire body surface. The responses of a high threshold neuron in the SI cortex are illustrated in

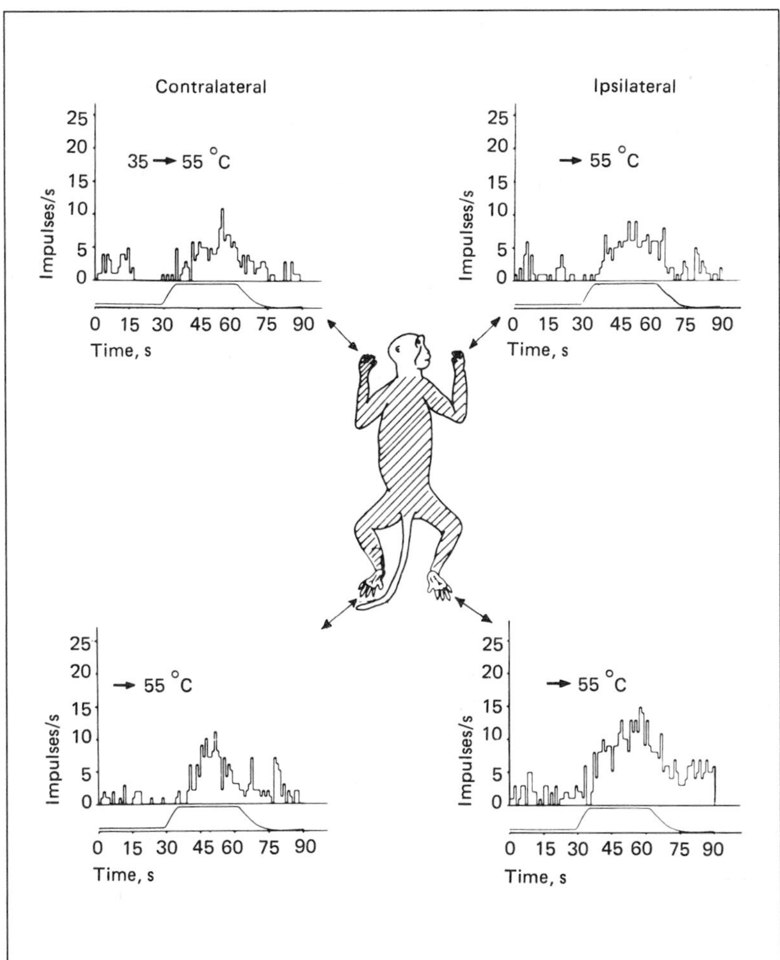

Fig. 6/25. Nociceptive neuron in the SI cortex that had a large receptive field. Noxious heat stimuli applied to any of the four extremities could excite the cell. [From *Kenshalo and Isensee,* 1983.]

figure 6/23. The receptive field was a small region on the glabrous skin of the contralateral hallux. The cell was located near the boundary between areas 3b and 1. The neuron did not respond to brushing the skin, but it did discharge following pressure and pinching applied to the receptive field and also to noxious heat stimulation. The cell responded in a graded fashion to graded noxious heat stimuli, as shown in figure 6/24. There was no response

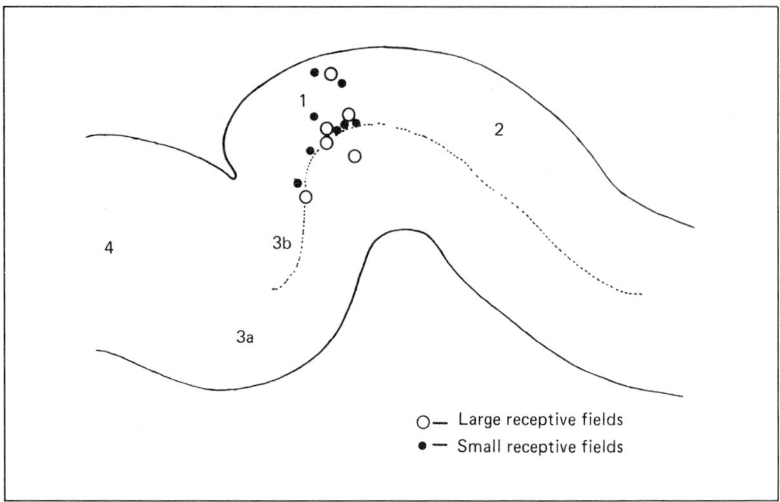

Fig. 6/26. Drawing of a parasagittal section through the SI cortex showing the locations of the recording sites for 15 nociceptive neurons. The dashed line is at the middle of layer 4. [From *Kenshalo and Isensee,* 1983.]

when noxious heat was applied to the opposite foot, indicating that the excitation of the cell was related to a specific receptive field and not to a generalized effect.

Some cortical nociceptive neurons had a small contralateral receptive field for tactile stimulation, but a high threshold receptive field over the rest of the body surface. The responses of one of these cells to noxious heat stimuli applied to each of the four extremities are shown in figure 6/25.

The locations of the nociceptive neurons found in the SI cortex by *Kenshalo and Isensee* [1983] are shown in figure 6/26. Most were near the boundary between areas 3b and 1 in the region described by *Kenshalo* et al. [1980] as the site from which most nociceptive VPL neurons in their sample could be antidromically activated.

The mean responses of a population of nociceptive cortical neurons to graded noxious stimuli were compared by *Kenshalo and Isensee* [1983] to the mean responses of spinothalamic tract cells and of nociceptive VPL neurons (fig. 6/27) [*Kenshalo* et al., 1979, 1980]. The threshold for activation of the cortical neurons was higher than for the cells in the spinal cord and thalamus. This was attributed to the effects of the anesthetic, since additional doses of barbiturate could be demonstrated to elevate the thresh-

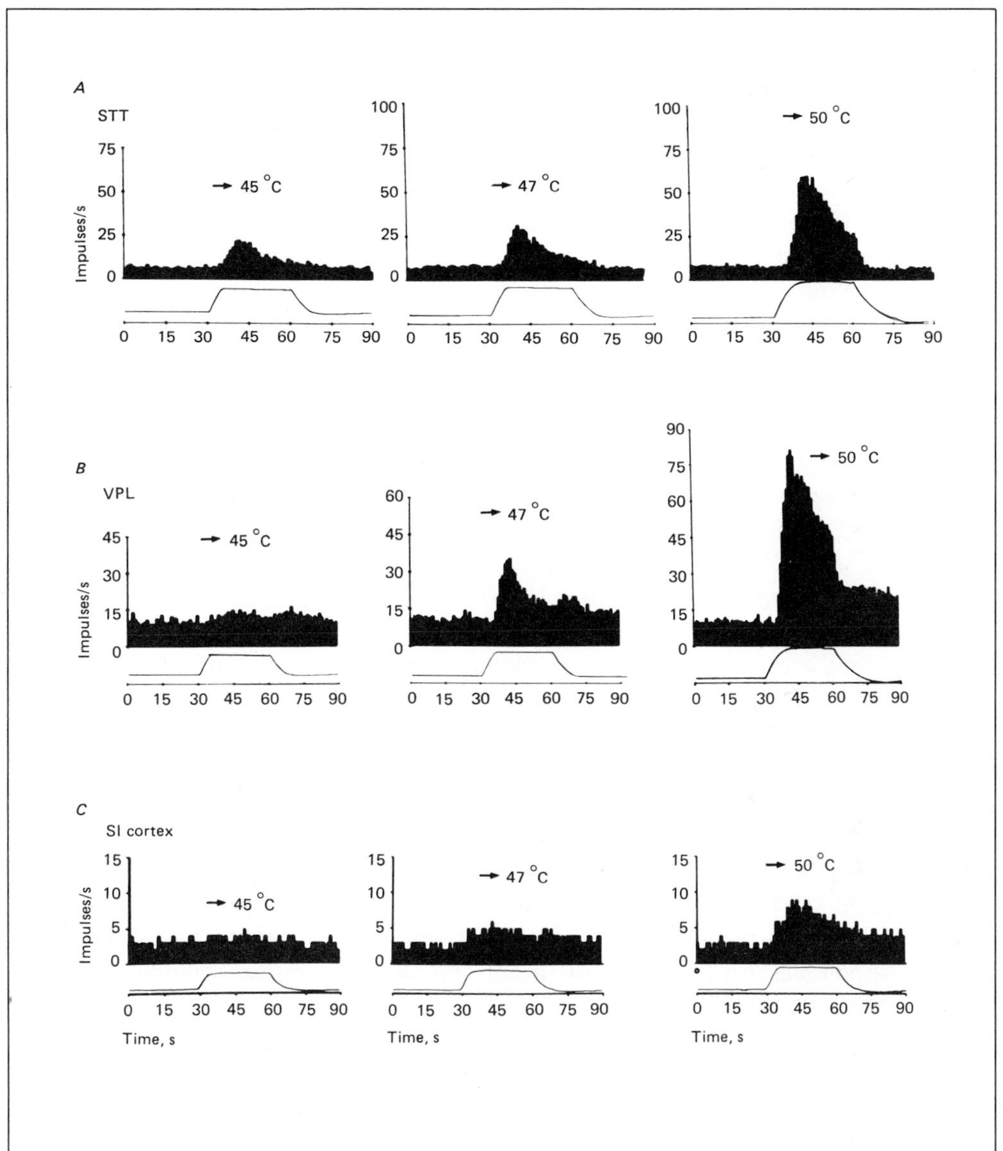

Fig. 6/27. Averaged histograms from populations of *(A)* spinothalamic tract cells (STT), nociceptive neurons of the *(B)* ventral posterior lateral (VPL) nucleus and nociceptive neurons of the *(C)* SI cortex showing responses to graded noxious heat stimulation. The number of cells were 16 STT cells, 10 VPL cells, and 29 SI neurons. [From *Kenshalo and Isensee,* 1983.]

old of cortical neurons to noxious heat stimuli. Another difference between the cortical neurons and those in the spinal cord and thalamus was that the responses adapted more slowly. (It is of interest that thalamic neurons had thresholds and adaptation rates intermediate between those of spinal and cortical nociceptive neurons.) The slow adaptation was suggested to bear relation to the fact that thermal pain in humans produced by stimuli over 47 °C does not adapt [*Hardy* et al., 1968; *LaMotte,* 1979]. Thus, the nociceptive cells in the SI cortex with restricted contralateral receptive fields have response properties that suit them well for a role in sensory discrimination of pain. However, the nociceptive cortical neurons with very large receptive fields did show adaptation of their responses to noxious heat. Therefore, it was suggested that these cells lacked properties suitable for a role in either stimulus localization or in recognition of maintained noxious heat pain, and it was suggested that such cells might contribute to the arousal mechanism rather than to sensory processing.

Rat SI Cortex

Nociceptive neurons have also been investigated in the rat SI cortex [*Lamour* et al., 1982, 1983a, c]. Of 292 cortical neurons that had peripheral receptive fields, 91 were activated by noxious stimuli [*Lamour* et al., 1983c]. Of the nociceptive cells, 56 were nociceptive-specific and 35 had wide dynamic range (multireceptive) properties. The nociceptive-specific cells had large receptive fields, located either on the contralateral side of the body or including an ipsilateral area (fig. 6/28). The wide dynamic range cells had restricted contralateral excitatory receptive fields (fig. 6/29), and they often also had high threshold inhibitory receptive fields. The cells could often be excited by noxious heat stimuli, as well as by noxious mechanical stimuli. There did not appear to be a segregation of cells responding just to innocuous stimuli into separate cortical columns from cells responding to noxious stimuli [*Lamour* et al., 1983a]. Instead, both

---

Fig. 6/28. A–C Responses of a nociceptive-specific unit in the rat SI cortex. The large receptive field is shown at the right. The unit did not respond to brushing *(B)*, but it did discharge when the contralateral hindfoot (PCHF), ipsilateral hindfoot (PIHF), scrotum (PS) or tail (PT) were pinched. Noxious heat (48.5 °C) applied to the CHF also excited the cell. [From *Lamour* et al., 1983c.]

Fig. 6/29. A Responses of a wide dynamic range (multireceptive) neuron in the rat SI cortex. The unit responded to light touch and tapping, as well as to pinch and to noxious heat. The receptive field in *B* was small. [From *Lamour* et al., 1983c.]

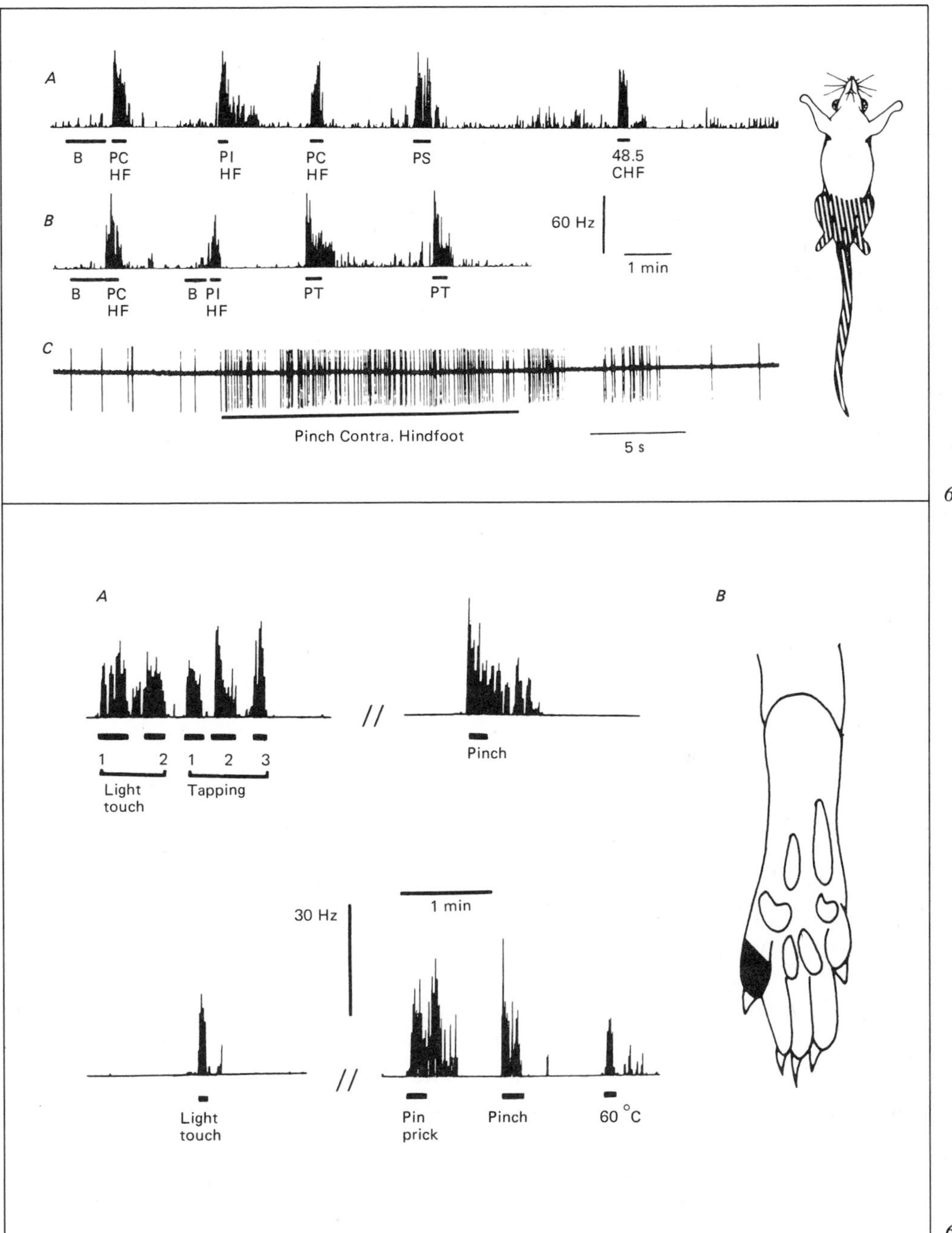
A
B
PC
HF
PI
HF
PC
HF
PS
48.5
CHF
B
60 Hz
1 min
B
PC
HF
B PI
HF
PT
PT
C
Pinch Contra. Hindfoot
5 s
A
1
2
1
2
3
Light
touch
Tapping
Pinch
B
30 Hz
1 min
Light
touch
Pin
prick
Pinch
60 °C

6/28

6/29

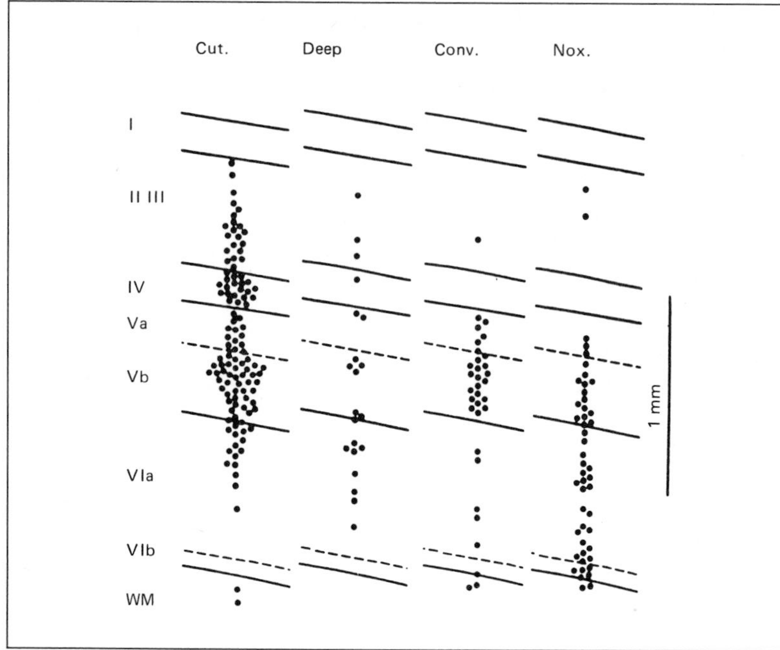

Fig. 6/30. Laminar distribution of neurons in rat SI cortex according to modalities. 'Cut'. refers to cells activated by innocuous skin stimulation. 'Deep' cells were excited by innocuous deep stimulation. 'Conv.' cells are wide dynamic range or multireceptive neurons and activated by innocuous and noxious stimuli. 'Nox.' are cells activated just by noxious stimuli. [From *Lamour* et al., 1983a.]

kinds of neuron could be found in the same penetrations. However, nociceptive neurons tended to be located in the deeper layers of the cortex (Vb and VI), whereas the cells responding just to innocuous stimuli were in more superficial layers (II–V). The nociceptive-specific cells were chiefly in layer VI and the wide dynamic range cells in layer Vb (fig. 6/30). This distribution matched the tendency for neurons with the largest receptive fields to be located in the deeper layers of the cortex. No nociceptive neurons were found in layer IV. The argument is made that nociceptive inputs from the VPL thalamus must be affecting the cells in layers V and VI directly, since the information could not be relayed through layer IV. Furthermore, it was noted that layers V and VI are output layers, and so nociceptive neurons might transmit information immediately to subcortical structures.

Monkey SII Cortex

Whitsel et al. [1969] recorded from the SII cortex in unanesthetized monkeys. They subdivided the SII region into caudal and rostral areas based on differences in the response properties of the neurons encountered. The caudal region contained neurons that responded to cutaneous, auditory and visual stimuli. The somatic inputs were from large, often bilateral areas of the body surface, and the stimuli often had to be noxious to activate these cells. The main emphasis of the study was on neurons found rostral to the 0.0 Horsley-Clarke coronal plane. Cells in this region (called SII/r) usually responded to tactile stimuli and not additionally to more intense stimuli. Some of the cells were activated better by moving than by stationary stimuli. Habituation was prominent when repeated stimuli were delivered. 90% of the cells had bilateral receptive fields. In some cases the receptive fields were continuous, in other cases disjoint. By taking into account the responses of populations of adjacent neurons, it was possible to determine that the SII cortex contained a map of the body surface, with the most rostral dermatomes being represented medially and the most caudal laterally. The response properties of neurons in the rostral part of SII did not suggest that this region is specialized for nociception. The responses of cells in the caudal region resembled those described by *Poggio and Mountcastle* [1960] for neurons in the posterior thalamic nuclei. However, these cells would not be suited to code sensory-discriminative aspects of pain sensation, since they had large bilateral receptive fields and a convergent input from the auditory and visual systems.

The organization and response properties of SII cortex in the monkey were reexamined by *Robinson and Burton* [1980a–c], because it was felt that the study by *Whitsel* et al. [1969] might have included neurons not only of the SII cortex proper (defined on the basis of an input from the ventrobasal complex only) but also of adjacent areas receiving input from other thalamic nuclei. Some of the monkeys were anesthetized, but most were awake behaving preparations. Neurons in the SII region proper generally had small to medium-sized receptive fields, and 86.5% of the 793 units examined responded to weak mechanical stimuli [*Robinson and Burton,* 1980a]. Over half (55.4%) of the units had strictly contralateral receptive fields; only 26% had bilaterally symmetrical receptive fields. The somatotopic organization found differed from that described by *Whitsel* et al. [1969]. The trigeminal representation was anterior and the hindlimb representation posterior. However, each body zone was found to occupy an obliquely oriented strip of cortex. Somatic responsiveness was also demonstrated for neurons in cortical regions located posterior to SII in area 7b, the retroinsular and postauditory fields and the granular part of the insula [*Robinson and Burton,* 1980b]. However, neurons in each field had response properties that distinguished them from neurons in other fields. Neurons in area 7b were often activated from bilateral receptive fields, and the receptive fields tended to diminish when the animal went to sleep. Some of the cells had a visual input. Neurons in the retroinsular, granular insular and postauditory fields were generally activated by innocuous stimuli and often had strictly contralateral receptive fields. Some of the cells in the postauditory and retroinsular fields responded to both somatic and auditory inputs. There was a topographic organization in the postauditory and retroinsular fields but not in the granular insular field.

In a detailed examination of the stimuli that activated neurons in the SII cortex and in the adjacent regions that receive somatic input, *Robinson and Burton* [1980c] found no region to be dominated by a nociceptive input. In SII proper, almost all of the neurons responded just to innocuous stimuli. Less than 3% of the neurons responded just to noxious stimuli. A higher percentage of cells in area 7b could be activated by noxious stimuli (7.5%), even though proportionally fewer of the neurons in this area were activated by somatic stimulation. A few nociceptive cells were also found in the retroinsular and granular insular cortex. *Robinson and Burton* [1980c] suggest that area 7b might be involved in attention to novel, including noxious, stimuli, and that its projection to the limbic cortex would be consistent with a role in affect and in learning.

Cat SII Cortex

Recordings were made from single units of cats in an area believed to be the SII cortex by *Andersson* [1962] and *Carreras and Andersson* [1963]. They reported that many neurons in the SII cortex resembled those in the SI cortex that have cutaneous receptive fields. However, no cells were seen in SII that were activated by joint movement, and more cells were found that had large receptive fields and that could be activated by strong stimuli. Pathways ascending in several parts of the spinal cord white matter contributed inputs to the SII cortex [*Andersson,* 1962]. It was suggested that the thalamic link was more likely to be the posterior complex than the ventrobasal complex [*Andersson,* 1962].

Burton et al. [1982] reexamined the activity of neurons in the SII cortex of the cat in the light of anatomical characterization of the SII cortex proper as a region that receives an input just from the ventrobasal complex (see previous section in SII cortex in the monkey). They distinguished between two portions of SII, a larger lateral part and a smaller medial part. The larger part had a detailed somatotopic representation of the body surface, whereas the smaller medial SII had a less distinct somatotopic organization. Adjacent cortex belonged to the SIV and the suprasylvian fringe areas. The SIV cortex appears to receive input from the posterior complex, and the suprasylvian fringe seems to be comparable to the postauditory cortex of the monkey and to receive auditory as well as somatic inputs. Because of the projection from the SIV cortex to the superior colliculus, *Burton* et al. [1982] propose that this region may be involved in alerting and orientation behavior.

Clearly, the somatic sensory properties of the SII cortex proper in the cat need further study, especially in unanesthetized preparations.

Role of the Thalamus and Cortex in Pain

It can be concluded from what is so far known about the responses of neurons within the VPL nucleus and in the SI cortex of primates and the rat that sensory-discriminative processing of pain sensation probably involves neural mechanisms similar to those utilized for other qualities of somatic sensation. However, there are a number of features of the neural organization that appear to underlie pain processing that deserve further attention. For example, the patch-like terminal zones formed by spinothalamic tract terminals in the primate (and human) VPL nucleus are suggestive of 'thalamic columns'. However, the nociceptive neurons that have been demonstrated in the SI cortex of the rat do not seem to have a columnar organization. Instead, these neurons are located in the output layers (V and VI) of the cortex, beneath neurons of layer IV that respond exclusively to tactile stimuli. It would be important to know if the nociceptive neurons in the primate SI cortex have a similar organization, or if instead they are clustered in cortical columns as would be suggested by the anatomy of the thalamic termination. The recording sites reported by *Kenshalo and Isensee* [1983] in the monkey SI cortex (fig. 6/26) indicate that these neurons are concentrated in layer IV or the supragranular layers, rather than in the deep layers. These observations would be consistent with a columnar organization. If there are indeed 'pain' columns in primate cortex, these might be scattered in proportion to the scattering of the 'thalamic columns' demonstrated by the pattern of termination of the spinothalamic tract in the VPL nucleus. Scattered 'pain' columns could account for the infrequency of pain sensations produced by electrical stimulation of the exposed human cortex [*Penfield and Boldrey,* 1937].

Many of the nociceptive neurons in the VPL nucleus and SI cortex, at least in the monkey, have restricted contralateral receptive fields, and these cells encode the intensity of noxious stimuli. Furthermore, in the case of maintained noxious heating, the responses of cortical neurons adapt very little, as expected from the observation that human pain sensation under these circumstances does not adapt. Evidently, there are neurons in the VPL nucleus and SI cortex that have properties suitable for a role in the sensory-discriminative processing of pain.

However, there are also neurons, at least in the SI cortex, that are nociceptive but that have receptive fields that are too large to permit stimulus localization by the neurons. It seems likely that such cells are involved in some other aspect of pain than sensory discrimination. Possibly these neurons participate in the arousal mechanism.

A similar statement can be made about the nociceptive neurons in the posterior complex and in the central lateral nucleus. Many of the nociceptive neurons in these nuclei do not respond in a manner that would allow them to encode stimulus location. They do respond in a graded fashion to graded intensities of noxious stimuli, but the discharges often outlast the stimulus, and so the discharge trains do not signal stimulus duration well. It is possible that these neurons are also involved in the arousal mechanism. The wide distribution of axonal projections from the central lateral nucleus to the cerebral cortex would be consistent with such a role. However, an alternative or additional role for the nociceptive neurons in the central lateral nucleus would be participation in motor activity in response to painful input. The central lateral nucleus appears to be a part of the motor system, since it receives a cerebellar projection and in turn connects with the striatum. Motor responses are, of course, a prominent component of the pain reaction.

The nucleus submedius has not been sufficiently well investigated to permit any conclusions about its possible role in pain mechanisms. The fact that it projects to the orbitofrontal cortex suggests the possibility of a role in affective responses.

The thalamic reticular nucleus is unlikely to be directly involved in sensation, since it does not project to the cerebral cortex. However, its neurons are likely to be inhibitory interneurons, and so the reticular nucleus could play an important part in shaping the responses of neurons within the main part of the dorsal thalamus, including the VPL nucleus. *Peschanski* et al. [1980a] proposed the possibility that a lesion preventing the inhibition of thalamic neurons by the reticular nucleus might underlie the thalamic syndrome of *Dejerine and Roussy* [1906]. This suggestion reminds one of the hypothesis of *Head and Holmes* [1911] that the thalamic syndrome results from a removal of inhibitory input to the thalamus from the cortex.

The role of the somatosensory areas of the cortex other than the SI cortex in nociception are not as yet clear. The SII cortex proper (an area defined on the basis of an input just from the ventrobasal complex) does not appear to be a major nociceptive-processing region, since only a small percentage of the neurons in this region in the monkey respond to noxious stimuli. More nociceptive neurons are found in the adjacent area 7b and in the retroinsular cortex. One proposal is that area 7b may be involved in attentional mechanisms, including attention to noxious stimuli, and that the connections of area 7b with the limbic system might allow this area to participate in affect and in learning mechanisms. The retroinsular cortex in the monkey and area SIV in the cat receive an input from the PO_m nucleus. Since this cortical region projects to the superior colliculus, it is possible that it is involved in alerting and orienting behavior.

Conclusions

1. Based on their clinical observations, *Head and Holmes* and also *Penfield* and his associates concluded that the cerebral cortex was not concerned with pain sensation. Rather, they believed that pain was felt in the thalamus.

2. However, several lines of evidence indicate that the cerebral cortex has an important role in the processing of pain sensation: (a) stimulation of the exposed cerebral cortex in man sometimes produces pain; (b) lesions of the postcentral gyrus may reduce pain; (c) pain is sometimes experienced in epileptic auras; (d) lesions of the cortex can produce a syndrome resembling 'thalamic' pain.

3. Since pain produces arousal, the activity of much of the cerebral hemisphere is altered by a painful stimulus. This makes it difficult to sort out cause and effect in the pain-processing circuitry of the brain. However, the observation that certain kinds of lesions of the thalamus or cortex can interfere in particular ways with nociceptive processing may give clues as to the organization of the system.

4. The spinothalamic tract ends in the ventral posterior lateral nucleus in primates, including humans, and in rats. There are fewer terminals in the VPL nucleus of the cat, although there are endings in a 'shell region' surrounding VPL. The projection to the VPL nucleus in monkeys and rats is somatotopically organized. In primates, fibers from the lower extremity end laterally and from the upper extremity medially in the nucleus.

5. Spinothalamic tract cells projecting to the VPL nucleus are concentrated in laminae I and V of the contralateral spinal cord.

6. In primates, spinothalamic tract endings occur as patch-like clusters in transverse sections or as rod-like zones in three dimensions.

7. The VPL nucleus projects to the SI and SII regions of the somatosensory cortex. The same neuron may project to both areas, and it is likely that a given neuron can branch and supply more than one cytoarchitectural area within SI. Rod-like arrays of thalamic neurons project to small regions of SI cortex that may be equivalent to cortical columns. The thalamic rod-like arrays could then be considered 'thalamic columns'.

8. The neurons receiving spinothalamic terminals in rod-like zones of the VPL nucleus might be regarded as 'spinothalamic columns'.

9. Small neurons in VPL project to layer I and large neurons to layers IIIb and IV. What functional significance this has is as yet unknown.

10. Individual spinothalamic tract axons can be tracked into the VPL nucleus, using antidromic activation by microstimulation. High threshold (nociceptive-specific) and wide dynamic range (multireceptive) spinothalamic tract cells could be shown to end in the VPL nucleus. Some branching axons that supplied both VPL and the central lateral nucleus were identified.

11. Although many investigations have demonstrated the occurrence of neurons in the VPL nucleus that respond selectively to innocuous mechanical stimuli, there have been a number of reports of nociceptive responses in this nucleus. Recently, the number of such reports has increased.

12. Nociceptive neurons have been observed in the primate VPL nucleus after interruption of the somatosensory pathways that ascend in the dorsal part of the spinal cord. The receptive fields were often restricted in size and located on the contralateral side of the body.

13. Nociceptive responses can also be recorded in monkeys that have an intact spinal cord, although many of the neighboring cells in the VPL nucleus are not responsive to noxious stimuli. Nociceptive VPL neurons include both nociceptive-specific and wide dynamic range classes of cells.

Noxious heat stimuli provide a particularly effective test for these cells, since the responsible receptors are well documented. The receptive fields of the nociceptive VPL neurons are restricted in size, contralaterally placed, and somatotopically related to the locations within the VPL nucleus of the neurons. The nociceptive VPL cells can often be activated antidromically from the SI cortex by stimulation in the region of the boundary between areas 3b and I.

14. Wide dynamic range (multireceptive) neurons have been recorded in awake behaving monkeys in the VPL nucleus.

15. Nociceptive-specific neurons have been demonstrated in the 'shell region' (and some in VPL) of the thalamus of the cat.

16. Nociceptive-specific and wide dynamic range (multireceptive) neurons have also been recorded in the VPL nucleus of the rat thalamus. Like the nociceptive neurons in the monkey VPL nucleus, these cells could often be shown to encode the intensity of noxious heat stimuli. They were also activated by intraperitoneal injections of bradykinin.

17. Neurons in the medial part of the posterior complex (PO_m) receive an input from the spinothalamic tract and project to the retroinsular cortex, a somatosensory receiving area that is separate from SI and SII. According to some investigators, many of the cells in the posterior region respond to noxious stimuli. However, in reports from other laboratories, most of the neurons in PO_m are tactile. It is possible that differences in anesthetic level account for much of the variation in different reports.

18. The central lateral nucleus receives a projection from the spinothalamic tract and another from the cerebellum. This nucleus projects both to the striatum and to broad areas of the cerebral cortex, including SI. The centre median-parafascicular complex receives a major input from the reticular formation and so could be influenced by activity in the spinoreticular tracts. Neurons in the intralaminar region, presumably including the central lateral nucleus, as well as the centre median-parafascicular complex, are often described as being nociceptive. The receptive fields are often very large and bilateral.

19. The nucleus submedius is a small nucleus that receives a projection from lamina I cells of the spinothalamic tract. The nucleus submedius projects to the orbitofrontal cortex. Preliminary evidence suggests that neurons in the nucleus submedius may be nociceptive.

20. The thalamic reticular nucleus receives collaterals from thalamocortical axons as they pass from the specific relay nuclei to their projection zones in the cerebral cortex. Reticular neurons then project back into the

main body of the thalamus. It is believed that the reticular neurons of the nucleus serve as inhibitory interneurons. There have been a number of reports that some neurons in the reticular nucleus respond to or are inhibited by noxious somatosensory stimuli.

21. Candidate nociceptive tracts other than the spinothalamic tract provide input to many of the same thalamic nuclei. The spinoreticular, spinomesencephalic, spinocervical and postsynaptic dorsal column pathways could influence activity in a fairly direct fashion in the ventrobasal complex, PO_m, and intralaminar complex.

22. As in the case of the VPL nucleus of the thalamus, most studies of the SI cerebral cortex have concentrated on neurons that respond to innocuous mechanical stimuli. However, there have been reports that occasional neurons in the SI cortex respond to noxious stimuli.

23. Recently, neurons have been found in the SI cortex of the monkey in the region apparently receiving a projection from nociceptive neurons of the VPL nucleus (boundary between areas 3b and 1) that can be classified as nociceptive-specific or as wide dynamic range (multireceptive) cortical neurons. Many of these cells have restricted contralateral receptive fields, but some have large, bilateral receptive fields. The cells can encode graded noxious heat stimuli. Their responses to noxious heat differ, however, from those of spinothalamic tract cells in that the threshold is higher (perhaps due to anesthesia) and the discharge adapts less during heating. Nociceptive cells in the VPL nucleus have properties in between those of cortical and spinothalamic tract nociceptive cells.

24. Nociceptive neurons have also recently been described in the SI cortex of the rat. However, the receptive fields of nociceptive-specific SI neurons in the rat are large and often include an ipsilateral region; these cells tend to be located in cortical layer VI. The wide dynamic range cells have more restricted receptive fields, and the cells are in layer Vb.

25. Neurons in the SII cortex of the monkey are generally tactile and not nociceptive. More nociceptive cells are found in the adjacent area 7b and the retroinsular cortex. The nociceptive neurons in area 7b might play a role in attention and learning.

26. Similarly, the SII cortex in the cat may not be a major center for nociceptive processing. The SIV cortex, like the retroinsular cortex in the monkey, receives an input from the PO_m nucleus. Since the SIV cortex projects to the superior colliculus, it has been suggested that this part of the cortex may be involved in alerting and orienting behavior.

27. Based on the response properties of neurons in the VPL nucleus

and SI cortex, it can be proposed that information leading to pain sensation is processed in much the same way as information underlying other somatic sensations.

28. It would be important to know if the nociceptive neurons in the SI cortex of the monkey have a columnar organization. The evidence from the anatomical configuration of the termination zones of the spinothalamic tract in primates and the recording sites in layer IV and above in the monkey SI cortex suggest that this might be so, despite contrary evidence with respect to the nociceptive neurons found in the deeper layers of the SI cortex of the rat.

29. Many of the nociceptive neurons in the SI cortex of the monkey have response and receptive field properties that suit them well for a role in the sensory-discriminative processing of pain.

30. However, other neurons in the SI cortex and also neurons in the posterior complex, the intralaminar region of the thalamus, the SII cortex and adjacent somatosensory areas (7b, retroinsular, granular insular, postauditory) do not seem well suited for a sensory-discriminative role.

31. It is suggested that many nociceptive neurons in the cortex and thalamus may be involved in the arousal mechanism. However, cells in the central lateral nucleus could play an additional or alternative role in motor responses to painful stimuli.

32. The nucleus submedius might contribute to affective responses, judging from its connection to the orbitofrontal cortex.

33. The thalamic reticular nucleus could serve an important function as the source of inhibitory modulation of nociceptive responses in the thalamus.

34. Area 7b may be involved in attention and learning related to pain.

35. The retroinsular cortex and PO_m may be concerned with alerting and orienting behavior.

Chapter 7

Overview and Future Directions

Nociceptors

Recent work has extended considerably our understanding of the operation of the nociceptive transmission system, especially of the initial stages of the nociceptive pathways. It is now clear that there are indeed distinct receptors that are responsive to stimuli that damage or threaten damage to tissue, as *Sherrington* [1906] postulated. The evidence is most complete for the cutaneous nociceptors, but similar findings are being obtained for nociceptors supplying muscle, joints, and certain of the viscera. It seems only a matter of time before we will have a rather complete survey of the nociceptor population.

Most of the nociceptors are innervated by afferent fibers of small diameter. Nociceptive afferent fibers are generally either finely myelinated axons (Aδ fibers) or unmyelinated axons (C fibers). The functional role of nociceptors supplied by Aδ or C fibers is often quite distinct. For example, in the skin Aδ fibers are commonly excited just by mechanical forms of noxious stimuli, whereas C fibers are usually activated by mechanical, thermal and chemical noxious stimuli [*Burgess and Perl,* 1973]. Furthermore, the sensory quality of the pain evoked by Aδ fibers is generally described as 'pricking', whereas that evoked by C fibers is 'burning' [*Torebjörk and Hallin,* 1973]. When a noxious stimulus is intense enough and placed sufficiently far from the neuraxis, stimulation may evoke two separate pains, the first due to activity in Aδ fibers and the second to activity in C fibers [*Lewis,* 1942].

An amazing technical development that has greatly assisted the analysis of the sensory functions of various types of afferent fibers has been the introduction by several investigators in Scandinavia [*Torebjörk and Ochoa,* 1980, 1983; *Vallbo,* 1981] of the combined technique of microneurography and microstimulation in human peripheral nerves. It has been possible to correlate the activity recorded from and evoked in individual afferent nerve fibers, including C fibers, with human sensation [*Van Hees and Gybels,*

1981]. Although many of the findings have corroborated functions already proposed for the various fiber types on the basis of cross-species correlations, nevertheless having the evidence directly from human subjects who can communicate their sensory experience is an invaluable addition to the evidence.

When the receptive fields of individual cutaneous nociceptive afferent fibers are mapped, they are found to consist of one or more localized spots on the skin [e.g. *Perl,* 1968]. These receptive fields may help account for the 'pain spots' described by *von Frey* and others near the turn of the century. However, there may be more nociceptive receptor fields than are necessary to account for the density of pain spots, and so the latter may actually be formed by a close association of several nociceptive terminals. At any rate, the general notion that the distribution of receptor terminals accounts for the presence of definable sensory spots in the skin appears to be well grounded, despite a number of disputes and contradictions in the literature since the time of *von Frey.*

The activity described for nociceptors in animals in response to noxious thermal and chemical stimuli seems to account in detail for the sensory experiences produced by comparable stimuli in man. Noxious heating of normal skin appears to activate chiefly the C polymodal nociceptors, and the population of these receptors behaves in a way that appears to provide adequate coding of stimulus intensity [*LaMotte and Campbell,* 1978; *LaMotte* et al., 1983]. The discharges of C polymodal nociceptors also seem to account for the pain produced by chemical stimuli. There is a separate population of these receptors that respond to itch-provoking stimuli and that cause an itch sensation when stimulated [*Torebjörk and Ochoa,* 1981]. It is not yet entirely clear what receptors account for cold pain, although there is nothing in conflict with the suggestion that receptors innervating blood vessels might be involved [*Wolf and Hardy,* 1941].

A topic of active recent investigation is hyperalgesia. To some extent, primary hyperalgesia can be accounted for by the sensitization of C polymodal nociceptors when a less intense thermal damage is used to cause the hyperalgesia [*LaMotte* et al., 1982, 1983] and of Aδ nociceptors when a more intense damaging stimulus is employed [*Meyer and Campbell,* 1981]. However, it appears that some of the changes produced by thermal damage may depend upon central nervous system mechanisms. Further work in this area is warranted.

A related matter is the question of chemical activation of nociceptive endings. Although a number of substances can excite nociceptors, it is still not certain that the hypothesis that nociceptors are really chemoreceptors is correct [*Lewis,* 1942; *Lim,* 1970]. It is reasonable to suppose that damaging stimuli cause the release of substances from cells adjacent to nociceptive terminals, but more work is needed to determine what the important substances actually are and their role in transduction. In addition to peptides, like bradykinin, monoamines, like serotonin, and potassium, a clarification of the role of the prostaglandins is needed. This is an active field of research, and so it is to be expected that major improvements in our knowledge of the mechanisms of nociceptive transduction will be forthcoming.

Similarly, the experimental use of capsaicin has already begun to add importantly to our understanding of peripheral events in nociception. The depletion of chemical agents like substance P [*Gamse* et al., 1980; *Jancsó* et al., 1981] and the blockage of conduction in C polymodal nociceptors but not other C fibers [*Petsche* et al., 1983] indicate a relative specificity of action that may lead to important insights through the use of this neurotoxin. Already it is becoming apparent that substance P is involved in the antidromic vasodilation that results from stimulation of dorsal root fibers, and that substance P depletion by capsaicin treatment can prevent the flare reaction and plasma extravasation that normally accompany axon reflexes in C fibers [*Carpenter and Lynn,* 1981; *Gamse* et al., 1980; *Jancsó* et al., 1977, 1980]. It would appear that a further detailed analysis of the behavior of nociceptors will provide information of value not only for an understanding of pain mechanisms but also of inflammation. A related subject is the secondary hyperalgesia that occurs in areas of the skin adjacent to a damaged area. At least a part of the mechanism of secondary hyperalgesia may relate to peripheral events that could involve axon reflexes [*Fitzgerald,* 1979], although central processes may also contribute.

There has been considerable recent research, particularly in the Federal Republic of Germany, on the functions of fine afferent fibers from muscle and joints. It is too early to relate all of the specific receptor types to particular sensations, but it is clear that there are nociceptors in muscle that are likely to account for the pain of ischemic muscle [*Mense and Stahnke,* 1983] and for that of inflammed joints [*Coggeshall* et al., 1983; *Schaible and Schmidt,* 1983a, b]. It remains to be seen if the microneurography technique can be applied to fine afferent fibers in muscle or joint nerves.

It appears that there are now at least two acceptable hypotheses to account for pain referral in visceral disease. In addition to convergence of cutaneous and visceral afferent inputs onto central neurons, some of which have been shown to belong to ascending tracts (including the spinothalamic and spinoreticular tracts), there has recently also been a demonstration that primary afferent fibers, some of which are nociceptors, can branch [*Bahr* et al., 1981; *Perl,* 1984; *Pierau* et al., 1982; *Taylor and Pierau,* 1982] and supply different tissues (skin and muscle or skin and viscera). It is not clear how frequent such branching nociceptors are, but they clearly cannot be disregarded. It seems a reasonable assumption that axon reflexes in such branched nociceptive afferents could account for referred tenderness.

Our knowledge of nociceptors in viscera is growing. A number of studies have demonstrated the presence of specific types of nociceptive endings in particular visceral organs. For example, there are both Aδ and C fibers supplying the heart whose activity is enhanced following occlusion of a coronary artery [*Brown,* 1967; *Uchida and Murao,* 1974c]. Similarly, the respiratory system is supplied by afferent fibers of Aδ (lung irritant and cough receptors) and C fiber (J receptors) caliber that may be nociceptive [*Coleridge* et al., 1965; *Mills* et al., 1969; *Sellick and Widdicombe,* 1969; *Paintal,* 1969]. Nociceptors have been difficult to demonstrate in the gastrointestinal tract, but recently it has been possible to distinguish nociceptors from mechanoreceptors supplying the biliary tree on the basis of graded distentions that do or do not evoke pseudoaffective responses [*Cervero,* 1982a]. Similarly, many of the studies of genitourinary tract afferents have not demonstrated a separate class of nociceptor, but recently it has been possible to show that polymodal nociceptors with both Aδ and C fibers innervate the testicle [*Kumazawa and Mizumura,* 1980a, b]. It seems highly likely that other gastrointestinal and genitourinary nociceptors will be found in future work in which the appropriate adequate stimulus is used, especially if the viscus studied is inflamed.

Input System

Nociceptive afferent fibers generally enter the spinal cord by way of the dorsal roots. Some nociceptive afferents may instead enter through the ventral roots [*Clifton* et al., 1976; *Coggeshall* et al., 1974; *Coggeshall and Ito,* 1977; *Light and Metz,* 1978; *Maynard* et al., 1977]. However, the number of such fibers may not be as great as was thought on the basis of counts of ventral root axons in which about 30% of the fibers were found to be afferent, because some of the ventral root afferents may loop back into the dorsal root [*Risling and Hildebrand,* 1982] or may arise from receptors supplying the pia mater underneath the spinal cord [*Dalsgaard* et al., 1982]. The looping fibers may account for the 'recurrent sensibility' first described by *Magendie* [1822b].

The fine afferent fibers tend to become segregated from the large fibers within each dorsal rootlet as they approach the spinal cord, at least in some monkey species and in man. This segregation does not seem to occur in the cat [*Synder,* 1977]. Although most of the fine afferents collect on the ventral aspect of the dorsal rootlet, some take a dorsomedial position [*Sindou* et al., 1974b]. There is at least one report that it is possible to produce pain relief in man by interruption of fibers in the ventrolateral aspect of dorsal rootlets [*Sindou* et al., 1974a]. However, postmortem evidence is needed to be sure that the blood supply was not interrupted, since it has also been shown that lesions of the dorsal root entry zone that impinge upon the dorsal horn are also effective in producing pain relief [*Nashold and Ostdahl,* 1979], and such a lesion could be produced secondary to vascular interruption.

Lissauer's tract has now been shown to consist both of primary afferent fibers of fine caliber and also of the axons of propriospinal neurons [*Chung and Coggeshall,* 1979; *Chung* et al., 1979; *Coggeshall* et al., 1981]. In the monkey, the dominant population consists of primary afferent fibers. Several studies have been done that suggest that interruption of Lissauer's tract can produce pain relief [*Hyndman,* 1942; *Rand,* 1960]. However, it is not clear to what extent this depends upon an intrusion into the superficial dorsal horn by the lesion. Presumably multiple lesions would be required if pain relief is needed over several dermatomes [cf. *Nashold and Ostdahl,* 1979].

The fine afferent fibers appear to synapse chiefly in the superficial layers of the dorsal horn, although some Aδ fibers also terminate in lamina V. In fact, in the sacral spinal cord, a nociceptive Aδ fiber may also end in lamina X and in the contralateral dorsal horn [*Light and Perl,* 1979b]. There is a controversy as to the precise destinations of Aδ and C fibers in the superficial dorsal horn. One view is that Aδ fibers end just in lamina I and C fibers in lamina II [*Light and Perl,* 1979a; *Réthyli,* 1977; *Réthyli* et al., 1979]; another opinion is that the converse pattern obtains [*Gobel,* 1979; *Gobel and Binck,* 1977]. Still another possibility is that both Aδ and C fibers end in lamina I and C fibers alone in lamina II. It is clear from experiments in which individual Aδ fibers are injected with horseradish peroxidase and traced to their terminations that Aδ fibers do project to lamina I and apparently not to lamina II [*Light and Perl,* 1979b]. Similar evidence is needed for the destinations of C fibers. This appears to be technically very difficult, since intracellular impalements of individual C fibers are unlikely for dimensional reasons and transport of horseradish peroxidase does not seem to spread far enough along axons to allow them to be traced from dorsal root ganglion cells, which can be impaled, all the way into the spinal cord. Nevertheless, it seems likely that future technical advances will allow a mapping of the termination zones of individually identified C fibers, including C polymodal nociceptors. The previous impression that many of the afferents to the substantia gelatinosa are recurrent terminals ('flame-shaped arbors') of axons that initially descend well into the dorsal horn is now believed to be incorrect and based upon examination of material derived from neonatal animals in which the substantia gelatinosa was incompletely developed. These terminals are now known to belong to hair follicle afferents and to end no more superficially than in lamina III [*Brown* et al., 1977b].

Several types of axon terminals can be identified in electron micrographs of sections of the dorsal horn. Primary afferent fibers typically form the central terminals of complex synaptic structures known as glomeruli. There are several distinctly different types of central terminals, based on synaptic vesicular content, cytoplasmic density, and the type of degeneration that results when the dorsal root is sectioned [*Knyihár-Csillik* et al., 1982; *Ralston,* 1979; *Ralston and Ralston,* 1979]. One type of synaptic terminal found in laminae I and II contains numerous large dense cored vesicles. Although it has been shown that some of these terminals contain substance P [*deLanerolle and LaMotte,* 1983; *DiFiglia* et al., 1982; *Pickel* et al., 1977], it seems likely that other terminals of the same type might contain other peptides known to be present in dorsal root ganglion cells: somatostatin, cholecystokinin, or vasoactive intestinal polypeptide. Another type of central terminal occurs in lamina II and is characterized by dense cytoplasm. These may contain (in rodents) fluoride-resistant acid phosphatase [*Knyihár* et al., 1974], and may prove to use ATP as a transmitter [*Dodd* et al., 1984]. Still another kind of ending, most prominent in the deeper layers of the dorsal horn, contains round vesicles. Presumably, these endings, which often belong to large myelinated fibers, contain still another neurotransmitter, perhaps glutamate. Terminals that do not degenerate following dorsal rhizotomy are of several types, including ones with flattened synaptic vesicles. Some can be shown to contain glutamic acid decarboxylase and so presumably utilize GABA as a transmitter [*McLaughlin* et al., 1975]. Other endings are associated with monoamines or contain enkephalin or other peptides found in neurons of the dorsal horn (neurotensin, avian pancreatic polypeptide). It will be a major task to sort out all of the various neurotransmitters in the dorsal horn and to assign them to particular functional categories of neurons, whether primary afferents, intrinsic neurons of the spinal cord, or axons descending from the brain. The fact that there are more candidate neurotransmitters than major classes of primary afferent fibers suggests the possibility that each separate receptor class could have its own transmitter(s). This suggests the further possibility of pharmacological manipulations that might interfere with or enhance transmission from each receptor population. With respect to pain, this would imply either interruption of the input from particular categories of nociceptors to produce analgesia directly or the enhancement of activity in afferents that produce inhibition of nociceptive pathways by a central action, and thus indirectly cause analgesia.

A great deal has been learned about the response properties of neurons in the dorsal horn of the spinal cord. A major advance was the recognition by *Perl's* group that many of the neurons in the marginal zone (and elsewhere) respond specifically to noxious stimuli [*Christensen and Perl,* 1970; *Kumazawa* et al., 1975]. Thus, there appear to be 'labeled line' pathways available to transmit purely nociceptive information.

Another type of nociceptive neuron that is commonly encountered in the dorsal horn, including lamina I, is the 'wide dynamic range' or multiconvergent neuron. These cells respond to innocuous mechanical stimuli, but they are most effectively activated by frankly noxious stimuli. Whereas it is clear that the nociceptive-specific neurons can provide unambiguous signals that noxious stimuli have occurred, it seems likely that the wide dynamic range (multireceptive) neurons also have an important contribution to make in nociception. The problem is to determine what this may be. One possibility is that the responses of these cells to innocuous stimuli are treated by the central nervous system as noise. Higher centers might ignore their activity unless it exceeds a threshold. A precedent for this is already known for nociceptive primary afferent fibers, since a few impulses in any one nociceptor are insufficient to evoke a sensation of pain [*Gybels* et al., 1979; *Van Hess and Gybels,* 1981; *Hallin* et al., 1982]. Summation is required. Another is that the activity of these cells is modulated by descending control systems, so that they act as detectors of tactile or of noxious stimuli at different times, depending upon the commands they receive from higher centers [see *Willis,* 1982]. This notion of switching depends upon the further postulate that the thalamic or cortical processing centers receiving input from wide dynamic range cells are also switched in some way so as to allow the ascending information to be interpreted differently, depending on whether the wide dynamic range cells are acting as tactile or nociceptive detectors. Evidence that something of this kind occurs in behaving animals comes from the study of *Casey* [1966] in which thalamic neurons responded like either nociceptive-specific or wide dynamic range cells, depending upon the stage of the sleep-wake cycle. Furthermore, it is also known that descending control systems are present that could cause such switching, at least at the spinal cord level. Another hypothesis is that wide dynamic range (multireceptive) cells code nociceptive inputs in part by virtue of the modulation of the population of such cells by the system of 'Diffuse Noxious Inhibitory Controls' (DNIC) described by *LeBars* et al. [1979a, b]. A set of neurons are excited by a given noxious stimulus, but all other wide dynamic range neurons whose excitatory receptive fields are outside of the area stimulated are inhibited. The higher processing centers receive a population input that reflects the changed activity of both the excited and the inhibited elements. Given tonic background activity in these neurons, the push-pull effect of a noxious input should enhance the contrast between excited and inhibitory components of the system and improve the signal-to-noise ratio within the population of wide dynamic range neurons. The tactile responses of the excited cells would tend to be suppressed by the inhibitory surround, since these responses are weak, and so the cells may behave more like nociceptive-specific neurons within the context of the system [*LeBars and Chitour,* 1983].

Another major advance in the study of dorsal horn sensory mechanisms has been the development of techniques for the study of neurons belonging to the substantia gelatinosa. Until recently, no data were available concerning the functional properties of these cells. Now we know that there are a number of functional types of neurons within the substantia gelatinosa, including some with properties quite different from those of adjacent laminae [*Cervero* et al., 1979b, c]. Particularly interesting are cells showing marked habituation or long-lasting discharges in response to peripheral stimulation [*Wall* et al., 1979a; *Fitzgerald,* 1981]. However, many neurons in the substantia gelatinosa have responses that resemble closely those of neurons in the neighboring laminae [*Bennett* et al., 1980; *Price* et al., 1979]. One of the problems for the future will be to establish the connectivity of functionally identified neurons of the substantia gelatinosa. A promising start in this direction has been the demonstration of the synaptic endings of many of the processes of several types of neurons in lamina II by the combined techniques of extra- and intracellular recordings (to show the functional properties of the cells) and histological reconstructions of the labeled dendrites and axons in electron micrographs following intracellular injection of horseradish peroxidase [*Gobel* et al., 1980]. Future studies will undoubtedly combine such labeling with a second labeling technique, such as degeneration, to establish the identity of both pre- and postsynaptic elements. Furthermore, it should be possible to use immunohistochemical markers to determine the neurotransmitter content of some of the neurons of the substantia gelatinosa. A beginning has also already been made in this direction [*Bennett* et al., 1982]. The goal will be a 'wiring diagram' of the dorsal horn, complete with connections of primary afferents, intrinsic neurons, and descending projections of functionally and chemically defined neurons. We are obviously far from such a goal at the moment, but much of the technology is already available for such an effort.

The role of neurons in deeper layers of the dorsal horn and also in the intermediate region and ventral horn needs to be clarified. Many of the neurons in these deeper layers are known to project axons through one or another of the ascending nociceptive tracts, and many of these cells have nociceptive-specific or wide dynamic range (multireceptive) types of responses. There is no reason to ignore their participation in signalling nociceptive information to the brain. In addition to nociceptive inputs from the skin, many of these cells receive nociceptive information from viscera or muscle. These convergent inputs suggest a role in pain referral.

Although there have been a number of studies demonstrating pharmacologic actions of various neurotransmitter candidates on nociceptive dorsal horn neurons, there has been little effort to relate these responses to particular functional systems. An important goal for future studies will be to analyze the connections and transmitters utilized by complete nociceptive systems, as well as of control systems. For example, if it proves to be the case that C polymodal nociceptors utilize substance P as an excitatory transmitter in the dorsal horn, it may eventually be possible to specify which neurons in the dorsal horn respond to an input from C polymodal nociceptors, using such information as which cells can be activated by iontophoretic application of substance P and by C polymodal nociceptors and which have their activity blocked by substance P depletion or by substance P antagonists. If somatostatin proves to be an inhibitory transmitter released by primary afferent fibers, it will be important to determine the receptor type(s) that contain somatostatin and the effects that such afferents have on nociceptive transmission.

Nociceptive Tracts

There are several ascending tracts that may well play a role in nociception. In addition, it is possible that important signals can ascend through multisynaptic pathways made up of propriospinal neurons. One of the important organizational features of the nociceptive system in humans and subhuman primates is that the most significant nociceptive system ascends in the anterolateral part of the spinal cord on the side opposite to the nociceptive input. The nociceptive tracts located in this region of the spinal cord white matter include the spinothalamic, spinoreticular and spinomesencephalic tracts. It can be presumed that the spinocerebellar and spino-olivary tracts are not important for pain sensation, since removal of the cerebellum does not impair sensation. However, it could not be excluded that the cerebellum plays a role in nociceptive motor and autonomic responses or in modulation of nociceptive transmission.

Good progress has been made in mapping the cells of origin of the various ascending tracts, especially using the horseradish peroxidase technique [see *Willis and Coggeshall,* 1978]. Furthermore, we are learning in more detail the distribution of the terminals of the various tracts within target regions of the brain. Evidence has been obtained that some spinal neurons participate in more than one ascending tract. The most important nuclei in the brain that should be considered as potential components of the nociceptive system include the rhombencephalic and midbrain reticular formation, the periaqueductal gray, and several thalamic nuclei, including the ventral posterior lateral nucleus, the medial part of the posterior complex, and several medial thalamic nuclei, including the intralaminar complex and the nucleus submedius. A given spinothalamic tract neuron may project to any or all of these. In addition, there are likely to be separate projections of neurons that terminate in just one or a few of these nuclei.

Spinothalamic tract cells are now among the best characterized of the ascending somatosensory tract neurons. Spinothalamic tract cells that project to the contralateral ventral posterior lateral nucleus can often be activated both by Aδ and C fibers, and often Aβ fibers, as well [*Beall* et al., 1977; *Chung* et al., 1979; *Foreman* et al., 1975]. They typically can be classified as nociceptive-specific or as wide dynamic range (multireceptive) cells with respect to mechanical stimulation of the skin [*Chung* et al., 1979; *Price* et al., 1978]. They tend to have restricted contralateral cutaneous receptive fields. They generally also respond in a graded fashion to graded noxious heat stimuli [*Kenshalo* et al., 1979], and they may be activated by noxious cold [*Kenshalo* et al., 1982]. They can also be excited by noxious chemical stimuli [*Foreman* et al., 1979b; *Levante* et al., 1975]. There is commonly a convergent input from fine muscle afferents and also nociceptive visceral afferents [*Foreman* et al., 1979a; *Milne* et al., 1981]. This convergent input may permit these cells to play a role in pain referral. It seems reasonable to speculate that spinothalamic tract cells projecting to the ventral posterior lateral nucleus participate in signalling the sensory-discriminative aspects of pain.

Other spinothalamic tract cells project just to the medial part of the thalamus, to the region of the central lateral nucleus [*Giesler* et al., 1981b]. Many of these neurons have very large receptive fields that depend in part upon a pathway that ascends to the brain and then descends into the spinal cord. They are generally nociceptive-specific cells, although some are multireceptive. The cells of origin of these spinothalamic neurons tend to be deeper in the spinal cord gray matter than are the cells that project to the lateral thalamus, often being in lamina VII. It seems likely that such cells participate in other aspects of pain than sensory discrimination. One hypothesis is that these cells are actually a component of the reticular formation and that they function as part of the ascending reticular activating system. They would thus contribute to arousal and other motivational-affective components of pain.

Other spinothalamic tract cells projecting to the medial thalamus are located in lamina I and have small, nociceptive-specific receptive fields [*Craig and Kniffki,* 1982]. These neurons may terminate in the nucleus submedius. Since the nucleus submedius projects to the frontal lobe, it is possible that this system participates in affective aspects of pain.

Spinothalamic tract cells often have large inhibitory receptive fields on much of the body surface, apart from the excitatory receptive field [*Gerhart* et al., 1981]. Usually the inhibition requires noxious intensities of stimulation, although occasionally a tactile input will produce inhibition from an area surrounding the excitatory field. The large, nociceptive inhibitory receptive fields are different from those described to be a part of the DNIC mechanism, since the inhibition of spinothalamic tract cells remains after spinal cord transection and so does not depend upon a supraspinal loop as DNIC does. Prolonged peripheral nerve stimulation, using stimulus intensities sufficient to excite nociceptive afferent fibers, can produce a prolonged inhibition of spinothalamic tract cells [*Chung* et al., 1984a, b]. This inhibition may be part of the mechanism of analgesia produced by transcutaneous electrical nerve stimulation and by acupuncture.

Spinothalamic tract cells can be used to test for the effects of candidate synaptic transmitters. A number of these have been shown to have effects on spinothalamic tract cells that are consistent with evidence from other studies of drug actions on nociceptive dorsal horn neurons in general [*Willcockson* et al., 1984a, b]. However, the use of spinothalamic tract cells offers the advantage of an identified set of sensory neurons, as opposed to unidentified elements that may play a role in reflex rather than in sensory mechanisms.

Studies of spinothalamic tract cells suggest that these neurons could well play a major role in a variety of pain phenomena. Future studies need to be designed to test whether or not these cells really do play such roles. A major need is the development of techniques for recording from spinothalamic tract cells in awake behaving animals. Comparable recordings are already possible in the trigeminal system [e.g. *Hoffman* et al., 1981]. Another important goal would be the identification of the transmitter(s) used by spinothalamic tract cells. Such a discovery might lead to a new pharmaceutical approach that could substitute for cordotomy, provided that the transmitter were uniquely different from that used by other neural pathways. Alternatively, a better understanding of the transmitters used in pathways that excite or inhibit spinothalamic tract cells might accomplish the same objective.

Much less is known about the spinoreticular tract than about the spinothalamic tract. One of the problems with experiments on the spinoreticular tract is the possibility that the responses of these cells are altered by anesthesia. Furthermore, recordings from individual spinoreticular cells fail to show what part of the brain is likely to be affected by the cell under observation, since it is unclear what the target cells in the reticular formation are, much less whether the target cells then project rostrally, caudally, or where. Thus, investigations of spinoreticular neurons are difficult to relate to pain mechanisms in any specific way. Furthermore, the receptive fields of spinoreticular neurons and of neurons in the reticular formation are often large [*Fields* et al., 1977; *Wolstencroft,* 1964], and so it seems unlikely that the spinoreticulo-thalamic pathway will provide information of much use for the sensory-discriminative aspects of pain. However, the arousal mechanism may be a major factor underlying the motivational-affective effects of stimulation within the reticular formation and with the neuroanatomical connectivity of the intralaminar complex, which is one of the main thalamic targets of the ascending reticular connections.

The spinomesencephalic tract can be regarded as a component of the spinoreticular system. However, the cells of origin of the spinomesencephalic tract are more similar in location to those cells that project to the ventral posterior lateral nucleus of the thalamus than to the intralaminar complex [*Trevino,* 1976; *Willis* et al., 1979], and some neurons have been shown to project both to the VPL nucleus and to the midbrain periaqueductal gray [*Price* et al., 1978]. Thus, the midbrain receives sensory-discriminative information from some of the same neurons that project to the VPL nucleus. There is a pathway from the periaqueductal gray to the VPL nucleus [*Barbaresi* et al., 1982], and so the periaqueductal gray could serve as a nociceptive relay. However, alternatively, the projection from the periaqueductal gray could be inhibitory. Descending inhibitory actions of the periaqueductal gray have been demonstrated [see *Willis,* 1982], and so it is not inconceivable that there are ascending inhibitory connections as well. Future work should be directed at a determination of the interactions of the periaqueductal gray and the VPL nucleus. The midbrain is clearly concerned with both nociceptive responses and with nociceptive modulation, and so much more attention should be placed on the connections of the midbrain and the response properties and connectivity of its neurons.

It seems unlikely that the spinocervical tract or the postsynaptic dorsal column pathway are very important for pain in humans or subhuman primates. However, these pathways could play a role in the return of pain after an initially successful cordotomy in humans or monkeys. This possibility shoud be investigated in monkeys that have been subjected to chronic cordotomies. The roles of these pathways in nociception in the cat merits further attention. Although the neurons belonging to the spinocervical tract include wide dynamic range (multireceptive) cells, it is not yet clear whether such cells occur in any numbers in the lateral cervical nucleus (as they do in the lateral cervical nucleus of the rat). The interrelations of the spinocervical and postsynaptic dorsal column pathway need further clarification. Double label studies are in order to determine what proportion of cells in laminae III–V of the dorsal horn belong to both of these pathways. The route taken by ascending fibers that make connections with the spinothalamic tract nucleus located in lamina VII at C1 and C2 studied by *Carstens and Trevino* [1978b] needs to be defined.

Thalamocortical Mechanisms

The ventral posterior lateral (VPL) nucleus of the thalamus is a major nucleus of termination of the spinothalamic tract in humans, monkeys and rats. In the cat, the spinothalamic tract appears to have only a minimal direct projection to the VPL nucleus, but it ends in a 'shell' region around the VPL nucleus [*Berkley,* 1980; *Boivie,* 1970; *Jones and Burton,* 1974]. The spinothalamic terminations in the human and monkey are made in burst-like patches when viewed in coronal sections, but in rod-like arrays when reconstructed in three dimensions [*Boivie,* 1979; *Mehler* et al., 1960]. These rod-like arrays may be in relation to 'thalamic columns' [*Jones,* 1983], which might then project to cortical columns that would presumably be concerned with nociceptive information.

The spinothalamic tract endings in the VPL nucleus of the monkey tend to be arranged around the periphery of the nucleus [*Berkley,* 1980]. If this is true, it may be that nociceptive processing occurs in a zone of the VPL nucleus that surrounds a centrally located core region engaged in tactile processing. This statement could apply both to primates and to carnivores. However, more attention needs to be paid to this point before one can judge its merits.

The notion that there may be nociceptive thalamic columns also needs to be pursued. For example, it should be possible to record from a large number of nociceptive neurons within the VPL nucleus along a properly oriented electrode track in the horizontal plane. Furthermore, nociceptive neurons in the SI cortex should receive connections from neurons in a nociceptive thalamic column, and so the latter should be labeled if horseradish peroxidase were injected into the appropriate cortical zone.

Additional work should be done following the course of individual nociceptive axons into the thalamus and also into the cortex, using either the antidromic activation technique [*Applebaum* et al., 1979] or better the injection of horseradish peroxidase into individual identified axons [*Landry and Deschenes,* 1981].

More recordings should be made from nociceptive neurons in the VPL thalamus and in the SI cerebral cortex. These cells should be characterized further in ways that allow judgements about the possible transformation of nociceptive signals from the level of peripheral nociceptive afferent fibers to ascending nociceptive tracts and then to higher levels. The objective would be to evaluate how well the information processing accounts for the properties of sensory experience presumably evoked by the activity.

In addition to experiments done on anesthetized animals, it will be necessary to have more work done on unanesthetized behaving animals. Anesthesia alters the responsiveness of neurons in the thalamus and cortex and so an evaluation of the response properties of cells at these levels requires recordings in the absence of anesthesia. However, it would not be appropriate to apply noxious stimuli to unanesthetized, paralyzed preparations, and so such experiments would best be done on awake behaving animals. It has been shown that at least mild noxious stimuli can be given to such animals, provided that care is taken to give the animals the means to terminate stimuli that they cannot tolerate.

The anesthesia problem becomes even more acute when attempts are made to evaluate the functional role of neurons in the medial thalamus and probably also in some of the areas of cerebral cortex that form ancillary somatosensory cortical processing zones. *Casey and Morrow* [pers. commun.] have shown that neurons in the medial thalamus that have somatosensory responsiveness in the awake behaving state become virtually silent when an anesthetic is given. Many of the discrepancies in the literature concerning the response properties of thalamocortical neurons may well derive from differences in levels or types of anesthesia employed by different groups of investigators or in different experiments.

As our knowledge of the connections made by different thalamic nuclei and neurons in different cortical areas grows, it will become easier to develop working hypotheses about the possible functional roles of various thalamocortical systems. Speculations have already been introduced about the possibility that the nociceptive neurons of area 7b in the monkey might be concerned with attention and learning in response to painful stimuli [*Robinson and Burton,* 1980c] and about a potential role for neurons in the SIV area in the cat in alerting and orientation behavior [*Burton* et al., 1982]. As such hypotheses are generated, methods should be developed to test them. Selective lesions might be produced and behavioral tests done to provide support for or to deny such hypotheses. For example, if the nucleus submedius is an important center for processing pain information and signalling this to the orbitofrontal cortex [*Craig and Burton,* 1981; *Craig* et al., 1982], what will be the behavioral effect of a lesion of the nucleus submedius? Another issue is the role of the inhibitory feedback that is presumably mediated by the thalamic reticular nucleus. Does elimination of this function result in the thalamic syndrome, as has been suggested [*Peschanski* et al., 1980a]?

One strategy that might be useful for investigators of the somatosensory system is an analysis of what has been learned of processing in the visual system to see what principles might apply to the somatosensory system. In the visual system, there now appear to be many different cortical areas that contribute to different aspects of visual processing. Each of these has a particular set of thalamic and cortical connections, as well as access to visual input and to a subcortical readout. The nociceptive system should presumably have a similar arrangement. Some of the common problems that the nervous system needs to solve that relate to both vision and pain include analysis of the quality, intensity, location, duration and form of the stimulus (sensory-discriminative processing); attention to novel or important stimuli; alerting and orienting to such stimuli; entering of significant information into memory; pattern recognition; emotional valuation (motivational-affective processing).

Plasticity

Essentially all of what has been stated so far has related to the nociceptive system as it exists in the normal individual. Little reference has been made to the nervous system that has been altered by disease, beyond pointing out, for example, that stimulation in the midbrain and medial thalamus can evoke the pain being suffered by a chronic pain patient, whereas comparable stimulation in normal (from the pain standpoint) subjects may evoke no sensation [*Tasker* et al., 1983]. However, evidence of this kind suggests strongly that the nociceptive system is plastic; it can change its character with experience of long-enduring pain. If this is true, presumably such plastic changes could account for the clinical differences between acute and chronic pain. Acute pain can generally be managed effectively. Narcotic and other nonnarcotic drugs are one satisfactory means, and there are a host of other techniques ranging up to full general anesthesia. However, drugs appear to be more a hindrance than a help in many forms of chronic pain, and the same can be said in most instances for surgery. Chronic pain appears to be a different problem than acute pain, and because it is so poorly understood, it is too often a refractory problem [*Casey* et al., 1979].

Chronic pain will be one of the most important challenges for research in the field of pain for the near future. One strategy that is being pursued is the development of animal models of chronic pain. It has been suggested that pain originates from neuromas following peripheral nerve section [*Wall* et al., 1979b; *Wiesenfeld and Lindblom,* 1980] and from changed activity of central neurons following dorsal rhizotomy [*Basbaum,* 1974; *Dennis and Melzack,* 1979; *Lombard* et al., 1979]. One indication of pain in these models is the occurrence of 'autotomy', self-mutilation resulting from biting the affected extremity. However, it is not certain that the animals that show autotomy do in fact invariably experience pain, since a form of autotomy occurs in children with congenital insensitivity to pain [*Sweet,* 1981a].

Another model of chronic pain that is being developed by *Guilbaud* and her associates is the rat made polyarthritic by injection of *Mycobacterium butyricus.* Altered responses have been recorded in the ventrobasal complex of the thalamus and in the SI cortex in these animals [*Gautron and Guilbaud,* 1982; *Lamour* et al., 1983b]. A disadvantage of this preparation is that the injection of the arthritogenic substance is systemic, and so it is not feasible to study the effects of afferent inputs from both normal and arthritic joints in the same animal. Another problem is that the inflammation may affect other tissue in addition to joints. Despite these reservations, the model appears to have considerable promise.

Another problem is that once valid models of chronic pain are established, there will be ethical constraints on the use of such a model. Clearly, a high ethical weight must be placed on solving the problem of chronic pain in humans. However, the judicious use of an animal model is a requirement as well [*Zimmermann,* 1983].

Clues about possible mechanisms of chronic pain may come from studies of nervous system plasticity. Such work on the somatosensory system should be a useful line to pursue. Already, a number of studies have been done in which peripheral nerve sections have produced neuromas, and the altered properties of afferent nerve fibers in the neuroma have been examined [*Devor and Jänig,* 1981; *Scadding,* 1982; *Wall and Gutnick,* 1974]. Peripheral nerve section also produces changes centrally. For example, the peptide and fluoride-resistant acid phosphatase content of the substantia gelatinosa is reduced in a manner similar to that caused by dorsal rhizotomy [*Barbut* et al., 1981; *Jessell* et al., 1979; *Knyihár and Csillik,* 1976]. Receptive fields of neurons in the dorsal horn may also be altered by peripheral nerve section [*Devor and Wall,* 1981a, b]. The substance P content of the dorsal horn is at first reduced by dorsal rhizotomy, but then it is restored, apparently by the growth of new terminals by intrinsic substance P interneurons [*Tessler* et al., 1980, 1981]. Changed receptive fields following peripheral nerve section are seen not only in the dorsal horn, but also in the dorsal column nuclei [*Dostrovsky* et al., 1976] and in the SI somatosensory cortex [*Kaas* et al., 1983]. Clearly, a lot more needs to be done in the area of plasticity of the somatosensory system, but there have been enough intriguing results already to indicate that this area will produce important new knowledge.

Descending Control Systems

As indicated in Chapter 1, the excitement in the area of pain mechanisms stems in part from the discovery of the opiate receptors, the intrinsic opioid substances, and the 'analgesia' pathways in the central nervous systems. This area was not part of the theme of the present review (see *Willis,* 1982], but future work on pain modulation will obviously be of great importance for furthering our understanding and enhancing our ability to control pain.

Conclusion

The 'puzzle of pain' [*Melzack,* 1973] has stimulated the interest of a large number of investigators, both because of its importance and its challenge, but also because many of the techniques needed to begin its solution are now available. The past decade or two have seen enormous advances in our understanding of pain mechanisms. Hopefully, the momentum can be maintained so that a similar statement can be made in another decade.

References

Abols, I.A.; Basbaum, A.I.: Afferent connections of the rostral medulla of the cat: a neural substrate for midbrain-medullary interactions in the modulation of pain. J. comp. Neurol. *201:* 285–297 (1981).

Adriaensen, H.; Gybels, J.; Handwerker, H.O.; Van Hees, J.: Latencies of chemically evoked discharges in human cutaneous nociceptors and of the concurrent subjective sensations. Neurosci. Lett. *20:* 55–59 (1980).

Adrian, E.D.: The impulses produced by sensory nerve endings. 1. J. Physiol. *61:* 49–72 (1926a).

Adrian, E.D.: The impulses produced by sensory nerve endings. 4. Impulses from pain receptors. J. Physiol. *62:* 33–51 (1926b).

Adrian, E.D.; Zotterman, Y.: The impulses produced by sensory nerve endings. 3. Impulses set up by touch and pressure. J. Physiol. *61:* 465–483 (1926).

Ainsworth, A.; Hall, P.; Wall, P.D.; Allt, G.; MacKensie, M.L.; Gibson, S.; Polak, J.M.: Effects of capsaicin applied locally to adult peripheral nerve. II. Anatomy and enzyme and peptide chemistry of peripheral nerve and spinal cord. Pain *11:* 379–388 (1981).

Albe-Fessard, D.; Boivie, J.; Grant, G.; Levante, A.: Labelling of cells in the medulla oblongata and the spinal cord of the monkey after injections of horseradish peroxidase in the thalamus. Neurosci. Lett. *1:* 75–80 (1975).

Albe-Fessard, D.; Kruger, L.: Duality of unit discharges from cat centrum medianum in response to natural and electrical stimulation. J. Neurophysiol. *25:* 3–20 (1962).

Albe-Fessard, D.; Levante, D.; Lamour, Y.: Origin of spino-thalamic tract in monkeys. Brain Res. *65:* 503–509 (1974).

Amassian, V.E.: Evoked single cortical unit activity in the somatic sensory areas. Electroenceph. clin. Neurophysiol. *5:* 415–438 (1953).

Ammons, W.S.; Foreman, R.D.: Responses of T2-T4 spinal neurons to stimulation of the greater splanchnic nerves of the cat. Expl Neurol. *83:* 288–301 (1984).

Anderson, F.D.: Distribution of dorsal root fibers in the cat spinal cord. Anat. Rec. *136:* 154–155 (1960).

Andersson, S.A.: Projection of different spinal pathways to the second somatic sensory area in cat. Acta physiol. scand. *56:* suppl. 194, pp. 1–74 (1962).

Andrew, B.L.: The sensory innervation of the medial ligament of the knee joint. J. Physiol. *123:* 241–250 (1954).

Andrezik, J.A.; Chan-Palay, V.; Palay, S.L.: The nucleus paragigantocellularis lateralis in the rat. Demonstration of afferents by the retrograde transport of horseradish peroxidase. Anat. Embryol. *161:* 373–390 (1981).

Angaut-Petit, D.: The dorsal column system. II. Functional properties and bulbar relay of the postsynaptic fibers of the cat's fasciculus gracilis. Exp. Brain Res. *22:* 471–493 (1975).

Angel, A.: The effect of peripheral stimulation on units located in the thalamic reticular nuclei. J. Physiol. *171:* 42–60 (1964).

Angel, A.; Clarke, K.A.: An analysis of the representation of the forelimb in the ventrobasal complex of the albino rat. J. Physiol. *249:* 399–423 (1975).

Applebaum, A.E.; Beall, J.E.; Foreman, R.D.; Willis, W.D.: Organization and receptive fields of primate spinothalamic tract neurons. J. Neurophysiol. *38:* 572–586 (1975).

Applebaum, A.E.; Leonard, R.B.; Kenshalo, D.R., Jr.; Martin, R.F.; Willis, W.D.: Nuclei in which functionally identified spinothalamic tract neurons terminate. J. comp. Neurol. *188:* 575–586 (1979).

Applebaum, M.L.; Clifton, G.L.; Coggeshall, R.E.; Coulter, J.D.; Vance, W.H.; Willis, W.D.: Unmyelinated fibres in the sacral 3 and caudal 1 ventral roots of the cat. J. Physiol. *256:* 557–572 (1976).

Armstrong-James, M.: The functional status and columnar organization of single cells responding to cutaneous stimulation in neonatal rat somatosensory cortex SI. J. Physiol. *246:* 501–538 (1975).

Aronin, N.; DiFiglia, M.; Liotta, A.S.; Martin, J.B.: Ultrastructural localization and biochemical features of immunoreactive leu-enkephalin in monkey dorsal horn. J. Neurosci. *1:* 561–577 (1981).

Bahr, R.; Blumberg, H.; Jänig, W.: Do dichotomizing afferent fibers exist which supply visceral organs as well as somatic structures? A contribution to the problem of referred pain. Neurosci. Lett. *24:* 25–28 (1981).

Baker, D.G.; Coleridge, H.M.; Coleridge, J.C.G.; Nerdrum, T.: A search for a cardiac nociceptor: stimulation by bradykinin of sympathetic afferent nerve endings in the heart of the cat. J. Physiol. *306:* 519–536 (1980).

Baker, M.A.: Spontaneous and evoked activity of neurones in the somatosensory thalamus of the waking cat. J. Physiol. *217:* 359–379 (1971).

Barbaresi, P.; Conti, F.; Manzoni, T.: Periaqueductal grey projection to the ventrobasal complex in the cat: a horseradish peroxidase study. Neurosci. Lett. *30:* 205–209 (1982).

Barber, R.P.; Vaughn, J.E.; Slemmon, J.R.; Salvaterra, P.M.; Roberts, E.; Leeman, S.E.: The origin, distribution and synaptic relationships of substance P axons in rat spinal cord. J. comp. Neurol. *184:* 331–352 (1979).

Barbut, D.; Polak, J.M.; Wall, P.D.: Substance P in spinal cord dorsal horn decreases following peripheral nerve injury. Brain Res. *205:* 289–298 (1981).

Basbaum, A.I.: Conduction of the effects of noxious stimulation by short-fiber multisynaptic systems of the spinal cord in the rat. Expl Neurol. *40:* 699–716 (1973).

Basbaum, A.I.: Effects of central lesions on disorders produced by multiple dorsal rhizotomy in rats. Expl Neurol. *42:* 490–501 (1974).

Basbaum, A.I.; Fields, H.L.: Endogenous pain control mechanisms: review and hypothesis. Ann. Neurol. *4:* 451–462 (1978).

Baxendale, R.H.; Ferrell, W.R.: Discharge characteristics of the elbow joint nerve of the cat. Brain Res. *261:* 195–203 (1983).

Baxter, D.W.; Olszewski, J.: Congenital universal insensitivity to pain. Brain *83:* 381–393 (1960).

Beal, J.A.: Serial reconstruction of Ramón y Cajal's large primary afferent complexes in laminae II and III of the adult monkey spinal cord: a Golgi study. Brain Res. *166:* 161–165 (1979a).

Beal, J.A.: The ventral dendritic arbor of marginal (lamina I) neurons in the adult primate spinal cord. Neurosci. Lett. *14:* 201–206 (1979b).

Beal, J.A.: Identification of presumptive long axon neurons in the substantia gelatinosa of the rat lumbosacral spinal cord: a Golgi study. Neurosci. Lett. *41:* 9–14 (1983).

Beal, J.A.; Bicknell, H.R.: Primary afferent distribution pattern in the marginal zone (lamina I) of adult monkey and cat lumbosacral spinal cord. J. comp. Neurol. *202:* 255–263 (1981).

Beal, J.A.; Cooper, M.H.: The neurons in the gelatinosal complex (laminae II and III) of the monkey (*Macaca mulatta*): a Golgi study. J. comp. Neurol. *179:* 89–122 (1978).

Beal, J.A.; Fox, C.A.: Afferent fibers in the substantia gelatinosa of the adult monkey (*Macaca mulatta*): a Golgi study. J. comp. Neurol. *168:* 113–144 (1976).

Beal, J.A.; Penny, J.E.; Bicknell, H.R.: Structural diversity of marginal (lamina I) neurons in the adult monkey (*Macaca mulatta*) lumbosacral spinal cord: a Golgi study. J. comp. Neurol. *202:* 237–254 (1981).

Beall, J.E.; Applebaum, A.E.; Foreman, R.D.; Willis, W.D.: Spinal cord potentials evoked by cutaneous afferents in the monkey. J. Neurophysiol. *40:* 199–211 (1977).

Beattie, M.S.; Bresnahan, J.C.; King, J.S.: Ultrastructural identification of dorsal root primary afferent terminals after anterograde filling with horseradish peroxidase. Brain Res. *153:* 127–134 (1978).

Becerra-Cabal, L.; LaMotte, R.H.; Ngeow, J.; Putterman, G.J.: Chemically induced itch, pain and hyperalgesia by intra-epidermal injection. Neurosci. Abstr. *9:* 1063 (1983).

Bechterew, W.: Über die hinteren Nervenwurzeln, ihre Endigung in der grauen Substanz des Rückenmarkes und ihre centrale Fortsetzung im letzteren. Arch. Anat. Physiol. 126–136 (1887).

Beck, P.W.; Handwerker, H.O.: Bradykinin and serotonin effects on various types of cutaneous nerve fibres. Pflügers Arch. *347:* 209–222 (1974).

Beck, P.W.; Handwerker, H.O.; Zimmermann, M.: Nervous outflow from the cat's foot during noxious radiant heat stimulation. Brain Res. *67:* 373–386 (1974).

Beecher, H.K.: Measurement of subjective responses; quantitative effects of drugs (Oxford University Press, New York 1959).

Beitel, R.E.; Dubner, R.: Fatigue and adaptation in unmyelinated (C) polymodal nociceptors to mechanical and thermal stimuli applied to the monkey's face. Brain Res. *112:* 402–406 (1976a).

Beitel, R.E.; Dubner, R.: The response of unmyelinated (C) polymodal nociceptors to thermal stimuli applied to the monkey's face. J. Neurophysiol. *39:* 1160–1175 (1976b).

Belcher, G.; Ryall, R.W.: Differential excitatory and inhibitory effects of opiates on non-nociceptive and nociceptive neurones in the spinal cord of the cat. Brain Res. *145:* 303–314 (1978).

Bennett, G.J.; Abdelmoumene, M.; Hayashi, H.; Dubner, R.: Physiology and morphology of substantia gelatinosa neurons intracellularly stained with horseradish peroxidase. J. comp. Neurol. *194:* 809–827 (1980).

Bennett, G.J.; Abdelmoumene, M.; Hayashi, H.; Hoffert, M.J.; Dubner, R.: Spinal cord layer I neurons with axon collaterals that generate local arbors. Brain Res. *209:* 421–426 (1981).

Bennett, G.J.; Hayashi, H.; Abdelmoumene, M.; Dubner, R.: Physiological properties of stalked cells of the substantia gelatinosa intracellularly stained with horseradish peroxidase. Brain Res. *164:* 285–289 (1979).

Bennett, G.J.; Ruda, M.A.; Gobel, S.; Dubner, R.: Enkephalin immunoreactive stalked cells and lamina IIb islet cells in cat substantia gelatinosa. Brain Res. *240:* 162–166 (1982).

Bennett, G.J.; Seltzer, Z.; Lu, G.W.; Nishikawa, N.; Dubner, R.: The cells of origin of the dorsal column postsynaptic projection in the lumbosacral enlargements of cats and monkeys. Somatosens. Res. *1:* 131–149 (1983).

Bennett, R.E.; Ferrington, D.G.; Rowe, M.: Tactile neuron classes within second somatosensory area (SII) of cat cerebral cortex. J. Neurophysiol. *43:* 292–309 (1980).

Bentivoglio, M.; Macchi, G.; Albanese, A.: The cortical projections of the thalamic intralaminar nuclei, as studied in cat and rat with the multiple fluorescent retrograde tracing technique. Neurosci. Lett. *26:* 5–10 (1981).

Berkley, K.J.: Response properties of cells in ventrobasal and posterior group nuclei of the cat. J. Neurophysiol. *36:* 940–952 (1973).

Berkley, K.J.: Spatial relationships between the terminations of somatic sensory and motor pathways in the rostral brainstem of cats and monkeys. I. Ascending somatic sensory inputs to lateral diencephalon. J. comp. Neurol. *193:* 283–317 (1980).

Besson, J.M.; Conseiller, C.; Hamann, K.F.; Maillard, M.C.: Modifications of dorsal horn cell activities in the spinal cord, after intra-arterial injection of bradykinin. J. Physiol. *221:* 189–205 (1972).

Besson, J.M.; Guilbaud, G.; Abdelmoumene, H.; Chaouch, A.: Physiologie de la nociception. J. Physiol., Paris *78:* 7–107 (1982).

Bessou, P.; Burgess, P.R.; Perl, E.R.; Taylor, C.B.: Dynamic properties of mechanoreceptors with unmyelinated (C) fibers. J. Neurophysiol. *34:* 116–131 (1971).

Bessou, P.; Laporte, Y.: Activation des fibres afférentes amyéliniques d'origine musculaire. C.r. Séanc. Soc. Biol. *152:* 1587–1590 (1958).

Bessou, P.; Laporte, Y.: Etude des récepteurs musculaires innervés par les fibres afférentes du Groupe III (fibres myelinisées fines) chez le chat. Arch. ital. Biol. *99:* 293–321 (1961).

Bessou, P.; Perl, E.R.: A movement receptor of the small intestine. J. Physiol. *182:* 404–426 (1966).

Bessou, P.; Perl, E.R.: Response of cutaneous sensory units with unmyelinated fibers to noxious stimuli. J. Neurophysiol. *32:* 1025–1043 (1969).

Bettag, W.; Yoshida, T.: Über stereotaktische Schmerzoperationen. Acta neurochir. *8:* 299–317 (1960).

Biemond, A.: The conduction of pain above the level of the thalamus opticus. Archs Neurol. Psychiat. *75:* 231–244 (1956).

Bigelow, N.; Harrison, I.; Goodell, H.; Wolff, H.G.: Studies on pain: quantitative measurements of two pain sensations of the skin, with reference to the nature of the 'hyperalgesia of peripheral neuritis'. J. clin. Invest. *24:* 503–512 (1945).

Blair, R.W.; Weber, R.N.; Foreman, R.D.: Characteristics of primate spinothalamic tract neurons receiving viscerosomatic convergent inputs in the T3–T5 segments. J. Neurophysiol. *46:* 797–811 (1981).

Blix, M.: Experimentelle Beiträge zur Lösung der Frage über die specifische Energie der Hautnerven. Z. Biol. *20:* 141–156 (1884).

Block, A.R.; Feinberg, H.; Herbaczynska-Cedro, K.; Vane, J.R.: Anoxia-induced release of prostaglandins in rabbit isolated hearts. Circulation Res. *36:* 34–42 (1975).

Blomqvist, A.; Flink, R.; Bowsher, D.; Griph, S.; Westman, J.: Tectal and thalamic projections of dorsal column and lateral cervical nuclei: a quantitative study in the cat. Brain Res. *141:* 335–341 (1978).

Blumgart, H.L.; Schlesinger, M.J.; Davis, D.: Studies on the relation of the clinical manifestations of angina pectoris, coronary thrombosis, and myocardial infarction to the pathological findings, with particular reference to the significance of the collateral circulation. Am. Heart J. *19:* 1–91 (1940).

Boivie, J.: The termination of the cervicothalamic tract in the cat. An experimental study with silver impregnation methods. Brain Res. *19:* 333–360 (1970).

Boivie, J.: The termination of the spinothalamic tract in the cat. An experimental study with silver impregnation methods. Exp. Brain Res. *12:* 331–353 (1971).

Boivie, J.: Anatomical observations on the dorsal column nuclei, their thalamic projection and the cytoarchitecture of some somatosensory thalamic nuclei in the monkey. J. comp. Neurol. *178:* 17–48 (1978).

Boivie, J.: An anatomical reinvestigation of the termination of the spinothalamic tract in the monkey. J. comp. Neurol. *186:* 343–370 (1979).

Boivie, J.: Thalamic projections from lateral cervical nucleus in monkey. A degeneration study. Brain Res. *198:* 13–26 (1980).

Bonica, J.J.: The management of pain (Lea & Febiger, Philadelphia 1953).

Bowsher, D.: Termination of the central pain pathway in man. The conscious appreciation of pain. Brain *80:* 606–622 (1957).

Bowsher, D.: The termination of secondary somatosensory neurons within the thalamus of *Macaca mulatta:* an experimental degeneration study. J. comp. Neurol. *117:* 213–227 (1961).

Bowsher, D.: Diencephalic projections from the midbrain reticular formation. Brain Res. *95:* 211–220 (1975).

Bowsher, D.: Role of the reticular formation in responses to noxious stimulation. Pain *2:* 361–378 (1976).

Bowsher, D.; Mallart, A.; Petit, D.; Albe-Fessard, D.: A bulbar relay to the centre median. J. Neurophysiol. *31:* 288–300 (1968).

Boyd, I.A.: The histological structure of the receptors in the knee-joint of the cat correlated with their physiological response. J. Physiol. *124:* 476–488 (1954).

Boyd, I.A.; Roberts, T.D.M.: Proprioceptive discharges from stretch-receptors in the knee-joint of the cat. J. Physiol. *122:* 38–58 (1953).

Breazile, J.E.; Kitchell, R.L.: A study of fiber systems within the spinal cord of the domestic pig that subserve pain. J. comp. Neurol. *133:* 373–382 (1968).

Brinkhus, H.B.; Carstens, E.; Zimmermann, M.: Encoding of graded noxious skin heating by neurons in posterior thalamus and adjacent areas in the cat. Neurosci. Lett. *15:* 37–42 (1979).

Brown, A.G.: Organization in the spinal cord: the anatomy and physiology of identified neurones (Springer, Berlin 1981).

Brown, A.G.; Brown, P.B.; Fyffe, R.E.W.; Pubols, L.M.: Receptive field organization and response properties of spinal neurones with axons ascending the dorsal columns in the cat. J. Physiol. *337:* 575–588 (1983).

Brown, A.G.; Franz, D.N.: Responses of spinocervical tract neurones to natural stimulation of identified cutaneous receptors. Exp. Brain Res. *7:* 231–249 (1969).

Brown, A.G.; Fyffe, R.E.W.: Form and function of dorsal horn neurones with axons ascending the dorsal columns in cat. J. Physiol. *321:* 31–47 (1981).

Brown, A.G.; Fyffe, R.E.W.; Noble, R.: Projections from Pacinian corpuscles and rapidly adapting mechanoreceptors of glabrous skin to the cat's spinal cord. J. Physiol. *307:* 385–400 (1980a).

Brown, A.G.; Fyffe, R.E.W.; Noble, R.; Rose, P.K.; Snow, P.J.: The density, distribution and topographical organization of spinocervical tract neurones in the cat. J. Physiol. *300:* 409–428 (1980b).

Brown, A.G.; Fyffe, R.E.W.; Rose, P.K.; Snow, P.J.: Spinal cord collaterals from axons of type II slowly adapting units in the cat. J. Physiol. *316:* 469–480 (1981).

Brown, A.G.; Hamann, W.C.; Martin, H.F.: Effects of activity in non-myelinated afferent fibres on the spinocervical tract. Brain Res. *98:* 243–259 (1975).

Brown, A.G.; House, C.R.; Rose, P.K.; Snow, P.J.: The morphology of spinocervical tract neurones in the cat. J. Physiol. *260:* 719–738 (1976).

Brown, A.G.; Rose, P.K.; Snow, P.J.: The morphology of spinocervical tract neurones revealed by intracellular injection of horseradish peroxidase. J. Physiol. *270:* 747–764 (1977a).

Brown, A.G.; Rose, P.K.; Snow, P.J.: The morphology of hair follicle afferent fibre collaterals in the spinal cord of the cat. J. Physiol. *272:* 779–797 (1977b).

Brown, A.G.; Rose, P.K.; Snow, P.J.: Morphology and organization of axon collaterals from afferent fibres of slowly adapting type I units in cat spinal cord. J. Physiol. *277:* 15–27 (1978).

Brown, A.M.: Excitation of afferent cardiac sympathetic fibres during myocardial ischaemia. J. Physiol. *190:* 35–53 (1967).

Brown, A.M.; Malliani, A.: Spinal sympathetic reflexes initiated by coronary receptors. J. Physiol. *212:* 685–705 (1971).

Brown-Séquard, C.E.: Course of lectures on the physiology and pathology of the central nervous system (Lippincott, Philadelphia 1860).

Bryan, R.N.; Coulter, J.D.; Willis, W.D.: Cells of origin of the spinocervical tract in the monkey. Expl Neurol. *42:* 574–586 (1974).

Bryan, R.N.; Trevino, D.L.; Coulter, J.D.; Willis, W.D.: Location and somatotopic organization of the cells of origin of the spino-cervical tract. Exp. Brain Res. *17:* 177–189 (1973).

Buck, S.H.; Deshmukh, P.P.; Yamamura, H.I.; Burks, T.F.: Thermal analgesia and substance-P depletion induced by capsaicin in guinea-pigs. Neuroscience *6:* 2217–2222 (1981).

Burgess, P.R.; Clark, F.J.: Characteristics of knee joint receptors in the cat. J. Physiol. *203:* 317–335 (1969).

Burgess, P.R.; Perl, E.R.: Myelinated afferent fibers responding specifically to noxious stimulation of the skin. J. Physiol. *190:* 541–562 (1967).

Burgess, P.R.; Perl, E.R.: Cutaneous mechanoreceptors and nociceptors; in Iggo, Handbook of sensory physiology, somatosensory system, vol. 2, pp. 29–78 (Springer, Heidelberg 1973).

Burgess, P.R.; Petit, D.; Warren, R.M.: Receptor types in cat hairy skin supplied by myelinated fibers. J. Neurophysiol. *31:* 833–848 (1968).

Burnweit, C.; Forssmann, W.G.: Somatostatinergic nerves in the cervical spinal cord of the monkey. Cell Tiss. Res. *200:* 83–90 (1979).

Burton, H.; Craig, A.D.: Spinothalamic projections in cat, raccoon and monkey: a study based on anterograde transport of horseradish peroxidase; in Macchi, Rustioni, Spreafico, Somatosensory integration in the thalamus, pp. 17–41 (Elsevier, Amsterdam 1983).

Burton, H.; Jones, E.G.: The posterior thalamic region and its cortical projection in New World and Old World monkeys. J. comp. Neurol. *168:* 249–302 (1976).

Burton, H.; Loewy, A.D.: Descending projections from the marginal cell layer and other regions of monkey spinal cord. Brain Res. *116:* 485–491 (1976).

Burton, H.; Mitchell, G.; Brent, D.: Second somatic sensory area in the cerebral cortex of cats: somatotopic organization and cytoarchitecture. J. comp. Neurol. *210:* 109–135 (1982).

Cadden, S.W.; Villanueva, L.; Chitour, D.; LeBars, D.: Depression of activities of dorsal horn convergent neurones by propriospinal mechanisms triggered by noxious inputs; comparison with diffuse noxious inhibitory controls (DNIC). Brain Res. *275:* 1–11 (1983).

Cajal, S.R. y: Histologie du système nerveux de l'homme et des vertébrés, vol. 1 (Maloine, Paris 1909) (reprinted by Consejo Superior de Investigaciones Científicas, Instituto Ramon y Cajal, Madrid 1952).

Calma, I.: The activity of the posterior group of thalamic nuclei in the cat. J. Physiol. *180:* 350–370 (1965).

Calvillo, O.; Henry, J.L.; Neuman, R.S.: Effects of morphine and naloxone on dorsal horn neurones in the cat. Can. J. Physiol. Pharmacol. *52:* 1207–1211 (1974).

Campbell, J.N.; LaMotte, R.H.: Latency to detection of first pain. Brain Res. *266:* 203–208 (1983).

Carli, G.; Farabollini, F.; Fontani, G.; Meucci, M.: Slowly adapting receptors in cat hip joint. J. Neurophysiol. *42:* 767–779 (1979).

Carpenter, M.B.; Stein, B.M.; Shriver, J.E.: Central projections of spinal dorsal roots in the monkey. II. Lower thoracic, lumbosacral and coccygeal dorsal roots. Am. J. Anat. *123:* 75–118 (1968).

Carpenter, S.E.; Lynn, B.: Vascular and sensory responses of human skin to mild injury after topical treatment with capsaicin. Br. J. Pharmacol. *73:* 755–758 (1981).

Carreras, M.; Andersson, S.A.: Functional properties of neurons of the anterior ectosylvian gyrus of the cat. J. Neurophysiol. *26:* 100–126 (1963).

Carstens, E.; Trevino, D.L.: Laminar origins of spinothalamic projections in the cat as determined by the retrograde transport of horseradish peroxidase. J. comp. Neurol. *182:* 151–166 (1978a).

Carstens, E.; Trevino, D.L.: Anatomical and physiological properties of ipsilaterally projecting spinothalamic neurons in the second cervical segment of the cat's spinal cord. J. comp. Neurol. *182:* 167–184 (1978b).

Casey, K.L.: Unit analysis of nociceptive mechanisms in the thalamus of the awake squirrel monkey. J. Neurophysiol. *29:* 727–750 (1966).

Casey, K.L.: Somatic stimuli, spinal pathways, and size of cutaneous fibers influencing unit activity in the medial medullary reticular formation. Expl Neurol. *25:* 35–56 (1969).

Casey, K.L.: Responses of bulboreticular units to somatic stimuli eliciting escape behavior in the cat. Int. J. Neurosci. *2:* 15–28 (1971a).

Casey, K.L.: Escape elicited by bulboreticular stimulation in the cat. Int. J. Neurosci. *2:* 29–34 (1971b).

Casey, K.L.; Bonica, J.J.; Dubner, R.; Fields, H.; Kerr, F.W.L.; Liebeskind, J.C.; Loeser, J.D.; Perl, E.R.; Willis, W.D.: Report of the panel on pain to the National Advisory Neurological and Communicative Disorders and Stroke Council. NIH Publ. No. 79-1912 (Department of Health, Education and Welfare, Washington 1979).

Casey, K.L.; Hall, B.R.; Morrow, T.J.: Effect of spinal cord lesions on responses of cats to thermal pulses. Pain, suppl. 1, p. 130 (1981).

Casey, K.L.; Morrow, T.J.: Ventral posterior thalamic neurons differentially responsive to noxious stimulation of the awake monkey. Science *221:* 675–677 (1983).

Cassinari, V.; Pagni, C.A.: Central pain. A neurosurgical survey (Harvard University Press, Cambridge 1969).

Cervero, F.: Afferent activity evoked by natural stimulation of the biliary system in the ferret. Pain *13:* 137–151 (1982a).

Cervero, F.: Noxious intensities of visceral stimulation are required to activate viscerosomatic multireceptive neurons in the thoracic spinal cord of the cat. Brain Res. *240:* 350–352 (1982b).

Cervero, F.: Somatic and visceral inputs to the thoracic spinal cord of the cat: effects of noxious stimulation of the biliary system. J. Physiol. *337:* 51–67 (1983a).

Cervero, F.: Mechanisms of visceral pain; in Persistent pain, vol. 4, pp. 1–19 (Grune & Stratton, New York 1983b).

Cervero, F.: Supraspinal connections of neurones in the thoracic spinal cord of the cat: ascending projections and effects of descending impulses. Brain Res. *275:* 251–261 (1983c).

Cervero, F.; Iggo, A.: Natural stimulation of urinary bladder afferents does not affect transmission through lumbosacral spinocervical tract neurones in the cat. Brain Res. *156:* 375–379 (1978).

Cervero, F.; Iggo, A.: The substantia gelatinosa of the spinal cord. A critical review. Brain *103:* 717–772 (1980).

Cervero, F.; Iggo, A.; Molony, V.: Responses of spinocervical tract neurones to noxious stimulation of the skin. J. Physiol. *267:* 537–558 (1977a).

Cervero, F.; Iggo, A.; Molony, V.: Ascending projections of nociceptor driven lamina I neurones in the cat. Exp. Brain Res. *35:* 135–149 (1979a).

Cervero, F.; Iggo, A.; Molony, V.: An electrophysiological study of neurones in the substantia gelatinosa Rolandi of the cat's spinal cord. Q. Jl exp. Physiol. *64:* 297–314 (1979b).

Cervero, F.; Iggo, A.; Molony, V.: Segmental and intersegmental organization of neurones in the substantia gelatinosa Rolandi of the cat's spinal cord. Q. Jl exp. Physiol. *64:* 315–326 (1979c).

Cervero, F.; Iggo, A.; Ogawa, H.: Nociceptor-driven dorsal horn neurones in the lumbar spinal cord of the cat. Pain *2:* 5–24 (1976).

Cervero, F.; McRitchie, H.A.: Neonatal capsaicin and thermal nociception: a paradox. Brain Res. *215:* 414–418 (1981).

Cervero, F.; McRitchie, H.A.: Neonatal capsaicin does not affect unmyelinated efferent fibers of autonomic nervous system: functional evidence. Brain Res. *239:* 283–288 (1982).

Cervero, F.; Molony, V.; Iggo, A.: Extracellular and intracellular recordings from neurones in the substantia gelatinosa Rolandi. Brain Res. *136:* 565–569 (1977b).

Chahl, L.A.; Iggo, A.: The effects of bradykinin and prostaglandin E_1 on rat cutaneous afferent nerve activity. Br. J. Pharmacol. *59:* 343–374 (1977).

Chambers, M.R.; Andres, K.H.; Duering, M.V.; Iggo, A.: The structure and function of the slowly adapting type II mechanoreceptor in hairy skin. Q. Jl exp. Physiol. *57:* 417–445 (1972).

Chang, H.T.: Integrative action of thalamus in the process of acupuncture for analgesia. Scientia sin. *16:* 25–60 (1973).

Chang, H.T.; Ruch, T.C.: Topographical distribution of spinothalamic fibres in the thalamus of the spider monkey. J. Anat. *81:* 150–164 (1947).

Chan-Palay, V.; Palay, S.L.: Immunocytochemical identification of substance P cells and their processes in rat sensory ganglia and their terminals in the spinal cord: light microscopic studies. Proc. natn. Acad. Sci. USA *74:* 3597–3601 (1977a).

Chan-Palay, V.; Palay, S.L.: Ultrastructural identification of substance P cells and their processes in rat sensory ganglia and their terminals in the spinal cord by immunocytochemistry. Proc. natn. Acad. Sci. USA *74:* 4050–4054 (1977b).

Chaouch, A.; Menétrey, D.; Binder, D.; Besson, J.M.: Neurons at the origin of the medial component of the bulbopontine spinoreticular tract in the rat: an anatomical study using horseradish peroxidase retrograde transport. J. comp. Neurol. *214:* 309–320 (1983).

Chéry-Croze, S.: Painful sensation induced by a thermal cutaneous stimulus. Pain *17:* 109–137 (1983).

Chi, C.C.: An experimental silver study of the ascending projections of the central gray substance and adjacent tegmentum in the rat with observation in the cat. J. comp. Neurol. *139:* 259–272 (1970).

Christensen, B.N.; Perl, E.R.: Spinal neurons specifically excited by noxious or thermal stimuli: marginal zone of the dorsal horn. J. Neurophysiol. *33:* 293–307 (1970).

Chung, J.M.; Fang, Z.R.; Hori, Y.; Lee, K.H.; Willis, W.D.: Prolonged inhibition of primate spinothalamic tract cells by peripheral nerve stimulation. Pain *19:* 259–275 (1984a).

Chung, J.M.; Kenshalo, D.R., Jr.; Gerhart, K.D.; Willis, W.D.: Excitation of primate spinothalamic neurons by cutaneous C-fiber volleys. J. Neurophysiol. *42:* 1354–1369 (1979).

Chung, J.M.; Lee, K.H.; Endo, K.; Coggeshall, R.E.: Activation of central neurons by ventral root afferents. Science *222:* 934–935 (1983).

Chung, J.M.; Lee, K.H.; Hori, Y.; Endo, K.; Willis, W.D.: Factors influencing peripheral nerve stimulation produced inhibition of primate spinothalamic tract cells. Pain *19:* 277–293 (1984b).

Chung, J.M.; Lee, K.H.; Hori, Y.; Willis, W.D.: Effects of capsaicin applied to a peripheral nerve on the responses of primate spinothalamic tract cells. Brain Res. (in press, 1984c).

Chung, K.; Coggeshall, R.E.: Primary afferent axons in the tract of Lissauer in the cat. J. comp. Neurol. *186:* 451–464 (1979).

Chung, K.; Langford, L.A.; Applebaum, A.E.; Coggeshall, R.E.: Primary afferent fibers in the tract of Lissauer in the rat. J. comp. Neurol. *184:* 587–598 (1979).

Clark, D.; Hughes, J.; Gasser, H.S.: Afferent function in the group of nerve fibers of slowest conduction velocity. Am. J. Physiol. *114:* 69–76 (1935).

Clark, F.J.: Information signaled by sensory fibers in medial articular nerve. J. Neurophysiol. *38:* 1464–1472 (1975).

Clark, F.J.; Burgess, P.R.: Slowly adapting receptors in cat knee joint: Can they signal joint angle? J. Neurophysiol. *38:* 1448–1463 (1975).

Clark, S.L.: Innervation of the pia mater of the spinal cord and medulla. J. comp. Neurol. *53:* 129–145 (1931).

Clark, W.E.L.: The termination of ascending tracts in the thalamus of the macaque monkey. J. Anat. *71:* 7–40 (1936).

Clarke, J.L.: Further researches on the grey substance of the spinal cord. Phil. Trans. R. Soc. *149:* 437–467 (1859).

Clifton, G.L.; Coggeshall, R.E.; Vance, W.H.; Willis, W.D.: Receptive fields of unmyelinated ventral root afferent fibres in the cat. J. Physiol. *256:* 573–600 (1976).

Coggeshall, R.E.: Afferent fibers in the ventral root. Neurosurgery *4:* 443–448 (1979).

Coggeshall, R.E.: Law of separation of function of the spinal roots. Physiol. Rev. *60:* 716–755 (1980).

Coggeshall, R.E.; Applebaum, M.L.; Fazen, M.; Stubbs, T.B.; Sykes, M.T.: Unmyelinated axons in human ventral roots, a possible explanation for the failure of dorsal rhizotomy to relieve pain. Brain *98:* 157–166 (1975).

Coggeshall, R.E.; Chung, K.; Chung, J.M.; Langford, L.A.: Primary afferent axons in the tract of Lissauer in the monkey. J. comp. Neurol. *196:* 431–442 (1981).

Coggeshall, R.E.; Coulter, J.D.; Willis, W.D.: Unmyelinated axons in the ventral roots of the cat lumbosacral enlargement. J. comp. Neurol. *153:* 39–58 (1974).

Coggeshall, R.E.; Hong, K.A.P.; Langford, L.A.; Schaible, H.G.; Schmidt, R.F.: Discharge characteristics of fine medial articular afferents at rest and during passive movements of inflamed knee joints. Brain Res. *272:* 185–188 (1983).

Coggeshall, R.E.; Ito, H.: Sensory fibres in ventral roots L7 and S1 in the cat. J. Physiol. *267:* 215–235 (1977).

Coimbra, A.; Magalhaes, M.M.; Sodré-Borges, B.P.: Ultrastructural localization of acid phosphatase in synaptic terminals of the rat substantia gelatinosa Rolandi. Brain Res. *22:* 142–146 (1970).

Coimbra, A.; Sodré-Borges, B.P.; Magalhaes, M.M.: The substantia gelatinosa Rolandi of the rat. Fine structure, cytochemistry (acid phosphatase) and changes after dorsal root section. J. Neurocytol. *3:* 199–217 (1974).

Coleridge, H.M.; Coleridge, J.C.G.; Luck, J.C.: Pulmonary afferent fibres of small diameter stimulated by capsaicin and by hyperinflation of the lungs. J. Physiol. *179:* 248–262 (1965).

Collins, W.F.; Nulsen, F.E.; Randt, C.T.: Relation of peripheral nerve fiber size and sensation in man. Archs Neurol. Psychiat. *3:* 381–385 (1960).

Comans, P.E.; Snow, P.J.: Ascending projections to nucleus parafascicularis of the cat. Brain Res. *230:* 337–341 (1981).

Corkin, S.; Twitchell, T.E.; Sullivan, E.V.: Safety and efficacy of cingulotomy for pain and psychiatric disorders; in Hitchcock et al., Modern concepts in psychiatric surgery, pp. 253–272 (Elsevier, Amsterdam 1979).

Cox, B.M.; Goldstein, A.; Li, C.H.: Opioid activity of a peptide, β-lipotropin-(61-91), derived from β-lipotropin. Proc. natn. Acad. Sci. USA *73:* 1821–1823 (1976).

Craig, A.D.: Spinocervical tract cells in cat and dog, labeled by the retrograde transport of horseradish peroxidase. Neurosci. Lett. *3:* 173–177 (1976).

Craig, A.D.: Spinal and medullary input to the lateral cervical nucleus. J. comp. Neurol. *181:* 729–744 (1978).

Craig, A.D.; Burton, H.: Spinal and medullary lamina I projection to nucleus submedius in medial thalamus: a possible pain center. J. Neurophysiol. *45:* 443–466 (1981).

Craig, A.D.; Kniffki, K.D.: Lumbosacral lamina I cells projecting to medial and/or lateral thalamus in the cat. Soc. Neurosci. Abstr. *8:* 95 (1982).

Craig, A.D.; Tapper, D.N.: Lateral cervical nucleus in the cat: functional organization and characteristics. J. Neurophysiol. *41:* 1511–1534 (1978).

Craig, A.D.; Wiegand, S.J.; Price, J.L.: The thalamo-cortical projection of the nucleus submedius in the cat. J. comp. Neurol. *206:* 28–48 (1982).

Cranefield, P.F.: The way in and the way out: Francois Magendie, Charles Bell and the roots of the spinal nerves (Futura, Mount Kisco 1974).

Croze, S.; Duclaux, R.: Thermal pain in humans: influence of the rate of stimulation. Brain Res. *157:* 418–421 (1978).

Croze, S.; Duclaux, R.; Kenshalo, D.R.: The thermal sensitivity of the polymodal nociceptors in the monkey. J. Physiol. *263:* 539–562 (1976).

Cuello, A.C.; Gamse, R.; Holzer, P.; Lembeck, F.: Substance P immunoreactive neurons following neonatal administration of capsaicin. Arch. Pharmacol. *315:* 185–194 (1981).

Curry, M.J.: The exteroceptive properties of neurones in the somatic part of the posterior group (PO). Brain Res. *44:* 439–462 (1972).

Curry, M.J.; Gordon, G.: The spinal input to the posterior group in the cat. An electrophysiological investigation. Brain Res. *44:* 417–437 (1972).

Curtis, D.R.; Phillis, J.W.; Watkins, J.C.: The chemical excitation of spinal neurones by certain acidic amino acids. J. Physiol. *150:* 656–682 (1960).

Curtis, D.R.; Watkins, J.C.: The excitation and depression of spinal neurones by structurally related amino acids. J. Neurochem. *6:* 117–141 (1960).

Dallenbach, K.M.: The temperature spots and end-organs. Am. J. Psychol. *39:* 402–427 (1927).

Dalsgaard, C.J.; Risling, M.; Cuello, C.: Immunohistochemical localization of substance P in the lumbosacral spinal pia mater and ventral roots of the cat. Brain Res. *246:* 168–171 (1982).

Darian-Smith, I.; Johnson, K.O.; Dykes, R.: 'Cold' fiber population innervating palmar and digital skin of the monkey: responses to cooling pulses. J. Neurophysiol. *36:* 325–346 (1973).

Darian-Smith, I.; Johnson, K.O.; LaMotte, C.; Shigenaga, Y.; Kenins, P.; Champness, P.: Warm fibers innervating palmar and digital skin of the monkey: responses to thermal stimuli. J. Neurophysiol. *42:* 1297–1315 (1979).

Davenport, H.A.; Ranson, S.W.: Ratios of cells to fibers and of myelinated to unmyelinated fibers in spinal nerve roots. Am. J. Anat. *49:* 193–207 (1931).

Davies, J.; Dray, A.: Depression and facilitation of synaptic responses in cat dorsal horn by substance P administered into substantia gelatinosa. Life Sci. *27:* 2037–2042 (1980).

Dejerine, J.; Mouzon, J.: Un nouveau type de syndrome sensitif cortical observé dans un cas de monoplégie corticale dissociée. Revue neurol. *28:* 1265–1271 (1915).

Dejerine, J.; Roussy, G.: Le syndrome thalamique. Revue neurol. *14:* 521–532 (1906).

deLanerolle, N.C.; LaMotte, C.C.: Ultrastructure of chemically defined neuron systems in the dorsal horn of the monkey. I. Substance P immunoreactivity. Brain Res. *274:* 31–49 (1983).

Dennis, S.G.; Melzack, R.: Self-mutilation after dorsal rhizotomy in rats: effects of prior pain and pattern of root lesions. Expl Neurol. *65:* 412–421 (1979).

Devor, M.; Wall, P.D.: Effect of peripheral nerve injury on receptive fields of cells in the cat spinal cord. J. comp. Neurol. *199:* 277–291 (1981a).

Devor, M.; Wall, P.D.: Plasticity in the spinal cord sensory map following peripheral nerve injury in rats. J. Neurosci. *1:* 679–684 (1981b).

Devor, M.; Jänig, W.: Activation of myelinated afferent endings in a neuroma by stimulation of the sympathetic supply in the rat. Neurosci. Lett. *24:* 43–47 (1981).

DiFiglia, M.; Aronin, N.; Leeman, S.E.: Light microscopic and ultrastructural localization of immunoreactive substance P in the dorsal horn of monkey spinal cord. Neuroscience *7:* 1127–1139 (1982).

Dilly, P.N.; Wall, P.D.; Webster, K.E.: Cells of origin of the spinothalamic tract in the cat and rat. Expl Neurol. *21:* 550–562 (1968).

Dimsdale, J.A.; Kemp, J.M.: Afferent fibres in ventral roots in the rat. J. Physiol. *187:* 25P–26P (1966).

Ditirro, F.J.; Ho, R.H.; Martin, C.F.: Immunohistochemical localization of substance P, somatostatin, and methionine-enkephalin in the spinal cord and dorsal root ganglia of the North American opossum, *Didelphis virginiana.* J. comp. Neurol. *198:* 351–363 (1981).

Dodd, J.; Jahr, C.E.; Jessell, T.M.: Neurotransmitters and neuronal markers at sensory synapses in the dorsal horn. Adv. Pain Res. Ther. *6:* 105–121 (1984).

Dodt, E.; Zotterman, Y.: The discharge of specific cold fibers at high temperatures (the paradoxical cold). Acta physiol. scand. *26:* 358–365 (1952).

Donaldson, H.H.: On the temperature sense. Mind *10:* 339–416 (1885).

Dong, W.K.; Ryu, H.; Wagman, I.H.: Nociceptive responses of neurons in medial thalamus and their relationship to spinothalamic pathways. J. Neurophysiol. *41:* 1592–1613 (1978).

Donoghue, J.P.; Kerman, K.L.; Ebner, F.F.: Evidence for two organizational plans within the somatic sensory-motor cortex of the rat. J. comp. Neurol. *183:* 647–664 (1979).

Dostrovsky, J.O.; Millar, J.; Wall, P.D.: The immediate shift of afferent drive of dorsal column nucleus cells following deafferentation: a comparison of acute and chronic deafferentation in gracile nucleus and spinal cord. Expl Neurol. *52:* 480–495 (1976).

Dostrovsky, J.O.; Pomeranz, B.: Interaction of iontophoretically applied morphine with responses of interneurons in cat spinal cord. Expl Neurol. *52:* 325–338 (1976).

Douglas, W.W.; Ritchie, J.M.: Non-medullated fibres in the saphenous nerve which signal touch. J. Physiol. *139:* 385–399 (1957).

Dreyer, D.A.; Schneider, R.J.; Metz, C.B.; Whitsel, B.L.: Differential contributions of spinal pathways to body representation in postcentral gyrus of *Macaca mulatta*. J. Neurophysiol. *37:* 119–145 (1974).

Dubner, R.; Sumino, R.; Wood, W.I.: A peripheral 'cold' fiber population responsive to innocuous and noxious thermal stimuli applied to the monkey's face. J. Neurophysiol. *38:* 1373–1389 (1975).

Dubuisson, D.; Fitzgerald, M.; Wall, P.D.: Ameboid receptive fields in laminae 1, 2 and 3. Brain Res. *177:* 376–378 (1979).

Duclaux, R.; Kenshalo, D.R.: Response characteristics of cutaneous warm receptors in the monkey. J. Neurophysiol. *43:* 1–15 (1980).

Duggan, A.W.; Griersmith, B.T.; Headley, P.M.; Hall, J.G.: Lack of effect by substance P at sites in the substantia gelatinosa where met-enkephalin reduces the transmission of nociceptive impulses. Neurosci. Lett. *12:* 313–317 (1979).

Duggan, A.W.; Hall, J.G.; Headley, P.M.: Suppression of transmission of nociceptive impulses by morphine: selective effects of morphine administered in the region of the substantia gelatinosa. Br. J. Pharmacol. *61:* 65–76 (1977a).

Duggan, A.W.; Hall, J.G.; Headley, P.M.: Enkephalins and dorsal horn neurones of the cat: effects on responses to noxious and innocuous skin stimuli. Br. J. Pharmacol. *61:* 399–408 (1977b).

Duggan, A.W.; Johnson, S.M.; Morton, C.R.: Differing distributions of receptors for morphine and met-5-enkephalinamide in the dorsal horn of the cat. Brain Res. *229:* 379–387 (1981).

Duncan, D.: A determination of the number of unmyelinated fibers in the ventral roots of the rat, cat, and rabbit. J. comp. Neurol. *55:* 459–471 (1932).

Duncan, D.; Keyser, L.L.: Some determinations of the ratio of nerve fibers to nerve cells in the thoracic dorsal roots and ganglia of the cat. J. comp. Neurol. *64:* 303–311 (1936).

Duncan, D.; Keyser, L.L.: Further determinations of the numbers of fibers and cells in the dorsal roots and ganglia of the cat. J. comp. Neurol. *68:* 479–490 (1938).

Duncan, D.; Morales, R.: Location of nerve cells producing the synaptic vesicles situated in the substantia gelatinosa of the spinal cord. Am. J. Anat. *138:* 139–144 (1973).

Duthie, H.L.; Gairns, F.W.: Sensory nerve endings and sensation in the anal region of man. Br. J. Surg. *47:* 585–595 (1960).

Dyck, P.S.; Lambert, E.H.: Numbers and diameters of nerve fibers and compound action potentials of sural nerve: controls and hereditary neuromuscular disorders. Trans. Am. neurol. Ass. *91:* 214–217 (1966).

Earle, K.M.: The tract of Lissauer and its possible relation to the pain pathway. J. comp. Neurol. *96:* 93–111 (1952).

Eccles, J.C.: The physiology of synapses (Springer, New York 1964).

Echols, D.H.; Colclough, J.A.: Abolition of painful phantom foot by resection of the sensory cortex. J. Am. med. Ass. *134:* 1476–1477 (1947).

Edinger, L.: Vergleichend-entwicklungsgeschichtliche und anatomische Studien im Bereiche des Centralnervensystems. 2. Über die Fortsetzung der hinteren Rückenmarkswurzeln zum Gehirn. Anat. Anz. *4:* 121–128 (1889).

Eklund, G.; Skoglund, S.: On the specificity of the Ruffini-like joint receptors. Acta physiol. scand. *49:* 184–191 (1960).

Erickson, T.C.; Bleckwenn, W.J.; Woolsey, C.N.: Observations on the post-central gyrus in relation to pain. Trans. Am. neurol. Ass. *77:* 57–59 (1952).

Ferrell, W.R.: The adequacy of stretch receptors in the cat knee joint for signalling joint angle throughout a full range of movement. J. Physiol. *299:* 85–99 (1980).

Ferrington, D.G.; Rowe, M.: Differential contributions to coding of cutaneous vibratory information by cortical somatosensory areas I and II. J. Neurophysiol. *43:* 310–331 (1980).

Fields, H.L.; Basbaum, A.I.: Brainstem control of spinal pain-transmission neurons. A. Rev. Physiol. *40:* 217–248 (1978).

Fields, H.L.; Clanton, C.H.; Anderson, S.D.: Somatosensory properties of spinoreticular neurons in the cat. Brain Res. *120:* 49–66 (1977).

Fields, H.L.; Meyer, G.A.; Partridge, L.D.: Convergence of visceral and somatic input onto spinal neurones. Expl Neurol. *26:* 36–52 (1970a).

Fields, H.L.; Partridge, L.D.; Winter, D.L.: Somatic and visceral receptive field properties of fibers in ventral quadrant white matter of the cat spinal cord. J. Neurophysiol. *33:* 827–837 (1970b).

Fields, H.L.; Wagner, G.M.; Anderson, S.D.: Some properties of spinal neurons projecting to the medial brain-stem reticular formation. Expl Neurol. *47:* 118–134 (1975).

Fillenz, M.; Widdicombe, J.G.: Receptors of the lungs and airways; in Neil, Handbook of sensory physiology, vol. III/1, pp. 81–112 (Springer, Berlin 1972).

Fitzgerald, M.: The spread of sensitization of polymodal nociceptors in the rabbit from nearby injury and by antidromic nerve stimulation. J. Physiol. *297:* 207–216 (1979).

Fitzgerald, M.: A study of the cutaneous afferent input to substantia gelatinosa. Neuroscience *6:* 2229–2237 (1981).

Fitzgerald, M.: The contralateral input to the dorsal horn of the spinal cord in the decerebrate spinal rat. Brain Res. *236:* 275–287 (1982).

Fitzgerald, M.: Capsaicin and sensory neurones – a review. Pain *15:* 109–130 (1983).

Fitzgerald, M.; Lynn, B.: The sensitization of high threshold mechanoreceptors with myelinated axons by repeated heating. J. Physiol. *265:* 549–563 (1977).

Fitzgerald, M.; Wall, P.D.: The laminar organization of dorsal horn cells responding to peripheral C fibre stimulation. Exp. Brain Res. *41:* 36–44 (1980).

Fjällbrant, N.; Iggo, A.: The effect of histamine, 5-hydroxytryptamine and acetylcholine on cutaneous afferent fibres. J. Physiol. *156:* 578–590 (1961).

Fleischer, E.; Handwerker, H.O.; Joukhadar, S.: Unmyelinated nociceptive units in two skin areas of the rat. Brain Res. *267:* 81–92 (1983).

Floyd, K.; Hick, V.E.; Koley, J.; Morrison, J.F.B.: The effects of bradykinin on afferent units in intra-abdominal sympathetic nerve trunks. Q. Jl exp. Physiol. *62:* 19–25 (1977).

Floyd, K.; Hick, V.E.; Morrison, J.F.B.: Mechanosensitive afferent units in the hypogastric nerve of the cat. J. Physiol. *259:* 457–471 (1976a).

Floyd, K.; Koley, J.; Morrison, J.F.B.: Afferent discharges in the sacral ventral roots of cats. J. Physiol. *259:* 37P–38P (1976b).

Floyd, K.; Morrison, J.F.B.: Splanchnic mechanoreceptors in the dog. Q. Jl Physiol. *59:* 361–366 (1974).

Fock, S.; Mense, S.: Excitatory effects of 5-hydroxytryptamine, histamine and potassium ions on muscular group IV afferent units: a comparison with bradykinin. Brain Res. *105:* 459–469 (1976).

Foerster, O.; Gagel, O.: Die Vorderseitenstrangdurchschneidung beim Menschen. Eine klinisch-patho-physiologisch-anatomische Studie. Z. ges. Neurol. Psychiat. *138:* 1–92 (1932).

Foltz, E.L.; White, L.E.: Pain 'relief' by frontal cingulotomy. J. Neurosurg. *19:* 89–99 (1962).

Foreman, R.D.: Viscerosomatic convergence onto spinal neurons responding to afferent fibers located in the inferior cardiac nerve. Brain Res. *137:* 164–168 (1977).

Foreman, R.D.; Applebaum, A.E.; Beall, J.E.; Trevino, D.L.; Willis, W.D.: Responses of primate spinothalamic tract neurons to electrical stimulation of hindlimb peripheral nerves. J. Neurophysiol. *38:* 132–145 (1975).

Foreman, R.D.; Hancock, M.B.; Willis, W.D.: Responses of spinothalamic tract cells in the thoracic spinal cord of the monkey to cutaneous and visceral inputs. Pain *11:* 149–162 (1981).

Foreman, R.D.; Kenshalo, D.R., Jr.; Schmidt, R.F.; Willis, W.D.: Field potentials and excitation of primate spinothalamic neurones in response to volleys in muscle afferents. J. Physiol. *286:* 197–213 (1979a).

Foreman, R.D.; Ohata, C.A.: Effects of coronary artery occlusion on thoracic spinal neurons receiving viscerosomatic inputs. Am. J. Physiol. *238:* H667–674 (1980).

Foreman, R.D.; Schmidt, R.F.; Willis, W.D.: Convergence of muscle and cutaneous input onto primate spinothalamic tract neurons. Brain Res. *124:* 555–560 (1977).

Foreman, R.D.; Schmidt, R.F.; Willis, W.D.: Effects of mechanical and chemical stimulation of fine muscle afferents upon primate spinothalamic tract cells. J. Physiol. *286:* 215–231 (1979b).

Foreman, R.D.; Weber, R.N.: Responses from neurons of the primate spinothalamic tract to electrical stimulation of afferents from the cardiopulmonary region and somatic structures. Brain Res. *186:* 463–468 (1980).

Forssmann, W.G.: A new somatostatinergic system in the mammalian spinal cord. Neurosci. Lett. *10:* 293–297 (1978).

Foster, R.W.; Ramage, A.G.: The action of some chemical irritants on somatosensory receptors of the cat. Neuropharmacology *20:* 191–198 (1981).

Fox, R.E.; Holloway, J.A.; Iggo, A.; Mokha, S.S.: Spinothalamic neurones in the cat: some electrophysiological observations. Brain Res. *182:* 186–190 (1980).

Franz, M.; Mense, S.: Muscle receptors with group IV afferent fibres responding to application of bradykinin. Brain Res. *92:* 369–383 (1975).

Freeman, M.A.R.; Wyke, B.: The innervation of the knee joint. An anatomical and histological study in the cat. J. Anat. *101:* 505–532 (1967).

Freeman, W.; Watts, J.W.: Psychosurgery in the treatment of mental disorders and intractable pain (Thomas, Springfield 1950).

Frey, M. von: Beiträge zur Physiologie des Schmerzsinns. Ber. sächs. ges. Wiss. Leipzig math.-Phys. Kl. *46:* 185–196, 283–296 (1894).

Frey, M. von: Beiträge zur Sinnesphysiologie der Haut. Ber. sächs. ges. Wiss. Leipzig math.-Phys. Kl. *47:* 166–184 (1895).

Frey, M. von: The distribution of afferent nerves in the skin. J. Am. med. Ass. *47:* 645–648 (1906).

Friedman, D.P.; Jones, E.G.: Focal projection of electrophysiologically defined groupings of thalamic cells on the monkey somatic sensory cortex. Brain Res. *191:* 249–252 (1980).

Friedman, D.P.; Jones, E.G.: Thalamic input to areas 3a and 2 in monkeys. J. Neurophysiol. *45:* 59–85 (1981).

Friedman, D.P.; Jones, E.G.; Burton, H.: Representation pattern in the second somatic sensory area of the monkey cerebral cortex. J. comp. Neurol. *192:* 21–41 (1980).

Frykholm, R.: Cervical nerve root compression resulting from disc degeneration and root-sleeve fibrosis. A clinical investigation. Acta chir. scand. *160:* suppl., pp. 1–149 (1951).

Frykholm, R.; Hyde, J.; Norlen, G.; Skoglund, C.R.: On pain sensations produced by stimulation of ventral roots in man. Acta physiol. scand. *29:* suppl. 106, pp. 455–469 (1953).

Fukushima, T.; Kerr, F.W.L.: Organization of trigeminothalamic tracts and other thalamic afferent systems of the brainstem in the rat: presence of gelatinosa neurons with thalamic connections. J. comp. Neurol. *183:* 169–184 (1979).

Gammon, G.D.; Bronk, D.M.: The discharge of impulses from Pacinian corpuscles in the mesentery and its relation to vascular changes. Am. J. Physiol. *114:* 77–84 (1935).

Gamse, R.: Capsaicin and nociception in the rat and mouse: possible role of substance P. Arch. Pharmacol. *320:* 205–216 (1982).

Gamse, R.; Holzer, P.; Lembeck, F.: Decrease of substance P in primary afferent neurones and impairment of neurogenic plasma extravasation by capsaicin. Br. J. Pharmacol. *68:* 207–213 (1980).

Gamse, R.; Lackner, D.; Gamse, G.; Leeman, S.E.: Effect of capsaicin pretreatment on capsaicin-evoked release of immunoreactive somatostatin and substance P from primary sensory neurons. Arch. Pharmacol. *316:* 38–41 (1981).

Gamse, R.; Petsche, U.; Lembeck, F.; Jancso, G.: Capsaicin applied to peripheral nerve inhibits axoplasmic transport of substance P and somatostatin. Brain Res. *239:* 447–462 (1982).

Garcin, R.: La douleur dans les affections organiques du système nerveux central. Revue neurol. *68:* 105–153 (1937).

Gardner, E.: The distribution and termination of nerves in the knee joint of the cat. J. comp. Neurol. *80:* 11–32 (1944).

Gardner, E.P.; Costanzo, R.M.: Spatial integration of multiple-point stimuli in primary somatosensory cortical receptive fields of alert monkeys. J. Neurophysiol. *43:* 420–443 (1980a).

Gardner, E.P.; Costanzo, R.M.: Temporal integration of multiple-point stimuli in primary somatosensory cortical receptive fields of alert monkeys. J. Neurophysiol. *43:* 444–468 (1980b).

Gasser, H.S.: Properties of dorsal root unmedullated fibers on the two sides of the ganglion. J. gen. Physiol. *38:* 709–728 (1955).

Gautron, M.; Guilbaud, G.: Somatic responses of ventrobasal thalamic neurones in polyarthritic rats. Brain Res. *237:* 459–471 (1982).

Gaze, R.M.; Gordon, G.: The representation of cutaneous sense in the thalamus of the cat and monkey. Q. Jl exp. Physiol. *39:* 279–304 (1954).

Georgopoulos, A.P.: Functional properties of primary afferent units probably related to pain mechanisms in primate glabrous skin. J. Neurophysiol. *39:* 71–83 (1976).

Georgopoulos, A.P.: Stimulus-response relations in high-threshold mechanothermal fibers innervating primate glabrous skin. Brain Res. *128:* 547–552 (1977).

Gerhart, K.D.; Yezierski, R.P.; Giesler, G.J.; Willis, W.D.: Inhibitory receptive fields of primate spinothalamic tract cells. J. Neurophysiol. *46:* 1309–1325 (1981).

Gernandt, B.; Zotterman, Y.: Intestinal pain: an electrophysiological investigation on mesenteric nerves. Acta physiol. scand. *12:* 56–72 (1946).

Gibson, S.J.; Polak, J.M.; Bloom, S.R.; Wall, P.D.: The distribution of nine peptides in rat spinal cord with special emphasis on the substantia gelatinosa and on the area around the central canal (lamina X). J. comp. Neurol. *201:* 65–79 (1981).

Giesler, G.J.; Cannon, J.T.; Urca, G.; Liebeskind, J.C.: Long ascending projections from substantia gelatinosa Rolandi and the subjacent dorsal horn in the rat. Science *202:* 984–986 (1978).

Giesler, G.J.; Menétrey, D.; Basbaum, A.I.: Differential origins of spinothalamic tract projections to medial and lateral thalamus in the rat. J. comp. Neurol. *184:* 107–126 (1979a).

Giesler, G.J.; Menétrey, D.; Guilbaud, G.; Besson, J.M.: Lumbar cord neurons at the origin of the spinothalamic tract in the rat. Brain Res. *118:* 320–324 (1976).

Giesler, G.J.; Nahin, R.L.; Madsen, A.M.: Postsynaptic dorsal column pathway of the rat. I. Anatomical studies. J. Neurophysiol. *51:* 260–275 (1984).

Giesler, G.J.; Speil, H.R.; Willis, W.D.: Organization of spinothalamic tract axons within the rat spinal cord. J. comp. Neurol. *195:* 243–252 (1981a).

Giesler, G.J.; Urca, G.; Cannon, J.T.; Liebeskind, J.C.: Response properties of neurons of the lateral cervical nucleus in the rat. J. comp. Neurol. *186:* 65–78 (1979b).

Giesler, G.J.; Yezierski, R.P.; Gerhart, K.D.; Willis, W.D.: Spinothalamic tract neurons that project to medial and/or lateral thalamic nuclei: evidence for a physiologically novel population of spinal cord neurons. J. Neurophysiol. *46:* 1285–1308 (1981b).

Glazer, E.J.; Basbaum, A.I.: Immunohistochemical localization of leucine-enkephalin in the spinal cord of the cat: enkephalin-containing marginal neurons and pain modulation. J. comp. Neurol. *196:* 377–389 (1981).

Glees, P.; Bailey, R.A.: Schichtung und Fasergrösse des Tractus spino-thalamicus des Menschen. Mschr. Psychiat. Neurol. *122:* 129–141 (1951).

Gobel, S.: Synaptic organization of the substantia gelatinosa glomeruli in the spinal trigeminal nucleus of the adult cat. J. Neurocytol. *3:* 219–243 (1974).

Gobel, S.: Golgi studies of the substantia gelatinosa neurons in the spinal trigeminal nucleus. J. comp. Neurol. *162:* 397–416 (1975).

Gobel, S.: Dendroaxonic synapses in the substantia gelatinosa glomeruli of the spinal trigeminal nucleus of the cat. J. comp. Neurol. *167:* 165–176 (1976).

Gobel, S.: Golgi studies of the neurons in layer I of the dorsal horn of the medulla (trigeminal nucleus caudalis). J. comp. Neurol. *180:* 375–394 (1978a).

Gobel, S.: Golgi studies of the neurons in layer II of the dorsal horn of the medulla (trigeminal nucleus caudalis). J. comp. Neurol. *180:* 395–414 (1978b).

Gobel, S.: Neural circuitry in the substantia gelatinosa of Rolando: anatomical insights. Adv. Pain Res. Ther. *3:* 175–195 (1979).

Gobel, S.; Binck, J.M.: Degenerative changes in primary trigeminal axons and in neurons in nucleus caudalis following tooth pulp extirpations in the cat. Brain Res. *132:* 347–354 (1977).

Gobel, S.; Falls, W.M.: Anatomical observations of horseradish peroxidase-filled terminal primary axonal arborizations in layer II of the substantia gelatinosa of Rolando. Brain Res. *175:* 335–340 (1979).

Gobel, S.; Falls, W.M.; Bennett, G.J.; Abdelmoumene, M.; Hayashi, H.; Humphrey, E.: An EM analysis of the synaptic connections of horseradish peroxidase-filled stalked cells and islet cells in the substantia gelatinosa of adult cat spinal cord. J. comp. Neurol. *194:* 781–807 (1980).

Godfraind, J.M.; Jessell, T.M.; Kelly, J.S.; McBurney, R.N.; Mudge, A.W.; Yamamoto, M.: Capsaicin prolongs action potential duration in cultured sensory neurones. J. Physiol. *312:* 32–33P (1981).

Gokin, A.P.; Kostyuk, P.G.; Preobrazhensky, N.N.: Neuronal mechanisms of interactions of high-threshold visceral and somatic afferent influences in spinal cord and medulla. J. Physiol., Paris *73:* 319–333 (1977).

Goldscheider, A.: Zur Lehre von den spezifischen Energien der Sinnesnerven (Schumacher, Berlin 1881).

Goldstein, A.: Opioid peptides (endorphins) in pituitary and brain. Science *193:* 1081–1086 (1976).

Goldstein, A.; Tachibana, S.; Lowney, L.I.; Hunkapiller, M.; Hood, L.: Dynorphin (1–13), an extraordinarily potent opioid peptide. Proc. natn. Acad. Sci. USA *76:* 6666–6670 (1979).

Gordon, G.: Nociceptive cells in ventroposterior thalamus of the cat (Letter to the editor). J. Neurophysiol. *50:* 1043 (1983).

Gordon, G.; Manson, J.R.: Cutaneous receptive fields of single nerve cells in the thalamus of the cat. Nature, Lond. *215:* 597–599 (1967).

Gowers, W.R.: A case of unilateral gunshot injury to the spinal cord. Trans. clin. Lond. *11:* 24–32 (1878).

Graham, L.T.; Shank, R.P.; Werman, R.; Aprison, M.H.: Distribution of some synaptic transmitter suspects in cat spinal cord: glutamic acid, aspartic acid, γ-aminobutyric acid, glycine, and glutamine. J. Neurochem. *14:* 465–472 (1967).

Graybiel, A.M.: The thalamo-cortical projection of the so-called posterior nuclear group: a study with anterograde degeneration methods in the cat. Brain Res. *49:* 229–244 (1973).

Grigg, P.: Mechanical factors influencing response of joint afferent neurons from cat knee. J. Neurophysiol. *38:* 1473–1484 (1975).

Guilbaud, G.; Benelli, G.; Besson, J.M.: Responses of thoracic dorsal horn interneurones to cutaneous stimulation and to the administration of algogenic substances into the mesenteric artery in the spinal cat. Brain Res. *124:* 437–448 (1977a).

Guilbaud, G.; Besson, J.M.; Oliveras, J.L.; Wyon-Maillard, M.C.: Modifications of the firing rate of bulbar reticular units (nucleus gigantocellularis) after intra-arterial injection of bradykinin into the limbs. Brain Res. *63:* 131–140 (1973).

Guilbaud, G.; Caille, D.; Besson, J.M.; Benelli, G.: Single unit activities in ventral posterior and posterior group thalamic nuclei during nociceptive and non-nociceptive stimulations in the cat. Arch. ital. Biol. *115:* 38–56 (1977b).

Guilbaud, G.; Peschanski, M.; Gautron, M.; Binder, D.: Neurones responding to noxious stimulation in VB complex and caudal adjacent regions in the thalamus of the rat. Pain *8:* 303–318 (1980).

Guz, A.; Noble, M.I.M.; Widdicombe, J.G.; Trenchard, D.; Mushin, W.W.; Makey, A.R.: The role of vagal and glossopharyngeal afferent nerves in respiratory sensation, control of breathing and arterial pressure regulation in conscious man. Clin. Sci. *30:* 161–170 (1966).

Guzman, F.; Braun, C.; Lim, R.K.S.: Visceral pain and the pseudoaffective response to intra-arterial injection of bradykinin and other algesic agents. Archs int. Pharmacodyn. Thér. *136:* 353–384 (1962).

Gybels, J.; Handwerker, H.O.; Van Hees, J.: A comparison between the discharges of human nociceptive fibres and the subject's ratings of his sensations. J. Physiol. *292:* 193–206 (1979).

Ha, H.: Cervicothalamic tract in the rhesus monkey. Expl Neurol. *33:* 205–212 (1971).

Ha, H.; Liu, C.N.: Organization of the spino-cervico-thalamic system. J. comp. Neurol. *127:* 445–470 (1966).

Haber, L.H.; Martin, R.F.; Chung, J.M.; Willis, W.D.: Inhibition and excitation of primate spinothalamic tract neurons by stimulation in region of nucleus reticularis gigantocellularis. J. Neurophysiol. *43:* 1578–1593 (1980).

Haber, L.H.; Moore, B.D.; Willis, W.D.: Electrophysiological response properties of spinoreticular neurons in the monkey. J. comp. Neurol. *207:* 75–84 (1982).

Hagbarth, K.E.; Vallbo, A.B.: Mechanoreceptor activity recorded percutaneously with semi-microelectrodes in human peripheral nerves. Acta physiol. scand. *69:* 121–122 (1967).

Halliday, A.M.; Logue, V.: Painful sensations evoked by electrical stimulation in the thalamus; in Somjen, Neurophysiology studied in man, pp. 221–230 (Excerpta Medica, Amsterdam 1972).

Hallin, R.G.; Torebjörk, H.E.: Studies on cutaneous A and C fibre afferents, skin nerve blocks and perception; in Zotterman, Sensory functions of the skin in primates, with special reference to man, pp. 137–148 (Pergamon Press, New York 1976).

Hallin, R.G.; Torebjörk, H.; Wiesenfeld, Z.: Nociceptors and warm receptors innervated by C fibres in human skin. J. Neurol. Neurosurg. Psychiat. *45:* 313–319 (1982).

Hamann, W.C.; Hong, S.K.; Kniffki, K.D.; Schmidt, R.F.: Projections of primary afferent fibres from muscle to neurones of the spinocervical tract of the cat. J. Physiol. *283:* 369–378 (1978).

Hamilton, B.L.: Projections of the nuclei of the periaqueductal gray matter in the cat. J. comp. Neurol. *152:* 45–58 (1973).

Hamilton, B.L.; Skultety, F.M.: Efferent connections of the periaqueductal gray matter in the cat. J. comp. Neurol. *139:* 105–114 (1970).

Hancock, M.B.; Foreman, R.D.; Willis, W.D.: Convergence of visceral and cutaneous input onto spinothalamic tract cells in the thoracic spinal cord of the cat. Expl Neurol. *47:* 240–248 (1975).

Hancock, M.B.; Rigamonti, D.D.; Bryan, R.N.: Convergence in the lumbar spinal cord of pathways activated by splanchnic nerve and hind limb cutaneous nerve stimulation. Expl Neurol. *38:* 337–348 (1973).

Hand, P.J.; Morrison, A.R.: Thalamocortical projections from the ventrobasal complex to somatic sensory areas I and II. Expl Neurol. *26:* 291–308 (1970).

Handwerker, H.O.: Influences of algogenic substances and prostaglandins on the discharges of unmyelinated cutaneous nerve fibers identified as nociceptors. Adv. Pain Res. Ther. *1:* 41–45 (1976).

Handwerker, H.O.; Iggo, A.; Zimmermann, M.: Segmental and supraspinal actions on dorsal horn neurons responding to noxious and non-noxious skin stimuli. Pain *1:* 147–165 (1975).

Handwerker, H.O.; Neher, K.D.: Characteristics of C-fibre receptors in the cat's foot responding to stepwise increase of skin temperature to noxious levels. Pflügers Arch. *365:* 221–229 (1976).

Hardy, J.D.; Stolwijk, J.A.J.; Hoffman, D.: Pain following step increase in skin temperature; in Kenshalo, The skin senses, pp. 444–454 (Thomas, Springfield 1968).

Hardy, J.D.; Wolff, H.G.; Goodell, H.: Pain sensations and reactions (Williams & Wilkins, New York 1952a/Hafner, New York 1967).

Hardy, J.D.; Wolff, H.G.; Goodell, H.: Pricking pain threshold in different body areas. Proc. Soc. exp. Biol. Med. *80:* 425–427 (1952b).

Harris, F.A.: Functional subsets of neurons in somatosensory thalamus of the cat. Expl Neurol. *58:* 149–170 (1978a).

Harris, F.A.: Regional variations of somatosensory input convergence in nucleus VPL of cat thalamus. Expl Neurol. *58:* 171–189 (1978b).

Hassler, R.: Die zentralen Systeme des Schmerzes. Acta neurochir. *8:* 353–423 (1960).

Hassler, R.: Dichotomy of facial pain conduction in the diencephalon; in Hassler, Walker, Trigeminal neuralgia, pp. 123–138 (Saunders, Philadelphia 1970).

Hassler, R.; Riechert, T.: Klinische und anatomische Befunde bei stereotaktischen Schmerzoperationen im Thalamus. Arch. Psychiat. NervKrankh. *200:* 93–122 (1959).

Hayes, A.G.; Scadding, J.W.; Skingle, M.; Tyers, M.B.: Effects of neonatal administration of capsaicin on nociceptive thresholds in the mouse and rat. J. Pharm. Pharmac. *33:* 183–185 (1981).

Hayes, A.G.; Tyers, M.B.: Effects of capsaicin on nociceptive heat, pressure and chemical thresholds and on substance P levels in the rat. Brain Res. *189:* 561–564 (1980).

Hayes, N.L.; Rustioni, A.: Spinothalamic and spinomedullary neurons in macaques: a single and double retrograde tracer study. Neuroscience *5:* 861–874 (1980).

Head, H.: On disturbances of sensation with especial reference to the pain of visceral disease. Brain *16:* 1–132 (1893).

Head, H.: Studies in neurology (Oxford University Press, London 1920).

Head, H.; Holmes, G.: Sensory disturbances from cerebral lesions. Brain *34:* 102–254 (1911).

Head, H.; Thompson, T.: The grouping of afferent impulses within the spinal cord. Brain *29:* 537–741 (1906).

Heath, C.J.; Jones, E.G.: An experimental study of ascending connections from the posterior group of thalamic nuclei in the cat. J. comp. Neurol. *141:* 397–426 (1971).

Heimer, L.; Wall, P.D.: The dorsal root distribution to the substantia gelatinosa of the rat with a note on the distribution in the cat. Exp. Brain Res. *6:* 89–99 (1968).

Heinbecker, P.; Bishop, G.H.; O'Leary, J.: Pain and touch fibers in peripheral nerves. Archs Neurol. Psychiat. *29:* 771–789 (1933).

Heinbecker, P.; Bishop, G.H.; O'Leary, J.: Analysis of sensation in terms of the nerve impulse. Archs Neurol. Psychiat. *31:* 34–54 (1934).

Helke, C.J.; DiMicco, J.A.; Jacobowitz, D.M.; Kopin, I.J.: Effect of capsaicin administration to neonatal rats on the substance P content of discrete CNS regions. Brain Res. *222:* 428–431 (1981).

Hellon, R.F.; Misra, N.K.: Neurones in the dorsal horn of the rat responding to scrotal skin temperature changes. J. Physiol. *232:* 375–388 (1973).

Henry, J.L.: Effects of substance P on functionally identified units in cat spinal cord. Brain Res. *114:* 439–451 (1976).

Henry, J.L.; Krnjevíc, K.; Morris, M.E.: Substance P and spinal neurones. Can. J. Physiol. Pharmacol. *53:* 423–432 (1975).

Henry, J.L.; Sessle, B.J.; Lucier, G.E.; Hu, J.W.: Effects of substance P on nociceptive and non-nociceptive trigeminal brain stem neurons. Pain *8:* 33–45 (1980).

Hensel, H.; Boman, K.K.A.: Afferent impulses in cutaneous sensory nerves in human subjects. J. Neurophysiol. *23:* 564–578 (1960).

Hensel, H.; Iggo, A.: Analysis of cutaneous warm and cold fibers in primates. Pflügers Arch. ges. Physiol. *329:* 1–8 (1971).

Hensel, H.; Kenshalo, D.R.: Warm receptors in the nasal region of cats. J. Physiol. *204:* 99–112 (1969).

Hentall, I.: A novel class of unit in the substantia gelatinosa of the spinal cat. Expl Neurol. *57:* 792–806 (1977).

Hertel, H.C.; Howaldt, B.; Mense, S.: Responses of group IV and group III muscle afferents to thermal stimuli. Brain Res. *113:* 201–205 (1976).

Hillman, P.; Wall, P.D.: Inhibitory and excitatory factors influencing the receptive fields of lamina 5 spinal cord cells. Exp. Brain Res. *9:* 284–306 (1969).

Hiss, E.; Mense, S.: Evidence for the existence of different receptor sites for algesic agents at the endings of muscular group IV afferent units. Pflügers Arch. *362:* 141–146 (1976).

Hockfield, S.; Gobel, S.: Neurons in and near nucleus caudalis with long ascending projection axons demonstrated by retrograde labeling with horseradish peroxidase. Brain Res. *139:* 333–339 (1978).

Hoffman, D.S.; Dubner, R.; Hayes, R.L.; Medlin, T.P.: Neuronal activity in medullary dorsal horn of awake monkeys trained in a thermal discrimination task. I. Responses to innocuous and noxious thermal stimuli. J. Neurophysiol. *46:* 409–427 (1981).

Hökfelt, T.; Elde, R.; Johansson, O.; Luft, R.; Nilsson, G.; Arimura, A.: Immunohistochemical evidence for separate populations of somatostatin-containing and substance P-containing primary afferent neurons in the rat. Neuroscience *1:* 131–136 (1976).

Hökfelt, T.; Kellerth, J.O.; Nilsson, G.; Pernow, B.: Experimental immunohistochemical studies on the localization and distribution of substance P in cat primary sensory neurons. Brain Res. *100:* 235–252 (1975).

Holloway, J.A.; Fox, R.E.; Iggo, A.: Projections of the spinothalamic tract to the thalamic nuclei of the cat. Brain Res. *157:* 336–340 (1978).

Holmes, F.W.; Davenport, H.A.: Cells and fibers in spinal nerves. J. comp. Neurol. *73:* 1–5 (1940).

Holmes, G.: Disorders of sensation produced by cortical lesions. Brain *50:* 413–427 (1927).

Holzer, P.; Jurna, I.; Gamse, R.; Lembeck, F.: Nociceptive threshold after neonatal capsaicin treatment. Eur. J. Pharmacol. *58:* 511–514 (1979).

Honda, C.N.; Mense, S.; Perl, E.R.: Neurons in ventrobasal region of cat thalamus selectively responsive to noxious mechanical stimulation. J. Neurophysiol. *49:* 662–673 (1983).

Honda, C.N.; Perl, E.R.: Properties of neurons in lamina X and the midline dorsal horn of the sacrococcygeal cord of the cat. Soc. Neurosci. Abstr. *7:* 610 (1981).

Hong, S.K.; Kniffki, K.D.; Mense, S.; Schmidt, R.F.; Wendisch, M.: Descending influences on the responses of spinocervical tract neurones to chemical stimulation of fine muscle afferents. J. Physiol. *290:* 129–140 (1979).

Hori, Y.; Lee, K.H.; Chung, J.M.; Willis, W.D.: The effects of small doses of barbiturate on the activity of primate nociceptive tract cells. Brain Res. *307:* 9–15 (1984).

Horrax, G.: Experiences with cordotomy. Archs Surg. *18:* 1140–1164 (1929).

Horrax, G.: Experiences with cortical excisions for the relief of intractable pain in the extremities. Surgery *20:* 593–602 (1946).

Horrobin, D.F.: The lateral cervical nucleus of the cat: an electrophysiological study. Q. Jl exp. Physiol. *51:* 351–371 (1966).

Howe, A.; Neil, E.: Arterial chemoreceptors; in Neil, Handbook of sensory physiology, vol. III/1: Enteroceptors, pp. 47–80 (Springer, Berlin 1972).

Hughes, J.: Isolation of an endogenous compound from the brain with pharmacological properties similar to morphine. Brain Res. *88:* 295–308 (1975).

Hughes, J.; Smith, T.W.; Kosterlitz, H.W.; Fothergill, L.A.; Morgan, B.A.; Morris, H.R.: Identification of two related pentapeptides from the brain with potent opiate agonist activity. Nature, Lond. *258:* 577–579 (1975).

Hunt, C.C.; McIntyre, A.K.: An analysis of fibre diameter and receptor characteristics of myelinated cutaneous afferent fibres in cat. J. Physiol. *153:* 99–112 (1960).

Hunt, S.P.; Kelly, J.S.; Emson, P.C.; Kimmel, J.R.; Miller, R.J.; Wu, J.Y.: An immunohistochemical study of neuronal populations containing neuropeptides or γ-aminobutyrate within the superficial layers of the rat dorsal horn. Neuroscience *6:* 1883–1898 (1981).

Hurt, R.W.; Ballantine, H.T.: Stereotactic anterior cingulate lesions for persistent pain: a report on 68 cases. Clin. Neurosurg. *21:* 334–351 (1974).

Hyndman, O.R.: Lissauer's tract section. A contribution to chordotomy for the relief of pain (preliminary report). J. int. Coll. Surg. *5:* 394–400 (1942).

Hyndman, O.R.; Van Epps, C.: Possibility of differential section of the spinothalamic tract. Archs Surg. *38:* 1036–1053 (1939).

Hyndman, O.R.; Wolkin, J.: Anterior chordotomy. Further observations on physiological results and optimum manner of performance. Archs Neurol. Psychiat. *50:* 129–148 (1943).

Hyvärinen, J.; Poranen, A.: Receptive field integration and submodality convergence in the hand area of the post-central gyrus of the alert monkey. J. Physiol. *283:* 539–556 (1978).

Iggo, A.: Tension receptors in the stomach and the urinary bladder. J. Physiol. *128:* 593–607 (1955).

Iggo, A.: Gastro-intestinal tension receptors with unmyelinated afferent fibres in the vagus of the cat. Q. Jl exp. Physiol. *42:* 130–143 (1957a).

Iggo, A.: Gastric mucosal chemoreceptors with vagal afferent fibres in the cat. Q. Jl exp. Physiol. *42:* 398–409 (1957b).

Iggo, A.: Cutaneous heat and cold receptors with slowly conducting (C) afferent fibres. Q. Jl exp. Physiol. *44:* 362–370 (1959).

Iggo, A.: Cutaneous mechanoreceptors with afferent C fibres. J. Physiol. *152:* 337–353 (1960).

Iggo, A.: Non-myelinated afferent fibres from mammalian skeletal muscle. J. Physiol. *155:* 52–53P (1961).

Iggo, A.: Physiology of visceral afferent systems. Acta neuroveg. *28:* 121–134 (1966).

Iggo, A.: Cutaneous thermoreceptors in primates and sub-primates. J. Physiol. *200:* 403–430 (1969).

Iggo, A.; Kornhuber, H.H.: A quantitative study of C-mechanoreceptors in hairy skin of the cat. J. Physiol. *271:* 549–565 (1977).

Iggo, A.; Ogawa, H.: Primate cutaneous thermal nociceptors. J. Physiol. *216:* 77P–78P (1971).

Iggo, A.; Ogawa, H.: Correlative physiological and morphological studies of rapidly adapting mechanoreceptors in cat's glabrous skin. J. Physiol. *266:* 275–296 (1977).

Iggo, A.; Ramsey, R.L.: Thermosensory mechanisms in the spinal cord of monkeys; in Zotterman, Sensory functions of the skin in primates, with special reference to man, pp. 285–302 (Pergamon Press, New York 1976).

Imai, Y.; Kusama, T.: Distribution of the dorsal root fibers in the cat. An experimental study with the Nauta method. Brain Res. *13:* 338–359 (1969).

Ingvar, S.: Zur Morphogenese der Tabes. Acta med. scand. *65:* 645–674 (1927).

Iriuchijima, J.; Zotterman, Y.: The specificity of afferent cutaneous C fibres in mammals. Acta physiol. scand. *49:* 267–278 (1960).

Ishijima, B.; Yoshimasu, N.; Fukushima, T.; Hori, T.; Sekino, H.; Sano, K.: Nociceptive neurons in the human thalamus. Confinia neurol. *37:* 99–106 (1975).

Jancsó, G.; Hökfelt, T.; Lundberg, J.M.; Király, E.; Halász, N.; Nilsson, G.; Terenius, L.; Rehfeld, J.; Steinbusch, H.; Verhofstad, A.; Elde, R.; Said, S.; Brown, M.: Immunohistochemical studies on the effect of capsaicin on spinal and medullary peptide and monoamine neurons using antisera to substance P, gastrin/CCK, somatostatin, VIP, enkephalin, neurotensin and 5-hydroxytryptamine. J. Neurol. *10:* 963–980 (1981).

Jancsó, G.; Király, E.: Distribution of chemosensitive primary sensory afferents in the central nervous system of the rat. J. comp. Neurol. *190:* 781–792 (1980).

Jancsó, G.; Király, E.: Sensory neurotoxins: chemically induced selective destruction of primary sensory neurons. Brain Res. *210:* 83–89 (1981).

Jancsó, G.; Király, E.; Jancsó-Gábor, A.: Pharmacologically induced selective degeneration of chemosensitive primary sensory neurones. Nature, Lond. *270:* 741–743 (1977).

Jancsó, G.; Király, E.; Jancsó-Gábor, A.: Direct evidence for an axonal site of action of capsaicin. Arch. Pharmacol. *313:* 91–94 (1980).

Jancsó, G.; Knyihár, E.: Functional linkage between nociception and fluoride resistant acid phosphatase activity in the Rolando substance. Neurobiology *5:* 42–43 (1979).

Jeftinija, S.; Miletić, V.; Randic, M.: Cholecystokinin octapeptide excites dorsal horn neurons both in vivo and in vitro. Brain Res. *213:* 231–236 (1981).

Jessell, T.M.; Iversen, L.L.: Opiate analgesics inhibit substance P release from rat trigeminal nucleus. Nature, Lond. *268:* 549–551 (1977).

Jessell, T.M.; Iversen, L.L.; Cuello, A.C.: Capsaicin-induced depletion of substance P from primary sensory neurones. Brain Res. *152:* 183–188 (1978).

Jessell, T.; Tsunoo, A.; Kanazawa, I.; Otsuka, M.: Substance P: depletion in the dorsal horn of rat spinal cord after section of the peripheral processes of primary sensory neurons. Brain Res. *168:* 247–259 (1979).

Jewsbury, E.C.O.: Insensitivity to pain. Brain *74:* 336–353 (1951).

Jones, E.G.: Some aspects of the organization of the thalamic reticular complex. J. comp. Neurol. *162:* 285–308 (1975a).

Jones, E.G.: Possible determinants of the degree of retrograde neuronal labeling with horseradish peroxidase. Brain Res. *85:* 249–253 (1975b).

Jones, E.G.: Thalamic basis for column-like input to monkey somatic sensory and motor cortex; in Macchi, Rustioni, Spreafico, Somatosensory integration in the thalamus, pp. 309–336 (Elsevier, Amsterdam 1983).

Jones, E.G.; Burton, H.: Cytoarchitecture and somatic sensory connectivity of thalamic nuclei other than the ventrobasal complex in the cat. J. comp. Neurol. *154:* 395–432 (1974).

Jones, E.G.; Burton, H.: Areal differences in the laminar distribution of thalamic afferents in cortical fields of the insular, parietal and temporal regions of primates. J. comp. Neurol. *168:* 197–248 (1976).

Jones, E.G.; Friedman, D.P.: Projection pattern of functional components of thalamic ventrobasal complex on monkey somatosensory cortex. J. Neurophysiol. *48:* 521–544 (1982).

Jones, E.G.; Friedman, D.P.; Hendry, S.H.C.: Thalamic basis of place- and modality-specific columns in monkey somatosensory cortex: a correlative anatomical and physiological study. J. Neurophysiol. *48:* 545–568 (1982).

Jones, E.G.; Leavitt, R.Y.: Demonstration of thalamo-cortical connectivity in the cat somato-sensory system by retrograde axonal transport of horseradish peroxidase. Brain Res. *63:* 414–418 (1973).

Jones, E.G.; Leavitt, R.Y.: Retrograde axonal transport and the demonstration of non-specific projections to the cerebral cortex and striatum from thalamic intralaminar nuclei in the rat, cat and monkey. J. comp. Neurol. *154:* 349–378 (1974).

Jones, E.G.; Powell, T.P.S.: The cortical projection of the ventroposterior nucleus of the thalamus in the cat. Brain Res. *13:* 298–318 (1969).

Jones, E.G.; Powell, T.P.S.: Connexions of the somatic sensory cortex of the rhesus monkey. III. Thalamic connexions. Brain *93:* 37–56 (1970).

Jones, E.G.; Wise, S.P.; Coulter, J.D.: Differential thalamic relationships of sensory-motor and parietal cortical fields in monkeys. J. comp. Neurol. *183:* 833–882 (1979).

Jordan, L.M.; Kenshalo, D.R., Jr.; Martin, R.F.; Haber, L.H.; Willis, W.D.: Depression of primate spinothalamic tract neurons by iontophoretic application of 5-hydroxytryptamine. Pain *5:* 135–142 (1978).

Kaas, J.H.; Merzenich, M.M.; Killackey, H.P.: The reorganization of somatosensory cortex following peripheral nerve damage in adult and developing mammals. Annu. Rev. Neurosci. *6:* 325–356 (1983).

Kato, M.; Hirata, Y.: Sensory neurons in the spinal ventral roots of the cat. Brain Res. *7:* 479–482 (1968).

Kato, M.; Tanji, J.: Physiological properties of sensory fibers in the spinal ventral roots in the cat. Jap. J. Physiol. *21:* 71–77 (1971).

Kawatani, M.; Lowe, I.P.; Nadelhaft, I.; Morgan, C.; DeGroat, W.C.: Vasoactive intestinal polypeptide in visceral afferent pathways to the sacral spinal cord of the cat. Neurosci. Lett. *42:* 311–316 (1983).

Kenins, P.: Responses of single nerve fibres to capsaicin applied to the skin. Neurosci. Lett. *29:* 83–88 (1982).

Kennard, M.A.: The course of ascending fibers in the spinal cord of the cat essential to the recognition of painful stimuli. J. comp. Neurol. *100:* 511–524 (1954).

Kenshalo, D.R.; Duclaux, R.: Response characteristics of cutaneous cold receptors in the monkey. J. Neurophysiol. *40:* 319–332 (1977).

Kenshalo, D.R., Jr.; Giesler, G.J.; Leonard, R.B.; Willis, W.D.: Responses of neurons in primate ventral posterior lateral nucleus to noxious stimuli. J. Neurophysiol. *43:* 1594–1614 (1980).

Kenshalo, D.R., Jr.; Isensee, O.: Responses of primate SI cortical neurons to noxious stimuli. J. Neurophysiol. *50:* 1479–1496 (1983).

Kenshalo, D.R., Jr.; Leonard, R.B.; Chung, J.M.; Willis, W.D.: Responses of primate spinothalamic neurons to graded and to repeated noxious heat stimuli. J. Neurophysiol. *42:* 1370–1389 (1979).

Kenshalo, D.R., Jr.; Leonard, R.B.; Chung, J.M.; Willis, W.D.: Facilitation of the responses of primate spinothalamic cells to cold and to tactile stimuli by noxious heating of the skin. Pain *12:* 141–152 (1982).

Kerr, F.W.L.: Neuroanatomical substrates of nociception in the spinal cord. Pain *1:* 325–356 (1975a).

Kerr, F.W.L.: The ventral spinothalamic tract and other ascending systems of the ventral funiculus of the spinal cord. J. comp. Neurol. *159:* 335–356 (1975b).

Kerr, F.W.L.; Lippman, H.H.: The primate spinothalamic tract as demonstrated by anterolateral cordotomy and commissural myelotomy. Adv. Neurol. *4:* 147–156 (1974).

Kevetter, G.A.; Haber, L.H.; Yezierski, R.P.; Chung, J.M.; Martin, R.F.; Willis, W.D.: Cells of origin of the spinoreticular tract in the monkey. J. comp. Neurol. *207:* 61–74 (1982).

Kevetter, G.A.; Willis, W.D.: Spinothalamic cells in the rat lumbar cord with collaterals to the medullary reticular formation. Brain Res. *238:* 181–185 (1982).

Kevetter, G.A.; Willis, W.D.: Collaterals of spinothalamic cells in the rat. J. comp. Neurol. *215:* 453–464 (1983).

King, H.E.; Clausen, J.; Scarff, J.E.: Cutaneous thresholds for pain before and after unilateral prefrontal lobotomy. J. nerv. ment. Dis. *112:* 93–96 (1950).

King, J.S.; Gallant, P.; Myerson, V.; Perl, E.R.: The effects of anti-inflammatory agents on the responses and the sensitization of unmyelinated (C) fiber polymodal nociceptors; in Zotterman, Sensory functions of the skin in primates, with special reference to man, pp. 441–461 (Pergamon Press, Oxford 1976).

Kircher, C.; Ha, H.: The nucleus cervicalis lateralis in primates, including the human. Anat. Rec. *160:* 376 (1968).

Kitai, S.T.; Ha, H.; Morin, F.: Lateral cervical nucleus of the dog: anatomical and microelectrode studies. Am. J. Physiol. *209:* 307–312 (1965).

Kniffki, K.D.; Mense, S.; Schmidt, R.F.: The spinocervical tract as a possible pathway for muscular nociception. J. Physiol., Paris *73:* 359–366 (1977).

Kniffki, K.D.; Mense, S.; Schmidt, R.F.: Responses of group IV afferent units from skeletal muscle to stretch, contraction and chemical stimulation. Exp. Brain Res. *31:* 511–522 (1978).

Kniffki, K.D.; Mizumura, K.: Responses of neurons in VPL and VPL-VL region of the cat to algesic stimulation of muscle and tendon. J. Neurophysiol. *49:* 649–661 (1983).

Knyihár, E.: Fluoride-resistant acid phosphatase system of nociceptive dorsal root afferents. Experientia *27:* 1205–1207 (1971).

Knyihár, E.; Csillik, B.: Effect of peripheral axotomy on the fine structure and histochemistry of the Rolando structure: degenerative atrophy of central processes of pseudounipolar cells. Exp. Brain Res. *26:* 73–87 (1976).

Knyihár, E.; Gerebtzoff, M.A.: Extra-lysosomal localization of acid phosphatase in the spinal cord of the rat. Exp. Brain Res. *18:* 383–395 (1973).

Knyihár, E.; László, I.; Tornyos, S.: Fine structure and fluoride-resistant acid phosphatase activity of electron dense sinusoid terminals in the substantia gelatinosa Rolandi of the rat after dorsal root transection. Exp. Brain Res. *19:* 529–544 (1974).

Knyihár-Csillik, E.; Csillik, B.; Rakic, P.: Ultrastructure of normal and degenerating glomerular terminals of dorsal root axons in the substantia gelatinosa of the rhesus monkey. J. comp. Neurol. *210:* 357–375 (1982).

Kolmodin, G.M.; Skoglund, C.R.: Analysis of spinal interneurons activated by tactile and nociceptive stimulation. Acta physiol. scand. *50:* 337–355 (1960).

Konietzny, F.; Perl, E.R.; Trevino, D.; Light, A.; Hensel, H.: Sensory experiences in man evoked by intraneural electrical stimulation of intact cutaneous afferent fibers. Exp. Brain Res. *42:* 219–222 (1981).

Krauthamer, G.; McGuinness, C.; Gottesman, L.: Unit responses in the ventrobasal thalamus (VPL) of the cat to bradykinin injected into somatic and visceral arteries. Brain Res. Bull. *2:* 299–306 (1977).

Kruger, L.; Saporta, S.; Feldman, S.G.: Axonal transport studies of the sensory trigeminal complex; in Anderson, Matthews, Pain in the trigeminal region, pp. 191–202 (Elsevier, Amsterdam 1977).

Kumazawa, T.; Mizumura, K.: The polymodal C-fiber receptor in the muscle of the dog. Brain Res. *101:* 589–593 (1976).

Kumazawa, T.; Mizumura, K.: Thin-fibre receptors responding to mechanical, chemical, and thermal stimulation in the skeletal muscle of the dog. J. Physiol. *273:* 179–194 (1977a).

Kumazawa, T.; Mizumura, K.: The polymodal receptors in the testis of dog. Brain Res. *136:* 553–558 (1977b).

Kumazawa, T.; Mizumura, K.: Effects of synthetic substance P on unit-discharge of testicular nociceptors of dogs. Brain Res. *170:* 553–557 (1979).

Kumazawa, T.; Mizumura, K.: Chemical responses of polymodal receptors of the scrotal contents in dogs. J. Physiol. *229:* 219–231 (1980a).

Kumazawa, T.; Mizumura, K.: Mechanical and thermal responses of polymodal receptors recorded from the superior spermatic nerve of dogs. J. Physiol. *299:* 233–245 (1980b).

Kumazawa, T.; Mizumura, K.: Temperature dependency of the chemical responses of the polymodal receptor units in vitro. Brain Res. *278:* 305–307 (1983).

Kumazawa, T.; Perl, E.R.: Differential excitation of dorsal horn substantia gelatinosa and marginal neurons by primary afferent units with fine (A and C) fibers; in Zotterman, Sensory functions of the skin in primates, pp. 67–88 (Pergamon, Oxford 1976).

Kumazawa, T.; Perl, E.R.: Primate cutaneous sensory units with unmyelinated (C) afferent fibers. J. Neurophysiol. *40:* 1325–1338 (1977).

Kumazawa, T.; Perl, E.R.: Excitation of marginal and substantia gelatinosa neurons in the primate spinal cord: indications of their place in dorsal horn functional organization. J. comp. Neurol. *177:* 417–434 (1978).

Kumazawa, T.; Perl, E.R.; Burgess, P.R.; Whitehorn, D.: Ascending projections from marginal zone (lamina I) neurons of the spinal dorsal horn. J. comp. Neurol. *162:* 1–12 (1975).

Kuru, M.: Sensory paths in the spinal cord and brain stem of man (Sogensya, Tokyo 1949).

LaMotte, C.: Distribution of the tract of Lissauer and the dorsal root fibers in the primate spinal cord. J. comp. Neurol. *172:* 529–562 (1977).

LaMotte, C.; Pert, C.B.; Snyder, S.H.: Opiate receptor binding in primate spinal cord: distribution and changes after dorsal root section. Brain Res. *112:* 407–412 (1976).

LaMotte, R.H.: Intensive and temporal determinants of thermal pain; in Kenshalo, Sensory functions of the skin of humans, pp. 327–358 (Plenum Press, New York 1979).

LaMotte, R.H.; Campbell, J.N.: Comparison of responses of warm and nociceptive C-fiber afferents in monkey with human judgments of thermal pain. J. Neurophysiol. *41:* 509–528 (1978).

LaMotte, R.H.; Thalhammer, J.G.: Response properties of high-threshold cutaneous cold receptors in the primate. Brain Res. *244:* 279–287 (1982).

LaMotte, R.H.; Thalhammer, J.G.; Robinson, C.J.: Peripheral neural correlates of magnitude of cutaneous pain and hyperalgesia: a comparison of neural events in monkey with sensory judgments in human. J. Neurophysiol. *50:* 1–26 (1983).

LaMotte, R.H.; Thalhammer, J.G.; Torebjörk, H.E.; Robinson, C.J.: Peripheral neural mechanisms of cutaneous hyperalgesia following mild injury by heat. J. Neurosci. *2:* 765–781 (1982).

Lamour, Y.; Guilbaud, G.; Willer, J.C.: Rat somatosensory (SmI) cortex. II. Laminar and columnar organization of noxious and non-noxious inputs. Exp. Brain Res. *49:* 46–54 (1983a).

Lamour, Y.; Guilbaud, G.; Willer, J.C.: Altered properties and laminar distribution of neuronal responses to peripheral stimulation in the SmI cortex of the arthritic rat. Brain Res. *273:* 183–187 (1983b).

Lamour, Y.; Willer, J.C.; Guilbaud, G.: Neuronal responses to noxious stimulation in rat somatosensory cortex. Neurosci. Lett. *29:* 35–40 (1982).

Lamour, Y.; Willer, J.C.; Guilbaud, G.: Rat somatosensory (SmI) cortex. I. Characteristics of neuronal responses to noxious stimulation and comparison with responses to non-noxious stimulation. Exp. Brain Res. *49:* 35–45 (1983c).

Landry, P.; Deschenes, M.: Intracortical arborizations and receptive fields of identified ventrobasal thalamocortical afferents to the primary somatic sensory cortex in the cat. J. comp. Neurol. *199:* 345–371 (1981).

Langford, L.A.: Unmyelinated axon ratios in cat motor, cutaneous and articular nerves. Neurosci. Lett. *40:* 19–22 (1983).

Langford, L.A.; Coggeshall, R.E.: Branching of sensory axons in the dorsal root and evidence for the absence of dorsal root efferent fibers. J. comp. Neurol. *184:* 193–204 (1979).

Langford, L.A.; Coggeshall, R.E.: Branching of sensory axons in the peripheral nerve of the rat. J. comp. Neurol. *203:* 745–750 (1981).

Langford, L.A.; Schmidt, R.F.: Afferent and efferent axons in the medial and posterior articular nerves of the cat. Anat. Rec. *296:* 71–78 (1983).

Larsson, L.I.; Rehfeld, J.F.: Localization and molecular heterogeneity of cholecystokinin in the central and peripheral nervous system. Brain Res. *165:* 201–218 (1979).

Lassen, N.A.; Ingvar, D.H.; Skinhoj, E.: Brain function and blood flow. Scient. Am. *239:* 62–71 (1978).

Lasson, S.N.; Nickels, S.M.: The use of morphometric techniques to analyse the effect of neonatal capsaicin treatment on rat dorsal root ganglia and dorsal roots. J. Physiol. *303:* 12P (1980).

Learmonth, J.R.: A contribution to the neurophysiology of the urinary bladder in man. Brain *54:* 147–176 (1931).

LeBars, D.; Chitour, D.: Do convergent neurones in the spinal dorsal horn discriminate nociceptive from non-nociceptive information? Pain *17:* 1–19 (1983).

LeBars, D.; Chitour, D.; Clot, A.M.: The encoding of thermal stimuli by diffuse noxious inhibitory controls (DNIC). Brain Res. *230:* 394–399 (1981).

LeBars, D.; Dickenson, A.H.; Besson, J.M.: Diffuse noxious inhibitory controls (DNIC). I. Effects on dorsal horn convergent neurones in the rat. Pain *6:* 283–304 (1979a).

LeBars, D.; Dickenson, A.H.; Besson, J.M.: Diffuse noxious inhibitory controls (DNIC). II. Lack of effect on non-convergent neurones, supraspinal involvement and theoretical implications. Pain *6:* 305–327 (1979b).

Leek, B.F.: Abdominal visceral receptors; in Neil, Handbook of sensory physiology, vol. III/1, pp. 113–160 (Springer, Berlin 1972).

Lele, P.P.; Weddell, G.: The relationship between neurohistology and corneal sensibility. Brain *79:* 119–154 (1956).

Lembeck, F.; Donnerer, J.: Time course of capsaicin-induced functional impairments in comparison with changes in neuronal substance P content. Arch. Pharmacol. *316:* 240–243 (1981).

Lende, R.A.; Kirsch, W.M.; Druckman, R.: Relief of facial pain after combined removal of precentral and postcentral cortex. J. Neurosurg. *34:* 537–543 (1971).

Levante, A.; Albe-Fessard, D.: Localisation dans les couches VII et VIII de Rexed des cellules d'origine d'un faisceau spino-réticulaire croisé. C.r. hebd. Séanc. Acad. Sci., Paris *274:* 3007–3010 (1972).

Levante, A.; Lamour, Y.; Guilbaud, G.; Besson, J.M.: Spinothalamic cell activity in the monkey during intense nociceptive stimulation: intra-arterial injection of bradykinin into the limbs. Brain Res. *88:* 560–564 (1975).

Lewin, W.; Phillips, C.G.: Observations on partial removal of the postcentral gyrus for pain. J. Neurol. Neurosurg. Psychiat. *15:* 143–147 (1952).

Lewis, T.: Pain (Macmillan, New York 1942).

Lewis, T.; Pochin, E.E.: The double pain response of the human skin to a single stimulus. Clin. Sci. *3:* 67–76 (1937).

Li, C.H.; Chung, D.: Isolation and structure of an untriakontapeptide with opiate activity from camel pituitary glands. Proc. natn. Acad. Sci. USA *73:* 1145–1148 (1976).

Light, A.R.; Metz, C.B.: The morphology of the spinal cord efferent and afferent neurons contributing to the ventral roots of the cat. J. comp. Neurol. *179:* 501–516 (1978).

Light, A.R.; Perl, E.R.: Differential termination of large-diameter and small-diameter primary afferent fibers in the spinal dorsal gray matter as indicated by labeling with horseradish peroxidase. Neurosci. Lett. *6:* 59–63 (1977).

Light, A.R.; Perl, E.R.: Reexamination of the dorsal root projection to the spinal dorsal horn including observations on the differential termination of coarse and fine fibers. J. comp. Neurol. *186:* 117–132 (1979a).

Light, A.R.; Perl, E.R.: Spinal termination of functionally identified primary afferent neurons with slowly conducting myelinated fibers. J. comp. Neurol. *186:* 133–150 (1979b).

Light, A.R.; Trevino, D.L.; Perl, E.R.: Morphological features of functionally defined neurons in the marginal zone and substantia gelatinosa of the spinal dorsal horn. J. comp. Neurol. *186:* 151–172 (1979).

Lim, R.K.S.: Pain. A. Rev. Physiol. *32:* 269–288 (1970).

Lim, R.K.S.; Liu, C.N.; Guzman, F.; Braun, C.: Visceral receptors concerned in visceral pain and the pseudoaffective response to intra-arterial injection of bradykinin and other algesic agents. J. comp. Neurol. *118:* 269–294 (1962).

Lin, C.S.; Merzenich, M.M.; Sur, M.; Kaas, J.H.: Connections of areas 3b and 1 of the parietal somatosensory strip with the ventroposterior nucleus in the owl monkey (*Aotus trivirgatus*). J. comp. Neurol. *185:* 355–372 (1979).

Lindgren, I.; Olivecrona, H.: Surgical treatment of angina pectoris. J. Neurosurg. *4:* 19–39 (1947).

Lissauer, H.: Beitrag zur pathologischen Anatomie der Tabes dorsalis und zum Faserverlauf im menschlichen Rückenmark. Neur. Zentbl. *4:* 245–246 (1885); English translation in Earle (1952).

Liu, R.P.C.: Laminar origins of spinal projection neurons to periaqueductal gray of the rat. Brain Res. *264:* 118–122 (1983).

Ljungdahl, A.; Hökfelt, T.; Nilsson, G.: Distribution of substance P-like immunoreactivity in the central nervous system of the rat. I. Cell bodies and nerve terminals. Neuroscience *3:* 861–943 (1978).

Loe, P.R.; Whitsel, B.L.; Dreyer, D.A.; Metz, C.B.: Body representation in ventrobasal thalamus of macaque: a single-unit analysis. J. Neurophysiol. *40:* 1339–1355 (1977).

Loeb, G.E.: Ventral root projections of myelinated dorsal root ganglion cells in the cat. Brain Res. *106:* 159–165 (1976).

Loh, H.H.; Tseng, L.F.; Wei, E.; Li, C.H.: β-Endorphin is a potent analgesic agent. Proc. natn. Acad. Sci. USA *73:* 2895–2898 (1976).

Lombard, M.C.; Nashold, B.S.; Albe-Fessard, D.: Deafferentation hypersensitivity in the rat after dorsal rhizotomy: a possible animal model of chronic pain. Pain *6:* 163–174 (1979).

Long, R.R.: Sensitivity of cutaneous cold fibers to noxious heat: paradoxical cold discharges. J. Neurophysiol. *40:* 489–502 (1977).

Longhurst, J.C.; Mitchell, J.H.; Moore, M.B.: The spinal cord ventral root: an afferent pathway of the hind-limb pressor reflex in cats. J. Physiol. *301:* 467–476 (1980).

Lu, G.W.; Bennett, G.J.; Nishikawa, N.; Dubner, R.: Spinal neurons with branched axons ascending both the dorsal and dorsolateral funiculi (in press, 1984).

Lu, G.W.; Bennett, G.J.; Nishikawa, N.; Hoffert, M.J.; Dubner, R.: Extra- and intracellular recordings from dorsal column postsynaptic spinomedullary neurons in the cat. Expl Neurol. *82:* 456–477 (1983).

Lund, R.D.; Webster, K.E.: Thalamic afferents from the spinal cord and trigeminal nuclei. An experimental anatomical study in the rat. J. comp. Neurol. *130:* 313–328 (1967).

Lundberg. A.; Oscarsson, O.: Three ascending spinal pathways in the dorsal part of the lateral funiculus. Acta physiol. scand. *51:* 1–16 (1961).

Lynn, B.: The heat sensitization of polymodal nociceptors in the rabbit and its independence of the local blood flow. J. Physiol. *287:* 493–507 (1979).

Lynn, B.; Carpenter, S.E.: Primary afferent units from the hairy skin of the rat hind limb. Brain Res. *238:* 29–43 (1982).

Macchi, G.; Angeleri, F.; Guazzi, G.: Thalamo-cortical connections of the first and second somatic sensory areas in the cat. J. comp. Neurol. *111:* 387–405 (1959).

MacDonald, R.L.; Nelson, P.G.: Specific-opiate–induced depression of transmitter release from dorsal root ganglion cells in culture. Science *199:* 1449–1451 (1978).

MacKenzie, J.: Some points bearing on the association of sensory disorders and visceral disease. Brain *16:* 321–354 (1893).

MacKenzie, J.: Symptoms and their interpretation (Shaw, London 1909).

Maderdrut, J.L.; Yaksh, T.L.; Petrusz, P.; Go, V.L.W.: Origin and distribution of cholecystokinin-containing nerve terminals in the lumbar dorsal horn and nucleus caudalis of the cat. Brain Res. *243:* 363–368 (1982).

Magendie, F.: Expériences sur les fonctions des racines des nerfs rachidiens. J. Physiol. exp. Path. *2:* 276–279 (1822a); reprinted in Cranefield (1974).

Magendie, F.: Expériences sur les fonctions des racines des nerfs qui naissent de la moelle épinière. J. Physiol. exp. Path. *2:* 366–371 (1822b); reprinted in Cranefield (1974).

Magoun, H.W.: The waking brain; 2nd ed. (Thomas, Springfield 1963).

Manson, J.: The somatosensory cortical projection of single nerve cells in the thalamus of the cat. Brain Res. *12:* 489–492 (1969).

Mantyh, P.W.: The terminations of the spinothalamic tract in the cat. Neurosci. Lett. *38:* 119–124 (1983).

Mark, V.H.; Ervin, F.R.; Hackett, T.P.: Clinical aspects of stereotactic thalamotomy in the human. I. The treatment of severe pain. Archs Neurol. *3:* 351–367 (1960).

Mark, V.H.; Ervin, F.R.; Yakovlev, P.I.: Stereotactic thalamotomy. III. The verification of anatomical lesion sites in the human thalamus. Archs Neurol. *8:* 528–538 (1963).

Marley, P.D.; Nagy, J.I.; Emson, P.C.; Rehfeld, J.F.: Cholecystokinin in the rat spinal cord: distribution and lack of effect of neonatal capsaicin treatment and rhizotomy. Brain Res. *238:* 494–498 (1982).

Marshall, J.: Sensory disturbances in cortical wounds with special reference to pain. J. Neurol. Neurosurg. Psychiat. *14:* 187–204 (1951).

Marshall, W.H.; Woolsey, C.N.; Bard, P.: Observations on cortical somatic sensory mechanisms of cat and monkey. J. Neurophysiol. *4:* 1–24 (1941).

Matsushita, M.; Ikeda, M.; Hosoya, Y.: The location of spinal neurons with long descending axons (long descending propriospinal tract neurons) in the cat: a study with the horseradish peroxidase technique. J. comp. Neurol. *184:* 63–80 (1979).

Matsushita, M.; Tanami, T.: Contralateral termination of primary afferent axons in the sacral and caudal segments of the cat, as studied by anterograde transport of horseradish peroxidase. J. comp. Neurol. *220:* 206–218 (1983).

Matthews, M.A.: An electron microscopic study of the relationship between axon diameter and the initiation of myelin production in the peripheral nervous system. Anat. Rec. *161:* 337–352 (1968).

Matthews, P.B.C.: Mammalian muscle receptors and their central actions (Williams & Wilkins, Baltimore 1972).

Maunz, R.A.; Pitts, N.G.; Peterson, B.W.: Cat spinoreticular neurons: locations, responses and changes in responses during repetitive stimulation. Brain Res. *148:* 365–379 (1978).

Mawe, G.M.; Bresnahan, J.C.; Beattie, M.S.: Primary afferent projections from dorsal and ventral roots to autonomic preganglionic neurons in the cat sacral spinal cord: light and electron microscopic observations. Brain Res. *290:* 152–157 (1984).

Maynard, C.W.; Leonard, R.B.; Coulter, J.D.; Coggeshall, R.E.: Central connections of ventral root afferents as demonstrated by the HRP method. J. comp. Neurol. *172:* 601–608 (1977).

Mayer, D.J.; Liebeskind, J.C.: Pain reduction by focal electrical stimulation of the brain: an anatomical and behavioral analysis. Brain Res. *68:* 73–93 (1974).

Mayer, D.J.; Price, D.D.: Central nervous system mechanisms of analgesia. Pain *2:* 379–404 (1976).

Mayer, D.J.; Price, D.D.; Becker, D.P.: Neurophysiological characterization of the anterolateral spinal cord neurons contributing to pain perception in man. Pain *1:* 51–58 (1975).

McCall, W.D.; Farias, M.C.; Williams, W.J.; BeMent, S.L.: Static and dynamic responses of slowly adapting joint receptors. Brain Res. *70:* 221–243 (1974).

McClung, J.R.; Castro, A.J.: Rexed's laminar scheme as it applies to the rat cervical spinal cord. Expl Neurol. *58:* 145–148 (1978).

McCreery, D.B.; Bloedel, J.R.: Reduction of the response of cat spinothalamic neurons to graded mechanical stimuli by electrical stimulation of the lower brain stem. Brain Res. *97:* 151–156 (1975).

McCreery, D.B.; Bloedel, J.R.: Effect of trigeminal stimulation on the excitability of cat spinothalamic neurons. Brain Res. *117:* 136–140 (1976).

McLaughlin, B.J.; Barber, R.; Saito, K.; Roberts, E.; Wu, J.Y.: Immunocytochemical localization of glutamate decarboxylase in rat spinal cord. J. comp. Neurol. *164:* 305–322 (1975).

McLellan, A.M.; Goodell, H.: Pain from the bladder, ureter, and kidney pelvis. Res. Publs Ass. Res. nerv. ment. Dis. *23:* 252–262 (1943).

McMahon, S.B.; Morrison, J.F.B.: Spinal neurones with long projections activated from the abdominal viscera of the cat. J. Physiol. *322:* 1–20 (1982a).

McMahon, S.B.; Morrison, J.F.B.: Two groups of spinal interneurones that respond to stimulation of the abdominal viscera of the cat. J. Physiol. *322:* 21–34 (1982b).

McMahon, S.B.; Wall, P.D.: A system of rat spinal cord lamina I cells projecting through the contralateral dorsolateral funiculus. J. comp. Neurol. *214:* 217–223 (1983).

McMurray, G.A.: Experimental study of a case of insensitivity to pain. Archs Neurol. Psychiat. *64:* 650–667 (1950).

Mehler, W.R.: The anatomy of the so-called 'pain tract' in man: an analysis of the course and distribution of the ascending fibers of the fasciculus anterolateralis; in French, Porter, Basic research in paraplegia, pp. 26–55 (Thomas, Spingfield 1962).

Mehler, W.R.: Some neurological species differences – a posteriori. Ann. N.Y. Acad. Sci. *167:* 424–468 (1969).

Mehler, W.R.; Feferman, M.E.; Nauta, W.J.H.: Ascending axon degeneration following anterolateral cordotomy. An experimental study in the monkey. Brain *83:* 718–751 (1960).

Mehler, W.R.; Nauta, W.J.H.: Connections of the basal ganglia and of the cerebellum. Confinia neurol. *36:* 205–222 (1974).

Mei, N.: Mécanorécepteurs vagaux digestifs chez le chat. Exp. Brain Res. *11:* 502–514 (1970).

Melzack, R.: The puzzle of pain (Basic Books, New York 1973).

Melzack, R.; Casey, K.L.: Sensory, motivational and central control determinants of pain; in Kenshalo, The skin senses, pp. 423–443 (Thomas, Spingfield 1968).

Melzack, R.; Stotler, W.A.; Livingston, W.K.: Effects of discrete brainstem lesions in cats on perception of noxious stimulation. J. Neurophysiol. *21:* 353–367 (1958).

Melzack, R.; Wall, P.D.: On the nature of cutaneous sensory mechanisms. Brain *85:* 331–356 (1962).

Melzack, R.; Wall, P.D.: Pain mechanisms: a new theory. Science *150:* 971–979 (1965).

Mendell, L.M.: Physiological properties of unmyelinated fiber projection to the spinal cord. Expl Neurol. *16:* 316–332 (1966).

Menétrey, D.; Chaouch, A.; Besson, J.M.: Location and properties of dorsal horn neurons at origin of spinoreticular tract in lumbar enlargement of the rat. J. Neurophysiol. *44:* 862–877 (1980).

Menétrey, D.; Chaouch, A.; Binder, D.; Besson, J.M.: The origin of the spinomesencephalic tract in the rat: an anatomical study using the retrograde transport of horseradish peroxidase. J. comp. Neurol. *206:* 193–207 (1982).

Menétrey, D.; Giesler, G.J.; Besson, J.M.: An analysis of response properties of spinal cord dorsal horn neurones to nonnoxious and noxious stimuli in the spinal rat. Exp. Brain Res. *27:* 15–33 (1977).

Mense, S.: Nervous outflow from skeletal muscle following chemical noxious stimulation. J. Physiol. *267:* 75–88 (1977).

Mense, S.: Sensitization of group IV muscle receptors to bradykinin by 5-hydroxytryptamine and prostaglandin E_2. Brain Res. *225:* 95–105 (1981).

Mense, S.; Schmidt, R.F.: Activation of group IV afferent units from muscle by algesic agents. Brain Res. *72:* 305–310 (1974).

Mense, S.; Schmidt, R.F.: Muscle pain: which receptors are responsible for the transmission of noxious stimuli? in Rose, Physiological aspects of clinical neurology, pp. 265–278 (Blackwell, Oxford 1977).

Mense, S.; Stahnke, M.: Responses in muscle afferent fibres of slow conduction velocity to contractions and ischaemia in the cat. J. Physiol. *342:* 383–397 (1983).

Meyer, R.A.; Campbell, J.N.: Myelinated nociceptive afferents account for the hyperalgesia that follows a burn to the hand. Science *213:* 1527–1529 (1981).

Meyers, D.E.R.; Snow, P.J.: The responses to somatic stimuli of deep spinothalamic tract cells in the lumbar spinal cord of the cat. J. Physiol. *329:* 355–371 (1982a).

Meyers, D.E.R.; Snow, P.J.: The morphology of physiologically identified deep spinothalamic tract cells in the lumbar spinal cord of the cat. J. Physiol. *329:* 373–388 (1982b).

Mikeladze, A.L.: Endings of afferent nerve fibers in lumbosacral region of spinal cord. Fed. Proc. Trans. *25:* suppl., pp. T211–T216 (1965).

Miletić, V.; Randić, M.: Neurotensin excites cat spinal neurones located in laminae I–III. Brain Res. *169:* 600–604 (1979).

Millar, J.: Joint afferent fibres responding to muscle stretch, vibration and contraction. Brain Res. *63:* 380–383 (1973a).

Millar, J.: The topography and receptive fields of ventroposterolateral thalamic neurons excited by afferents projecting through the dorsolateral funiculus of the spinal cord. Expl Neurol. *41:* 303–313 (1973b).

Millar, J.; Armstrong-James, M.: The responses of neurones of the superficial dorsal horn to iontophoretically applied glutamate ion. Brain Res. *231:* 267–277 (1982).

Mills, J.E.; Sellick, H.; Widdicombe, J.G.: Activity of lung irritant receptors in pulmonary micro-embolism, anaphylaxis, and drug-induced bronchoconstrictions. J. Physiol. *203:* 337–357 (1969).

Milne, R.J.; Foreman, R.D.; Giesler, G.J.; Willis, W.D.: Convergence of cutaneous and pelvic visceral nociceptive inputs onto primate spinothalamic neurons. Pain *11:* 163–183 (1981).

Milne, R.J.; Foreman, R.D.; Willis, W.D.: Responses of primate spinothalamic neurons located in the sacral intermediomedial gray (Stilling's nucleus) to proprioceptive input from the tail. Brain Res. *234:* 227–236 (1982).

Minderhoud, J.M.: An anatomical study of the efferent connections of the thalamic reticular nucleus. Exp. Brain Res. *12:* 435–446 (1971).

Mitchell, D.; Hellon, R.F.: Neuronal and behavioural responses in rats during noxious stimulation of the tail. Proc. R. Soc. *197:* 169–194 (1977).

Molenaar, I.; Kuypers, H.G.J.M.: Cells of origin of propriospinal fibers and of fibers ascending to supraspinal levels. A HRP study in the cat and rhesus monkey. Brain Res. *152:* 429–450 (1978).

Molinari, H.H.: The cutaneous sensitivity of units in laminae VII and VIII of the cat. Brain Res. *234:* 165–169 (1982).

Molony, V.; Steedman, W.M.; Cervero, F.; Iggo, A.: Intracellular marking of identified neurones in the superficial dorsal horn of the cat spinal cord. Q. Jl exp. Physiol. *66:* 211–223 (1981).

Moncada, S.; Ferreira, S.H.; Vane, J.R.: Inhibition of prostaglandin biosynthesis as the mechanism of analgesia of aspirin-like drugs in the dog knee joint. Eur. J. Pharmacol. *31:* 250–260 (1975).

Morgan, C.; Nadelhaft, I.; DeGroat, W.C.: The distribution of visceral primary afferents from the pelvic nerve to Lissauer's tract and the spinal gray matter and its relationship to the sacral parasympathetic nucleus. J. comp. Neurol. *201:* 415–440 (1981).

Morin, F.: A new spinal pathway for cutaneous impulses. Am. J. Physiol. *183:* 245–252 (1955).

Morin, F.; Schwartz, H.G.; O'Leary, J.L.: Experimental study of the spinothalamic and related tracts. Acta psychiat. neurol. *26:* 371–396 (1951).

Morrison, A.R.; Hand, P.J.; O'Donoghue, J.: Contrasting projections from the posterior and ventrobasal thalamic nuclear complexes to the anterior ectosylvian gyrus of the cat. Brain Res. *21:* 115–121 (1970).

Morrison, J.F.B.: Splanchnic slowly adapting mechanoreceptors with punctate receptive fields in the mesentery and gastrointestinal tract of the cat. J. Physiol. *233:* 349–361 (1973).

Morrison, J.F.B.: The afferent innervation of the gastrointestinal tract; in Brooks, Evers, Nerves and the gut, pp. 297–322 (Slack, Thorofare 1977).

Mosso, J.A.; Kruger, L.: Spinal trigeminal neurons excited by noxious and thermal stimuli. Brain Res. *38:* 206–210 (1972).

Mosso, J.A.; Kruger, L.: Receptor categories represented in spinal trigeminal nucleus caudalis. J. Neurophysiol. *36:* 472–488 (1973).

Mott, F.W.: Experimental enquiry upon the afferent tracts of the central nervous system of the monkey. Brain *18:* 1–20 (1895).

Mountcastle, V.B.: Modality and topographic properties of single neurons of cat's somatic sensory cortex. J. Neurophysiol. *20:* 408–434 (1957).

Mountcastle, V.B.; Davies, P.W.; Berman, A.L.: Response properties of neurons of cat's somatic sensory cortex to peripheral stimuli. J. Neurophysiol. *20:* 374–407 (1957).

Mountcastle, V.B.; Henneman, E.: The representation of tactile sensibility in the thalamus of the monkey. J. comp. Neurol. *97:* 409–440 (1952).

Mountcastle, V.B.; Poggio, G.F.; Werner, G.: The relation of thalamic cell response to peripheral stimuli varied over an intensive continuum. J. Neurophysiol. *26:* 807–834 (1963).

Mountcastle, V.B.; Powell, T.P.S.: Neural mechanisms subserving cutaneous sensibility, with special reference to the role of afferent inhibition in sensory perception and discrimination. Bull. Johns Hopkins Hosp. *105:* 201–232 (1959).

Mountcastle, V.B.; Talbot, W.H.; Sakata, H.; Hyvarinen, J.: Cortical neuronal mechanisms in flutter-vibration studied in unanesthetized monkeys. Neuronal periodicity and frequency discrimination. J. Neurophysiol. *32:* 452–484 (1969).

Mudge, A.W.; Leeman, S.E.; Fischbach, G.D.: Enkephalin inhibits release of substance P from sensory neurons in culture and decreases action potential duration. Proc. natn. Acad. Sci. USA *76:* 526–530 (1979).

Nafe, J.P.: A quantitative theory of feeling. J. gen. Psychol. *2:* 199–210 (1929).

Nagy, J.I.; Hunt, S.P.; Iversen, L.L.; Emson, P.C.: Biochemical and anatomical observations on the degeneration of peptide-containing primary afferent neurons after neonatal capsaicin. Neuroscience *6:* 1923–1934 (1981).

Nagy, J.I.; Vincent, S.R.; Staines, W.A.; Fibiger, H.C.; Reisine, T.D.; Yamamura, H.I.: Neurotoxic action of capsaicin on spinal substance P neurons. Brain Res. *186:* 435–444 (1980).

Nahin, R.L.; Madsen, A.M.; Giesler, G.J.: Anatomical and physiological studies of the gray matter surrounding the spinal cord central canal. J. comp. Neurol. *220:* 321–335 (1983).

Narotzky, R.A.; Kerr, F.W.L.: Marginal neurons of the spinal cord: types, afferent synaptology and functional considerations. Brain Res. *139:* 1–20 (1978).

Nashold, B.S.; Ostdahl, R.H.: Dorsal root entry zone lesions for pain relief. J. Neurosurg. *51:* 59–69 (1979).

Nashold, B.S.; Wilson, W.P.; Slaughter, D.G.: Sensations evoked by stimulation in the midbrain of man. J. Neurosurg. *30:* 14–24 (1969).

Nashold, B.S.; Wilson, W.P.; Slaughter, G.: The midbrain and pain. Adv. Neurol. *4:* 157–166 (1974).

Nathan, P.W.; Smith, M.C.: Fasciculi proprii of the spinal cord in man: review of present knowledge. Brain *82:* 610–668 (1959).

Nauta, W.J.H.; Kuypers, H.G.J.M.: Some ascending pathways in the brainstem reticular formation; in Jasper et al., Reticular formation of the brain, pp. 3–30 (Little, Brown, Boston 1958).

Newman, P.P.: Visceral afferent functions of the nervous system (Arnold, London 1974).

Noordenbos, W.; Wall, P.D.: Diverse sensory functions with an almost totally divided spinal cord. A case of spinal cord transection with preservation of part of one anterolateral quadrant. Pain *2:* 185–195 (1976).

Nyquist, J.K.; Greenhoot, J.H.: Unit analysis of nonspecific thalamic responses to high-intensity cutaneous input in the cat. Expl Neurol. *42:* 609–622 (1974).

Ochoa, J.; Mair, W.G.P.: The normal sural nerve in man. I. Ultrastructure and numbers of fibres and cells. Acta neuropath. *13:* 197–216 (1969).

O'Leary, J.L.; Heinbecker, P.; Bishop, G.H.: Dorsal root fibers which contribute to the tract of Lissauer. Proc. Soc. exp. Biol. Med. *30:* 302–303 (1932).

Olszewski, J.: The thalamus of *Macaca mulatta* (Karger, New York 1952).

Oswaldo-Cruz, E.; Kidd, C.: Functional properties of neurons in the lateral cervical nucleus of the cat. J. Neurophysiol. *27:* 1–14 (1964).

Paintal, A.S.: A study of gastric stretch receptors. Their role in the peripheral mechanism of satiation of hunger and thirst. J. Physiol. *126:* 255–270 (1954a).

Paintal, A.S.: The response of gastric stretch receptors and certain other abdominal and thoracic vagal receptors to some drugs. J. Physiol. *126:* 271–285 (1954b).

Paintal, A.S.: Impulses in vagal afferent fibres from specific pulmonary deflation receptors. The response of these receptors to phenyl diguanide, potato starch, 5-hydroxytryptamine and nicotine, and their role in respiratory and cardiovascular reflexes. Q. Jl exp. Physiol. *40:* 89–111 (1955).

Paintal, A.S.: Responses from mucosal mechanoreceptors in the small intestine of the cat. J. Physiol. *139:* 353–368 (1957).

Paintal, A.S.: Functional analysis of group III afferent fibres of mammalian muscles. J. Physiol. *152:* 250–270 (1960).

Paintal, A.S.: Mechanism of stimulation of type J pulmonary receptors. J. Physiol. *203:* 511–532 (1969).

Paintal, A.S.: Cardiovascular receptors; in Neil, Handbook of sensory physiology, vol. III/1: Enteroceptors, pp. 1–45 (Springer, Heidelberg 1972).

Paintal, A.S.: Vagal sensory receptors and their reflex effects. Physiol. Rev. *53:* 159–227 (1973).

Palermo, N.N.; Brown, H.K.; Smith, D.L.: Selective neurotoxic action of capsaicin on glomerular C-type terminals in rat substantia gelatinosa. Brain Res. *208:* 506–510 (1981).

Pasternak, G.W.; Goodman, R.; Snyder, S.H.: An endogenous morphine-like factor in mammalian brain. Life Sci. *16:* 1765–1769 (1975).

Pattle, R.E.; Weddell, G.: Observations on electrical stimulation of pain fibres in an exposed human sensory nerve. J. Neurophysiol. *11:* 93–98 (1948).

Paul, R.L.; Merzenich, M.; Goodman, H.: Representation of slowly and rapidly adapting cutaneous mechanoreceptors of the hand in Brodmann's areas 3 and 1 of *Macaca mulatta.* Brain Res. *36:* 229–249 (1972).

Pearson, J.C.; Haines, D.E.: Somatosensory thalamus of a prosimian primate *(Galago senegalensis).* I. Configuration of nuclei and termination of spinothalamic fibers. J. comp. Neurol. *190:* 533–558 (1980).

Penfield, W.; Boldrey, E.: Somatic motor and sensory representation in the cerebral cortex of man as studied by electrical stimulation. Brain *60:* 389–443 (1937).

Penfield, W.; Jasper, H.: Epilepsy and the functional anatomy of the human brain (Little, Brown, Boston 1954).

Penny, G.R.; Itoh, K.; Diamond, I.T.: Cells of different sizes in the ventral nuclei project to different layers of the somatic cortex in the cat. Brain Res. *242:* 55–65 (1982).

Perl, E.R.: Myelinated afferent fibres innervating the primate skin and their response to noxious stimuli. J. Physiol. *197:* 593–615 (1968).

Perl, E.R.: Is pain a specific sensation? J. psychiat. Res. *8:* 273–287 (1971).

Perl, E.R.: Sensitization of nociceptors and its relation to sensation. Adv. Pain Res. Ther. *1:* 17–28 (1976).

Perl, E.R.: Characterization of nociceptors and their activation of neurons in the superficial dorsal horn: first steps for the sensation of pain. Adv. Pain Res. Ther. *6:* 23–51 (1984a).

Perl, E.R.: Pain and nociception; in Darian-Smith, Handbook of physiology, section 1, vol. III: Sensory processes, part 2, pp. 915–975 (American Physiological Society, Bethesda 1984b).

Perl, E.R.; Whitlock, D.G.: Somatic stimuli exciting spinothalamic projections to thalamic neurons in cat and monkey. Expl Neurol. *3:* 256–296 (1961).

Pert, C.B.; Snyder, S.H.: Opiate receptor: demonstration in nervous tissue. Science *179:* 1011–1014 (1973).

Peschanski, M.; Guilbaud, G.; Gautron, M.: Neuronal responses to cutaneous electrical and noxious mechanical stimuli in the nucleus reticularis thalami of the rat. Neurosci. Lett. *20:* 165–170 (1980a).

Peschanski, M.; Guilbaud, G.; Gautron, M.: Posterior intralaminar region in rat: neuronal responses to noxious and nonnoxious cutaneous stimuli. Expl Neurol. *72:* 226–238 (1981).

Peschanski, M.; Guilbaud, G.; Gautron, M.; Besson, J.M.: Encoding of noxious heat messages in neurons of the ventrobasal thalamic complex of the rat. Brain Res. *197:* 401–413 (1980b).

Peschanski, M.; Mantyh, P.W.; Besson, J.M.: Spinal afferents to the ventrobasal thalamic complex in the rat: an anatomical study using wheatgerm agglutinin conjugated to horseradish peroxidase. Brain Res. *278:* 240–244 (1983).

Peterson, D.F.; Brown, A.M.: Functional afferent innervation of testis. J. Neurophysiol. *36:* 425–433 (1973).

Petras, J.M.: Afferent peripheral nerve fibers to the spinal cord and dorsal column nuclei in the cat. An analysis and comparison with the distribution of terminal efferent brain fibers to the spinal cord. Anat. Rec. *151:* 399–400 (1965).

Petras, J.M.: The substantia gelatinosa of Rolando. Experientia *24:* 1045–1047 (1968).

Petrén, K.: Ein Beitrag zur Frage vom Verlaufe der Bahnen der Hautsinne im Rückenmarke. Skand. Arch. Physiol. *13:* 9–98 (1902).

Petsche, U.; Fleischer, E.; Lembeck, F.; Handwerker, H.O.: The effect of capsaicin application to a peripheral nerve on impulse conduction in functionally identified afferent nerve fibres. Brain Res. *265:* 233–240 (1983).

Pickel, V.M.; Reis, D.J.; Leeman, S.E.: Ultrastructural localization of substance P in neurons of rat spinal cord. Brain Res. *122:* 534–540 (1977).

Pierau, F.K.; Taylor, D.C.M.; Abel, W.; Friedrich, B.: Dichotomizing peripheral fibres revealed by intracellular recording from rat sensory neurones. Neurosci. Lett. *31:* 123–128 (1982).

Poggio, G.F.; Mountcastle, V.B.: A study of the functional contributions of the lemniscal and spinothalamic systems to somatic sensibility. Bull. Johns Hopkins Hosp. *106:* 266–316 (1960).

Poggio, G.F.; Mountcastle, V.B.: The functional properties of ventrobasal thalamic neurons studied in unanesthetized monkeys. J. Neurophysiol. *26:* 775–806 (1963).

Poirier, L.J.; Bertrand, C.: Experimental and anatomical investigation of the lateral spino-thalamic and spino-tectal tracts. J. comp. Neurol. *102:* 745–757 (1955).

Pollin, B.; Albe-Fessard, D.: Organization of somatic thalamus in monkeys with and without section of dorsal spinal tracts. Brain Res. *173:* 431–449 (1979).

Pomeranz, B.; Wall, P.D.; Weber, W.V.: Cord cells responding to fine myelinated afferents from viscera, muscle and skin. J. Physiol. *199:* 511–532 (1968).

Pórszász, J.; Jancsó, N.: Studies on the action potentials of sensory nerves in animals desensitized with capsaicin. Acta physiol. hung. *16:* 299–306 (1959).

Poulos, D.A.; Molt, J.T.: Response of central trigeminal neurons to cutaneous thermal stimulation; in Zotterman, Sensory functions of the skin in primates, with special reference to man, pp. 263–282 (Pergamon Press, Oxford 1976).

Powell, T.P.S.; Mountcastle, V.B.: Some aspects of the functional organization of the cortex of the postcentral gyrus of the monkey: a correlation of findings obtained in a single unit analysis with cytoarchitecture. Bull. Johns Hopkins Hosp. *105:* 133–162 (1959).

Price, D.D.; Browe, A.C.: Responses of spinal cord neurons to graded noxious and non-noxious stimuli. Brain Res. *64:* 425–429 (1973).

Price, D.D.; Dubner, R.: Neurons that subserve the sensory-discriminative aspects of pain. Pain *3:* 307–338 (1977).

Price, D.D.; Dubner, R.; Hu, J.W.: Trigeminothalamic neurons in nucleus caudalis responsive to tactile, thermal, and nociceptive stimulation of monkey's face. J. Neurophysiol. *39:* 936–953 (1976).

Price, D.D.; Hayashi, H.; Dubner, R.; Ruda, M.A.: Functional relationships between neurons of marginal and substantia gelatinosa layers of primate dorsal horn. J. Neurophysiol. *42:* 1590–1608 (1979).

Price, D.D.; Hayes, R.L.; Ruda, M.A.; Dubner, R.: Spatial and temporal transformations of input to spinothalamic tract neurons and their relation to somatic sensations. J. Neurophysiol. *41:* 933–947 (1978).

Price, D.D.; Mayer, D.J.: Physiological laminar organization of the dorsal horn of *M. mulatta.* Brain Res. *79:* 321–325 (1974).

Price, D.D.; Mayer, D.J.: Neurophysiological characterization of the anterolateral quadrant neurons subserving pain in *M. mulatta.* Pain *1:* 59–72 (1975).

Privy Council, Medical Research Council: Aids to the investigation of peripheral nerve injuries; 2nd ed. (Her Majesty's Stationery Office, London 1958).

Proshansky, E.; Egger, M.D.: Staining of the dorsal root projection to the cat's dorsal horn by anterograde movement of horseradish peroxidase. Neurosci. Lett. *5:* 103–110 (1977).

Puil, E.: S-glutamate: its interactions with spinal neurons. Brain Res. Rev. *3:* 299–322 (1981).

Ralston, H.J.: The organization of the substantia gelatinosa Rolandi in the cat lumbosacral spinal cord. Z. Zellforsch. *67:* 1–23 (1965).

Ralston, H.J.: The fine structure of neurons in the dorsal horn of the cat spinal cord. J. comp. Neurol. *132:* 275–302 (1968a).

Ralston, H.J.: Dorsal root projections to dorsal horn neurons in the cat spinal cord. J. comp. Neurol. *132:* 303–330 (1968b).

Ralston, H.J.: The fine structure of laminae I, II and III of the macaque spinal cord. J. comp. Neurol. *184:* 619–642 (1979).

Ralston, H.J.: The fine structure of laminae IV, V and VI of the macaque spinal cord. J. comp. Neurol. *212:* 425–434 (1982).

Ralston, H.J.: Synaptic organization of spinothalamic tract projections to the thalamus, with special reference to pain. Adv. Pain Res. Ther. *6:* 183–195 (1984).

Ralston, H.J.; Ralston, D.D.: The distribution of dorsal root axons in laminae I, II and III of the macaque spinal cord: a quantitative electron microscopic study. J. comp. Neurol. *184:* 643–684 (1979).

Ralston, H.J.; Ralston, D.D.: The distribution of dorsal root axons to laminae IV, V and VI of the macaque spinal cord: a quantitative electron microscopic study. J. comp. Neurol. *212:* 435–448 (1982).

Rand, R.W.: Further observations on Lissauer tractolysis. Neurochirurgia *3:* 151–168 (1960).

Randić, M.; Miletić, V.: Effect of substance P in cat dorsal horn neurones activated by noxious stimuli. Brain Res. *128:* 164–169 (1977).

Randić, M.; Miletić, V.: Depressant actions of methionine-enkephalin and somatostatin in cat dorsal horn neurones activated by noxious stimuli. Brain Res. *152:* 196–202 (1978).

Randić, M.; Yu, H.H.: Effects of 5-hydroxytryptamine and bradykinin in cat dorsal horn neurones activated by noxious stimuli. Brain Res. *111:* 197–203 (1976).

Ranieri, F.; Mei, N.; Crousillat, J.: Les afférences splanchniques provenant des mécanorécepteurs gastro-intestinaux et péritonéaux. Exp. Brain Res. *16:* 276–290 (1973).

Ranson, S.W.: The course within the spinal cord of the non-medullated fibers of the spinal dorsal roots: a study of Lissauer's tract in the cat. J. comp. Neurol. *23:* 259–274 (1913).

Ranson, S.W.: The tract of Lissauer and the substantia gelatinosa Rolandi. Am. J. Anat. *16:* 97–126 (1914a).

Ranson, S.W.: An experimental study of Lissauer's tract and the dorsal roots. J. comp. Neurol. *24:* 531–545 (1914b).

Ranson, S.W.; Billingsley, P.R.: The conduction of painful afferent impulses in the spinal nerves. Studies in vasomotor reflex arc II. Am. J. Physiol. *40:* 571–584 (1916).

Ranson, S.W.; Hess, C.L. von: The conduction within the spinal cord of the afferent impulses producing pain and the vasomotor reflexes. Am. J. Physiol. *38:* 128–152 (1915).

Rasmussen, A.T.; Peyton, W.T.: The course and termination of the medial lemniscus in man. J. comp. Neurol. *88:* 411–424 (1948).

Réthelyi, M.: Preterminal and terminal axon arborizations in the substantia gelatinosa of cat's spinal cord. J. comp. Neurol. *172:* 511–528 (1977).

Réthelyi, M.; Capowski, J.J.: The terminal arborization pattern of primary afferent fibers in the substantia gelatinosa of the spinal cord in the cat. J. Physiol., Paris *73:* 269–277 (1977).

Réthelyi, M.; Szentágothai, J.: The large synaptic complexes of the substantia gelatinosa. Exp. Brain Res. *7:* 258–274 (1969).

Réthelyi, M.; Szentágothai, J.: Distribution and connections of afferent fibres in the spinal cord; in Iggo, Handbook of sensory physiology, pp. 207–252 (Springer, Berlin 1973).

Réthelyi, M.; Trevino, D.L.; Perl, E.R.: Distribution of primary afferent fibers within the sacrococcygeal dorsal horn: an autoradiographic study. J. comp. Neurol. *185:* 603–622 (1979).

Rexed, B.: The cytoarchitectonic organization of the spinal cord in the cat. J. comp. Neurol. *96:* 415–494 (1952).

Rexed, B.: A cytoarchitectonic atlas of the spinal cord in the cat. J. comp. Neurol. *100:* 297–380 (1954).

Rexed, B.: Some aspects of the cytoarchitectonics and synaptology of the spinal cord; in Eccles, Schadé, Progress in brain research, organization of the spinal cord, vol. 11, pp. 58–90 (Elsevier, Amsterdam 1964).

Reynolds, D.V.: Surgery in the rat during electrical analgesia induced by focal brain stimulation. Science *164:* 444–445 (1969).

Ribeiro-da-Silva, A.; Coimbra, A.: Two types of synaptic glomeruli and their distribution in laminae I–III of the rat spinal cord. J. comp. Neurol. *209:* 176–186 (1982).

Rinvik, E.: A re-evaluation of the cytoarchitecture of the ventral nuclear complex of the cat's thalamus on the basis of corticothalamic connections. Brain Res. *8:* 237–254 (1968).

Rioch, D. McK.: Studies on the diencephalon of carnivora. I. The nuclear configuration of the thalamus, epithalamus, and hypothalamus of the dog and cat. J. comp. Neurol. *49:* 1–120 (1929).

Risling, M.; Hildebrand, C.: Occurrence of unmyelinated axon profiles at distal, middle and proximal levels in the ventral root L7 of cats and kittens. J. neurol. Sci. *56:* 219–231 (1982).

Robertson, R.T.; Lynch, G.S.; Thompson, R.F.: Diencephalic distributions of ascending reticular systems. Brain Res. *55:* 309–322 (1973).

Robinson, C.J.; Burton, H.: Somatotopic organization in the second somatosensory area of *M. fascicularis.* J. comp. Neurol. *192:* 43–67 (1980a).

Robinson, C.J.; Burton, H.: Organization of somatosensory receptive fields in cortical areas 7b, retroinsula, postauditory and granular insula of *M. fascicularis.* J. comp. Neurol. *192:* 69–92 (1980b).

Robinson, C.J.; Burton, H.: Somatic submodality distribution within the second somatosensory (SII), 7b, retroinsular, postauditory, and granular insular cortical areas of *M. fascicularis.* J. comp. Neurol. *192:* 93–108 (1980c).

Rolando, L.: Ricerche anatomiche sulla struttura del midollo spinale (Stamperia Reale, Torino 1824).

Rose, J.E.: The thalamus of the sheep: cellular and fibrous structure and comparison with pig, rabbit and cat. J. comp. Neurol. *77:* 469–524 (1942).

Rose, J.E.; Mountcastle, V.B.: The thalamic tactile region in rabbit and cat. J. comp. Neurol. *97:* 441–490 (1952).

Rose, J.E.; Mountcastle, V.B.: Activity of single neurons in the tactile thalamic region of the cat in response to a transient peripheral stimulus. Bull. Johns Hopkins Hosp. *94:* 238–282 (1954).

Rose, J.E.; Woolsey, C.N.: A study of thalamo-cortical relations in the rabbit. Bull. Johns Hopkins Hosp. *73:* 65–128 (1943).

Rossi, A.; Grigg, P.: Characteristics of hip joint mechanoreceptors in the cat. J. Neurophysiol. *47:* 1029–1042 (1982).

Ruch, T.C.: Visceral sensation and referred pain; in Fulton, Howell's textbook of physiology; 15th ed., pp. 385–401 (Saunders, Philadelphia 1946).

Rucker, H.K.; Holloway, J.A.: Viscerosomatic convergence onto spinothalamic tract neurons in the cat. Brain Res. *243:* 155–157 (1982).

Russell, W.R.: Transient disturbances following gunshot wounds of the head. Brain *68:* 79–97 (1945).

Rustioni, A.: Non-primary afferents to the nucleus gracilis from the lumbar cord of the cat. Brain Res. *51:* 81–95 (1973).

Rustioni, A.: Non-primary afferents to the cuneate nucleus in the brachial dorsal funiculus of the cat. Brain Res. *75:* 247–259 (1974).

Rustioni, A.; Hayes, N.L.; O'Neill, S.: Dorsal column nuclei and ascending spinal afferents in macaques. Brain *102:* 95–125 (1979).

Rustioni, A.; Kaufman, A.B.: Identification of cells of origin of non-primary afferents to the dorsal column nuclei of the cat. Exp. Brain Res. *27:* 1–14 (1977).

Ryall, R.W.; Piercey, M.F.: Visceral afferent and efferent fibers in sacral ventral roots in cats. Brain Res. *23:* 57–65 (1970).

Samuel, E.P.: The autonomic and somatic innervation of the articular capsule. Anat. Rec. *113:* 53–70 (1952).

Sano, K.; Yoshioka, M.; Ogashiwa, M.; Ishijima, B.; Ohye, C.: Thalamolaminotomy. Confinia neurol. *27:* 63–66 (1966).

Saporta, S.; Kruger, L.: The organization of thalamocortical relay neurons in the rat ventrobasal complex studied by the retrograde transport of horseradish peroxidase. J. comp. Neurol. *174:* 187–208 (1977).

Sastry, B.R.: Substance P effects on spinal nociceptive neurones. Life Sci. *24:* 2169–2178 (1979).

Scadding, J.W.: The permanent anatomical effects of neonatal capsaicin on somatosensory nerves. J. Anat. *131:* 473–484 (1980).

Scadding, J.W.: Ectopic impulse generation in experimental neuromas: behavioral, physiological and anatomical correlates; in Culp, Ochoa, Abnormal nerves and muscles as impulse generators, pp. 533–552 (Oxford University Press, New York 1982).

Schady, W.J.L.; Torebjörk, H.E.; Ochoa, J.L.: Peripheral projections of nerve fibres in the human median nerve. Brain Res. *277:* 249–261 (1983).

Schaible, H.G.; Schmidt, R.F.: Activation of groups III and IV sensory units in medial articular nerve by local mechanical stimulation of knee joint. J. Neurophysiol. *49:* 35–44 (1983a).

Schaible, H.G.; Schmidt, R.F.: Responses of fine medial articular nerve afferents to passive movements of knee joint. J. Neurophysiol. *49:* 1118–1126 (1983b).

Scharf, B.; Hyvärinen, J.; Poranen, A.; Merzenich, M.M.: Electrical stimulation of human hair follicles via microelectrodes. Percept. Psychophys. *14:* 273–276 (1973).

Scheibel, M.E.; Scheibel, A.B.: Structural substrates for integrative patterns in the brain stem reticular core; in Jasper et al., Reticular formation of the brain, pp. 31–55 (Little, Brown, Boston 1958).

Scheibel, M.E.; Scheibel, A.B.: The organization of the nucleus reticularis thalami: a Golgi study. Brain Res. *1:* 43–62 (1966).

Scheibel, M.E.; Scheibel, A.B.: Terminal axonal patterns in cat spinal cord. II. The dorsal horn. Brain Res. *9:* 32–58 (1968).

Schofield, G.C.: Experimental studies on the innervation of the mucous membrane of the gut. Brain *83:* 490–514 (1960).

Schultzberg, M.; Dockray, G.J.; Williams, R.G.: Capsaicin depletes CCK-like immunoreactivity detected by immunohistochemistry, but not that measured by radioimmunoassay in rat dorsal spinal cord. Brain Res. *235:* 198–204 (1982).

Sellick, H.; Widdicombe, J.G.: The activity of lung irritant receptors during pneumothorax, hyperpnoea and pulmonary vascular congestion. J. Physiol. *203:* 359–381 (1969).

Selzer, M.; Spencer, W.A.: Convergence of visceral and cutaneous afferent pathways in the lumbar spinal cord. Brain Res. *14:* 331–348 (1969).

Seybold, V.; Elde, R.: Immunohistochemical studies of peptidergic neurons in the dorsal horn of the spinal cord. J. Histochem. Cytochem. *28:* 367–370 (1980).

Seybold, V.S.; Elde, R.P.: Neurotensin immunoreactivity in the superficial laminae of the dorsal horn of the rat. I. Light microscopic studies of cell bodies and proximal dendrites. J. comp. Neurol. *205:* 89–100 (1982).

Sheehan, D.: The afferent nerve supply of the mesentery and its significance in the causation of abdominal pain. J. Anat. *67:* 233–249 (1932).

Sherrington, C.S.: On the anatomical constitution of nerves of skeletal muscles; with remarks on recurrent fibers in the ventral spinal nerve root. J. Physiol. *17:* 211–258 (1894).

Sherrington, C.S.: The integrative action of the nervous system (Yale University Press, New Haven 1906); 2nd ed. 1947; Yale paperbound, 1981.

Shigenaga, Y.; Sakai, A.; Okada, K.: Effects of tooth pulp stimulation in trigeminal nucleus caudalis and adjacent reticular formation in rat. Brain Res. *103:* 400–406 (1976).

Shriver, J.E.; Stein, B.M.; Carpenter, M.B.: Central projections of spinal dorsal roots in the monkey. I. Cervical and upper thoracic dorsal roots. Am. J. Anat. *123:* 27–74 (1968).

Simantov, R.; Snyder, S.H.: Morphine-like peptides in mammalian brain: isolation, structure elucidation and interactions with the opiate receptor. Proc. natn. Acad. Sci. USA *73:* 2515–2519 (1976).

Simon, E.J.; Hiller, J.M.; Edelman, I.: Stereospecific binding of the potent narcotic analgesic [^{3}H]-etorphine to rat-brain homogenate. Proc. natn. Acad. Sci. USA *70:* 1947–1949 (1973).

Sinclair, D.C.: Cutaneous sensation and the doctrine of specific energy. Brain *78:* 584–614 (1955).

Sinclair, D.: Cutaneous sensation (Oxford University Press, London 1967).

Sinclair, D.: Mechanisms of cutaneous sensation (Oxford University Press, London 1981).

Sinclair, D.C.; Stokes, B.A.R.: The production and characteristics of 'second pain'. Brain *87:* 609–618 (1964).

Sinclair, D.C.; Weddell, G.; Feindel, W.H.: Referred pain and associated phenomena. Brain *71:* 184–211 (1948).

Sindou, M.; Fischer, G.; Goutelle, A.; Mansury, L.: La radicellotomie postérieure sélective. Premiers résultats dans la chirurgie de la douleur. Neurochirurgie *20:* 391–408 (1974a).

Sindou, M.; Quoex, C.; Baleydier, C.: Fiber organization at the posterior spinal cord-rootlet junction in man. J. comp. Neurol. *153:* 15–26 (1974b).

Skinner, R.D.; Coulter, J.D.; Adams, R.J.; Remmel, R.S.: Cells of origin of long descending propriospinal fibers connecting the spinal enlargements in cat and monkey determined by horseradish peroxidase and electrophysiological techniques. J. comp. Neurol. *188:* 443–454 (1979).

Skoglund, R.W.; McRoberts, J.W.; Radge, H.: Torsion of the spermatic cord: a review of the literature and an analysis of 70 new cases. J. Urol. *104:* 604–607 (1970).

Skoglund, S.: Anatomical and physiological studies of knee joint innervation in the cat. Acta physiol. scand. *36:* suppl. 124, pp. 1–101 (1956).

Skultety, F.M.: The behavioral effects of destructive lesions of the periaqueductal gray matter in adult cats. J. comp. Neurol. *110:* 337–365 (1958).

Skultety, F.M.: Stimulation of periaqueductal gray and hypothalamus. Archs Neurol. *8:* 608–620 (1963).

Smith, M.C.: Retrograde cell changes in human spinal cord after anterolateral cordotomies. Location and identification after different periods of survival. Adv. Pain Res. Ther. *1:* 91–98 (1976).

Snyder, R.L.: The organization of the dorsal root entry zone in cats and monkeys. J. comp. Neurol. *174:* 47–70 (1977).

Snyder, R.L.: Light and electron microscopic autoradiographic study of the dorsal root projections to the cat dorsal horn. Neuroscience *7:* 1417–1437 (1982).

Spiegel, E.A.; Kletzkin, M.; Szekely, E.G.: Pain reactions upon stimulation of the tectum mesencephali. J. Neuropath. exp. Neurol. *13:* 212–220 (1954).

Spiegel, E.A.; Wycis, H.T.: In Anniversary volume for O. Poetzl (Vienna), p. 438 (1948): quoted in Nashold et al. (1974).

Spiegel, E.A.; Wycis, H.T.; Szekely, E.G.; Gildenberg, P.L.: Medial and basal thalamotomy in so-called intractable pain; in Knighton, Dumke, Pain, pp. 503–517 (Little, Brown, Boston 1966).

Spiller, W.G.: The occasional clinical resemblance between caries of the vertebrae and lumbothoracic syringomyelia, and the location within the spinal cord of the fibres for the sensations of pain and temperature. Univ. Pa. med. Bull. *18:* 147–154 (1905).

Spiller, W.G.; Martin, E.: The treatment of persistent pain of organic origin in the lower part of the body by division of the anterolateral column of the spinal cord. J. Am. med. Ass. *58:* 1489–1490 (1912).

Spivy, D.F.; Metcalf, J.S.: Differential effect of medial and lateral dorsal root sections upon subcortical evoked potentials. J. Neurophysiol. *22:* 367–373 (1959).

Sprague, J.M.; Ha, H.: The terminal fields of dorsal root fibers in the lumbosacral spinal cord of the cat, and the dendritic organization of the motor nuclei. Organization of the spinal cord. Prog. Brain Res. *11:* 120–152 (1964).

Spreafico, R.; Hayes, N.L.; Rustioni, A.: Thalamic projections to the primary and secondary somatosensory cortices in cat: single and double retrograde tracer studies. J. comp. Neurol. *203:* 67–90 (1981).

Stacey, M.J.: Free nerve endings in skeletal muscle of the cat. J. Anat. *105:* 231–254 (1969).

Stanzione, P.; Zieglgänsberger, W.: Action of neurotensin on spinal cord neurons in the rat. Brain Res. *268:* 111–118 (1983).

Staszewska-Barczak, J.; Ferreira, S.H.; Vane, J.R.: An excitatory nociceptive cardiac reflex elicited by bradykinin and potentiated by prostaglandins and myocardial ischaemia. Cardiovasc. Res. *10:* 314–327 (1976).

Steriade, M.; Glenn, L.L.: Neocortical and caudate projections of intralaminar thalamic neurons and their synaptic excitation from midbrain reticular core. J. Neurophysiol. *48:* 352–371 (1982).

Sterling, P.; Kuypers, H.G.J.M.: Anatomical organization of the brachial spinal cord of the cat. I. The distribution of dorsal root fibers. Brain Res. *4:* 1–15 (1967).

Stone, T.T.: Phantom limb pain and central pain; relief by ablation of a portion of posterior central cerebral convolution. Archs Neurol. Psychiat. *63:* 739–748 (1950).

Sugimoto, T.; Gobel, S.: Primary neurons maintain their central axonal arbors in the spinal dorsal horn following peripheral nerve injury: an anatomical analysis using transganglionic transport of horseradish peroxidase. Brain Res. *248:* 377–381 (1982).

Sugita, K.; Mutsuga, N.; Takaoaka, Y.; Doi, T.: Results of stereotaxic thalamotomy for pain. Confinia neurol. *34:* 265–274 (1972).

Sugitani, M.: Electrophysiological and sensory properties of the thalamic reticular neurones related to somatic sensation in rats. J. Physiol. *290:* 79–95 (1979).

Sumino, R.; Dubner, R.; Starkman, S.: Responses of small myelinated 'warm' fibers to noxious heat stimuli applied to the monkey's face. Brain Res. *62:* 260–263 (1973).

Sur, M.; Nelson, R.J.; Kaas, J.H.: Representations of the body surface in cortical areas 3b and 1 of squirrel monkeys: comparisons with other primates. J. comp. Neurol. *211:* 177–192 (1982).

Swanson, A.G.: Congenital insensitivity to pain with anhidrosis. Archs Neurol. *8:* 299–306 (1963).

Swanson, A.G.; Buchan, G.C.; Alvord, E.C.: Anatomic changes in congenital insensitivity to pain. Archs Neurol. *12:* 12–18 (1965).

Sweet, W.H.: Animal models of chronic pain: their possible validation from human experience with posterior rhizotomy and congenital analgesia (Part I of the Second John J. Bonica Lecture). Pain *10:* 275–295 (1981a).

Sweet, W.H.: Cerebral localization of pain; in Thompson, New perspectives in cerebral localization, pp. 205–240 (Raven Press, New York 1981b).

Sweet, W.H.; White, J.C.; Selverstone, B.; Nilges, R.: Sensory responses from anterior roots and from surface and interior of spinal cord in man. Trans. Am. neurol. Ass. 165–169 (1950).

Szentágothai, J.: Neuronal and synaptic arrangement in the substantia gelatinosa Rolandi. J. comp. Neurol. *122:* 219–239 (1964).

Szolcsányi, J.: A pharmacological approach to elucidation of the role of different nerve fibres and receptor endings in mediation of pain. J. Physiol., Paris *73:* 251–259 (1977).

Szolcsányi, J.; Jancsó-Gábor, A.; Joó, F.: Functional and fine structural characteristics of the sensory neuron blocking effect of capsaicin. Arch. Pharmacol. *287:* 157–169 (1975).

Takahashi, M.; Yokota, T.: Convergence of cardiac and cutaneous afferents onto neurons in the dorsal horn of the spinal cord in the cat. Neurosci. Lett. *38:* 251–256 (1983).

Takahashi, T.; Otsuka, M.: Regional distribution of substance P in the spinal cord and nerve roots of the cat and the effect of dorsal root section. Brain Res. *87:* 1–11 (1975).

Talaat, M.: Afferent impulses in the nerves supplying the urinary bladder. J. Physiol. *89:* 1–13 (1937).

Talairach, J.; Tournoux, P.; Bancaud, J.: Chirurgie pariétale de la douleur. Acta. neurochir. *8:* 153–250 (1960).

Tasker, R.R.; Organ, L.W.; Rowe, I.H.; Hawrylyshyn, P.: Human spinothalamic tract-stimulation mapping in the spinal cord and brainstem. Adv. Pain Res. Ther. *1:* 251–257 (1976).

Tasker, R.R.; Tsuda, T.; Hawrylyshyn, P.: Clinical neurophysiological investigation of deafferentation pain. Adv. Pain Res. Ther. *5:* 713–738 (1983).

Taylor, D.C.M.; Pierau, F.K.: Double fluorescence labelling supports electrophysiological evidence for dichotomizing peripheral sensory nerve fibres in rats. Neurosci. Lett. *33:* 1–6 (1982).

Terenius, L.: Stereospecific interaction between narcotic analgesics and a synaptic plasma membrane fraction of rat cerebral cortex. Acta pharmac. tox. *32:* 317–320 (1973).

Terenius, L.; Wahlström, A.: Morphine-like ligand for opiate receptors in human CSF. Life Sci. *16:* 1759–1764 (1975).

Tessler, A.; Glazer, E.; Artymyshyn, R.; Murray, M.; Goldberger, M.E.: Recovery of substance P in the cat spinal cord after unilateral lumbosacral deafferentation. Brain Res. *191:* 459–470 (1980).

Tessler, A.; Himes, B.T.; Artymyshyn, R.; Murray, M.; Goldberger, M.E.: Spinal neurons mediate return of substance P following deafferentation of cat spinal cord. Brain Res. *230:* 263–281 (1981).

Thalhammer, J.G.; LaMotte, R.H.: Spatial properties of nociceptor sensitization following heat injury of the skin. Brain Res. *231:* 257–265 (1982).

Theriault, E.; Otsuka, M.; Jessell, T.: Capsaicin-evoked release of substance P from primary sensory neurons. Brain Res. *170:* 209–213 (1979).

Thunberg, T.: Untersuchungen über die bei einer einzelnen momentaren Hautreizung auftretenden zwei stechenden Empfindungen. Skand. Arch. Physiol. *12:* 394–442 (1901).

Todd, J.K.: Afferent impulses in the pudendal nerves of the cat. Q. Jl exp. Physiol. *49:* 258–267 (1964).

Torebjörk, H.E.: Afferent C units responding to mechanical, thermal and chemical stimuli in human non-glabrous skin. Acta physiol. scand. *92:* 374–390 (1974).

Torebjörk, H.E.; Hallin, R.G.: Activity in C fibres correlated to perception in man; in Hirsch, Zotterman, Cervical pain, pp. 171–177 (Pergamon Press, Oxford 1972).

Torebjörk, H.E.; Hallin, R.G.: Perceptual changes accompanying controlled preferential blocking of A and C fibre responses in intact human skin nerves. Exp. Brain Res. *16:* 321–332 (1973).

Torebjörk, H.E.; Hallin, R.G.: Identification of afferent C units in intact human skin nerves. Brain Res. *67:* 387–403 (1974).

Torebjörk, H.E.; Hallin, R.G.: Skin receptors supplied by unmyelinated (C) fibres in man; in Zotterman, Sensory functions of the skin in primates, with special reference to man, pp. 475–485 (Pergamon Press, Oxford 1976).

Torebjörk, H.E.; Hallin, R.G.: Microneurographic studies of peripheral pain mechanisms in man. Adv. Pain Res. Ther. *3:* 121–131 (1979).

Torebjörk, H.E.; LaMotte, R.H.; Robinson, C.J.: Peripheral neural correlates of magnitude of cutaneous pain and hyperalgesia: simultaneous recordings in humans of sensory judgments of pain and evoked responses in nociceptors with C-fibers. J. Neurophysiol. *51:* 325–339 (1984).

Torebjörk, H.E.; Ochoa, J.L.: Specific sensations evoked by activity in single identified sensory units in man. Acta physiol. scand. *110:* 445–447 (1980).

Torebjörk, H.E.; Ochoa, J.L.: Pain and itch from C fiber stimulation. Neurosci. Abstr. *7:* 228 (1981).

Torebjörk, H.E.; Ochoa, J.L.: Selective stimulation of sensory units in man. Adv. Pain Res. Ther. *5:* 99–104 (1983).

Towe, A.L.; Amassian, V.E.: Patterns of activity in single cortical units following stimulation of the digits in monkeys. J. Neurophysiol. *21:* 292–311 (1958).
Trevino, D.L.: The origin and projections of a spinal nociceptive and thermoreceptive pathway; in Zotterman, Sensory functions of the skin in primates, with special reference to man, pp. 367–376 (Pergamon Press, New York 1976).
Trevino, D.L.; Carstens, E.: Confirmation of the location of spinothalamic neurons in the cat and monkey by the retrograde transport of horseradish peroxidase. Brain Res. *98:* 177–182 (1975).
Trevino, D.L.; Coulter, J.D.; Willis, W.D.: Location of cells of origin of spinothalamic tract in lumbar enlargement of the monkey. J. Neurophysiol. *36:* 750–761 (1973).
Trevino, D.L.; Maunz, R.A.; Bryan, R.N.; Willis, W.D.: Location of cells of origin of the spinothalamic tract in the lumbar enlargement of cat. Expl Neurol. *34:* 64–77 (1972).
Truex, R.C.; Taylor, M.J.; Smythe, M.Q.; Gildenberg, P.L.: The lateral cervical nucleus of cat, dog and man. J. comp. Neurol. *139:* 93–104 (1965).
Tsubokawa, T.: The correlation of pain relief, neurological signs, EEG and anatomical lesion sites in pain patients treated by stereotaxic thalamotomy. Folia psychiat. neurol. jap. *21:* 41–51 (1967).
Tsubokawa, T.; Katayama, Y.; Ueno, Y.; Moriyasu, N.: Evidence for involvement of the frontal cortex in pain-related cerebral events in cats: increase in local cerebral blood flow by noxious stimuli. Brain Res. *217:* 179–185 (1981).
Tsubokawa, T.; Moriyasu, N.: Follow-up results of centre median thalamotomy for relief of intractable pain. Confinia neurol. *37:* 280–284 (1975).
Tsumoto, T.: Characteristics of the thalamic ventrobasal relay neurons as a function of conduction velocities of medial lemniscal fibers. Exp. Brain Res. *21:* 211–224 (1974).
Tsumoto, T.; Nakamura, S.: Inhibitory organization of the thalamic ventrobasal neurons with different peripheral representations. Exp. Brain Res. *21:* 195–210 (1974).
Tyers, M.B.; Haywood, H.: Effects of prostaglandins on peripheral nociceptors in acute inflammation. Agents Actions *6:* suppl., pp. 65–78 (1979).
Uchida, Y.; Murao, S.: Bradykinin-induced excitation of afferent cardiac sympathetic fibers. Jap. Heart J. *15:* 84–91 (1974a).
Uchida, Y.; Murao, S.: Potassium-induced excitation of afferent cardiac sympathetic nerve fibers. Am. J. Physiol. *226:* 603–607 (1974b).
Uchida, Y.; Murao, S.: Excitation of afferent cardiac sympathetic nerve fibers during coronary occlusion. Am. J. Physiol. *226:* 1094–1099 (1974c).
Uchida, Y.; Murao, S.: Acid-induced excitation of afferent cardiac sympathetic nerve fibers. Am. J. Physiol. *228:* 27–33 (1975).
Uddenberg, N.: Differential localization in dorsal funiculus of fibres originating from different receptors. Exp. Brain Res. *4:* 367–376 (1968a).
Uddenberg, N.: Functional organization of long, second-order afferents in the dorsal funiculus. Exp. Brain Res. *4:* 377–382 (1968b).
Urabe, M.; Tsubokawa, T.: Stereotaxic thalamotomy for the relief of intractable pain – CEM thalamotomy. Tohoku J. exp. Med. *85:* 286–298 (1965).
Urabe, M.; Tsubokawa, T.; Watanabe, Y.: Alteration of activity of single neurons in the nucleus centrum medianum following stimulation of the peripheral nerve and application of noxious stimuli. Jap. J. Physiol. *16:* 421–435 (1966).

Vallbo, A.B.: Sensations evoked from the glabrous skin of the human hand by electrical stimulation of unitary mechanosensitive afferents. Brain Res. *215:* 359–363 (1981).

Vallbo, A.B.; Hagbarth, K.E.: Activity from skin mechanoreceptors recorded percutaneously in awake human subjects. Expl Neurol. *21:* 270–289 (1968).

Van Hees, J.: Human C-fiber input during painful and nonpainful skin stimulation with radiant heat. Adv. Pain Res. Ther. *1:* 35–40 (1976).

Van Hees, J.; Gybels, J.M.: Pain related to single afferent C fibers from human skin. Brain Res. *48:* 397–400 (1972).

Van Hees, J.; Gybels, J.: C nociceptor activity in human nerve during painful and nonpainful skin stimulation. J. Neurol. Neurosurg. Psychiat. *44:* 600–607 (1981).

Vierck, C.J.; Luck, M.M.: Loss and recovery of reactivity to noxious stimuli in monkeys with primary spinothalamic cordotomies, followed by secondary and tertiary lesions of other cord sectors. Brain *102:* 233–248 (1979).

Voorhoeve, P.E.; Nauta, J.: Do nociceptive ventral root afferents exert central somatic affects? Adv. Pain Res. Ther. *5:* 105–110 (1983).

Wagman, I.H.; Price, D.D.: Responses of dorsal horn cells of *M. mulatta* to cutaneous and sural nerve A and C fiber stimuli. J. Neurophysiol. *32:* 803–817 (1969).

Waldeyer, W.: Das Gorilla-Rückenmark. Abh. königl. Akad. Wiss., Berlin 1–147 (1888).

Walker, A.E.: The thalamus of the chimpanzee. I. Terminations of the somatic afferent systems. Confinia neurol. *1:* 99–127 (1938).

Walker, A.E.: The spinothalamic tract in man. Archs Neurol. Psychiat. *43:* 284–298 (1940).

Walker, A.E.: Relief of pain by mesencephalic tractotomy. Archs Neurol. Psychiat. *48:* 865–880 (1942).

Wall, P.D.: The origin of a spinal-cord slow potential. J. Physiol. *164:* 508–526 (1962).

Wall, P.D.: The laminar organization of dorsal horn and effects of descending impulses. J. Physiol. *188:* 403–423 (1967).

Wall, P.D.; Fitzgerald, M.: Effects of capsaicin applied locally to adult peripheral nerve. I. Physiology of peripheral nerve and spinal cord. Pain *11:* 363–377 (1981).

Wall, P.D.; Gutnick, M.: Ongoing activity in peripheral nerves: the physiology and pharmacology of impulses originating from a neuroma. Expl Neurol. *43:* 580–593 (1974).

Wall, P.D.; Merrill, E.G.; Yaksh, T.L.: Responses of single units in lamina 2 and 3 of cat spinal cord. Brain Res. *160:* 245–260 (1979a).

Wall, P.D.; Scadding, J.W.; Tomkiewicz, M.M.: The production and prevention of experimental anesthesia dolorosa. Pain *6:* 175–182 (1979b).

Watkins, J.C.; Evans, R.H.: Excitatory amino acid transmitters. Annu. Rev. Pharmacol. Toxicol. *21:* 165–204 (1981).

Weaver, T.A.; Walker, A.E.: Topical arrangement within the spinothalamic tract of the monkey. Archs Neurol. Psychiat. *46:* 877–883 (1941).

Webber, R.H.; Wemett, A.: Distribution of fibers from nerve cell bodies in ventral roots of spinal nerves. Acta anat. *65:* 579–583 (1966).

Weddell, G.; Miller, S.: Cutaneous sensibility. A. Rev. Physiol. *24:* 199–222 (1962).

Welk, E.; Petsche, U.; Fleischer, E.; Handwerker, H.O.: Altered excitability of afferent C-fibres of the rat distal to a nerve site exposed to capsaicin. Neurosci. Lett. *38:* 245–250 (1983).

Welker, C.: Microelectrode delineation of fine grain somatotopic organization of SmI cerebral neocortex in albino rat. Brain Res. *26:* 259–275 (1971).

Wennmalm, A.; Chanh, P.H.; Junstad, M.: Hypoxia causes prostaglandin release from perfused rabbit hearts. Acta physiol. scand. *91:* 133–135 (1974).

White, J.C.; Bland, E.F.: The surgical relief of severe angina pectoris: methods employed and end results in 83 patients. Medicine *27:* 1–42 (1948).

White, J.C.; Sweet, W.H.: Pain, its mechanisms and neurosurgical control (Thomas, Springfield 1955).

White, J.C.; Sweet, W.H.: Pain and the neurosurgeon (Thomas, Springfield 1969).

Whitlock, D.G.; Perl, E.R.: Afferent projections through ventrolateral funiculi to thalamus of cat. J. Neurophysiol. *22:* 133–148 (1959).

Whitlock, D.G.; Perl, E.R.: Thalamic projections of spinothalamic pathways in monkey. Expl Neurol. *3:* 240–255 (1961).

Whitsel, B.L.; Dreyer, D.A.; Roppolo, J.R.: Determinants of body representation in postcentral gyrus of macaques. J. Neurophysiol. *34:* 1018–1034 (1971).

Whitsel, B.L.; Petrucelli, L.M.; Werner, G.: Symmetry and connectivity in the map of the body surface in somatosensory area II of primates. J. Neurophysiol. *32:* 170–183 (1969).

Whitsel, B.L.; Rustioni, A.; Dreyer, D.A.; Loe, P.R.; Allen, E.E.; Metz, C.B.: Thalamic projections to S-I in macaque monkey. J. comp. Neurol. *178:* 385–410 (1978).

Wiberg, M.; Blomquist, A.: Cells of origin of the feline spinotectal tract (Abstract). Neurosci. Lett. *7:* suppl., p. 134 (1981).

Widdicombe, J.G.: Receptors in the trachea and bronchi of the cat. J. Physiol. *123:* 71–104 (1954).

Wiesenfeld, Z.; Lindblom, U.: Behavioral and electrophysiological effects of various types of peripheral nerve lesions in the rat: a comparison of possible models for chronic pain. Pain *8:* 285–298 (1980).

Willcockson, W.S.; Chung, J.M.; Hori, Y.; Lee, K.H.; Willis, W.D.: Effects of iontophoretically released amino acids and amines on primate spinothalamic tract cells. J. Neurosci. *4:* 732–740 (1984a).

Willcockson, W.S.; Chung, J.M.; Hori, Y.; Lee, K.H.; Willis, W.D.: Effects of iontophoretically released peptides on primate spinothalamic tract cells. J. Neurosci. *4:* 741–750 (1984b).

Williams, J.T.; Zieglgänsberger, W.: The acute effects of capsaicin on rat primary afferents and spinal neurons. Brain Res. *253:* 125–131 (1982).

Willis, W.D.: Ascending pathways from the dorsal horn; in Brown, Réthelyi, Spinal cord sensation – sensory processing in the dorsal horn, pp. 169–178 (Scottish Academic Press, Edinburgh 1981).

Willis, W.D.: Control of nociceptive transmission in the spinal cord; in Ottoson, Progress in sensory physiology, vol. 3 (Springer, Berlin 1982).

Willis, W.D.: The spinothalamic tract; in Rosenberg, The clinical neurosciences, vol. 5: Neurobiology, pp. V-325–V-356 (Churchill Livingstone, New York 1983).

Willis, W.D.; Coggeshall, R.E.: Sensory mechanisms of the spinal cord (Plenum Press, New York 1978).

Willis, W.D.; Kenshalo, D.R., Jr.; Leonard, R.B.: The cells of origin of the primate spinothalamic tract. J. comp. Neurol. *188:* 543–574 (1979).

Willis, W.D.; Leonard, R.B.; Kenshalo, D.R., Jr.: Spinothalamic tract neurons in the substantia gelatinosa. Science *202:* 986–988 (1978).

Willis, W.D.; Trevino, D.L.; Coulter, J.D.; Maunz, R.A.: Responses of primate spinothalamic tract neurons to natural stimulation of hindlimb. J. Neurophysiol. *37:* 358–372 (1974).

Windle, W.F.: Neurones of the sensory type in the ventral roots of man and of other mammals. Archs Neurol. Psychiat. *26:* 791–800 (1931).

Winter, D.L.: Receptor characteristics and conduction velocities in bladder afferents. J. psychiat. Res. *8:* 225–235 (1971).

Wolf, S.: The stomach (Oxford University Press, New York 1965).

Wolf, S.; Hardy, J.D.: Studies on pain. Observations on pain due to local cooling and on factors involved in the 'cold pressor' effect. J. clin. Invest. *20:* 521–533 (1941).

Wolstencroft, J.H.: Reticulospinal neurones. J. Physiol. *174:* 91–108 (1964).

Wood, K.M.: The use of phenol as a neurolytic agent: a review. Pain *5:* 205–229 (1978).

Woollard, H.H.; Carmichael, E.A.: The testis and referred pain. Brain *56:* 293–303 (1933).

Woolf, C.J.; Fitzgerald, M.: The properties of neurones recorded in the superficial dorsal horn of the rat spinal cord. J. comp. Neurol. *221:* 313–328 (1983).

Woolsey, C.N.; Erickson, T.S.; Gilson, W.E.: Localization in somatic sensory and motor areas of human cerebral cortex as determined by direct recording of evoked potentials and electrical stimulation. J. Neurosurg. *51:* 476–506 (1979).

Woolsey, C.N.; Fairman, D.: Contralateral, ipsilateral, and bilateral representation of cutaneous receptors in somatic areas I and II of the cerebral cortex of pig, sheep, and other mammals. Surgery *19:* 684–702 (1946).

Woolsey, C.N.; Marshall, W.H.; Bard, P.: Representation of cutaneous tactile sensibility in the cerebral cortex of the monkey as indicated by evoked potentials. Bull. Johns Hopkins Hosp. *70:* 399–441 (1942).

Yaksh, T.L.; Abay, E.O.; Go, V.L.W.: Studies on the location and release of cholecystokinin and vasoactive intestinal peptide in rat and cat spinal cord. Brain Res. *242:* 279–290 (1982).

Yaksh, T.L.; Farb, D.H.; Leeman, S.E.; Jessell, T.M.: Intrathecal capsaicin depletes substance P in the rat spinal cord and produces prolonged thermal analgesia. Science *206:* 481–483 (1979).

Yaksh, T.L.; Hammond, D.L.: Peripheral and central substrates in the rostral transmission of nociceptive information. Pain *13:* 1–85 (1982).

Yamamota, T.; Takahashi, K.; Satomi, H.; Ise, H.: Origins of primary afferent fibers in the spinal ventral roots in the cat as demonstrated by the horseradish peroxidase method. Brain Res. *126:* 350–354 (1977).

Yezierski, R.P.; Culberson, J.L.; Brown, P.B.: Cells of origin of propriospinal connections to cat lumbosacral gray as determined with horseradish peroxidase. Expl Neurol. *69:* 493–512 (1980).

Yezierski, R.P.; Gerhart, K.D.; Schrock, B.J.; Willis, W.D.: A further examination of effects of cortical stimulation on primate spinothalamic tract cells. J. Neurophysiol. *49:* 424–441 (1983).

Ygge, J.; Grant, G.: The organization of the thoracic spinal nerve projection in the rat dorsal horn demonstrated with transganglionic transport of horseradish peroxidase. J. comp. Neurol. *216:* 1–9 (1983).

Yokota, T.: Excitation of units in marginal rim of trigeminal subnucleus caudalis elicited by tooth pulp stimulation. Brain Res. *95:* 154–158 (1975).

Yokota, T.; Matsumoto, N.: Somatotopic distribution of trigeminal nociceptive specific neurons within the caudal somatosensory thalamus of cat. Neurosci. Lett. *39:* 125–130 (1983a).

Yokota, T.; Matsumoto, N.: Location and functional organization of trigeminal wide dynamic range neurons within the nucleus ventralis posteromedialis of the cat. Neurosci. Lett. *39:* 231–236 (1983b).

Yoss, R.E.: Studies of the spinal cord. 3. Pathways for deep pain within the spinal cord and brain. Neurology *3:* 163–175 (1953).

Zemlan, F.P.; Leonard, C.M.; Kow, L.M.; Pfaff, D.W.: Ascending tracts of the lateral columns of the rat spinal cord: a study using the silver impregnation and horseradish peroxidase techniques. Expl Neurol. *62:* 298–334 (1978).

Zhu, C.G.; Sandro, C.; Akert, K.: Morphological identification of axo-axonic and dendro-dendritic synapses in the rat substantia gelatinosa. Brain Res. *230:* 25–40 (1981).

Zieglgänsberger, W.; Bayerl, H.: The mechanism of inhibition of neuronal activity by opiates in the spinal cord of cat. Brain Res. *115:* 111–128 (1976).

Zieglgänsberger, W.; Puil, E.A.: Actions of glutamic acid on spinal neurones. Exp. Brain Res. *17:* 35–49 (1973).

Zieglgänsberger, W.; Tulloch, I.F.: Effects of substance P on neurones in the dorsal horn of the spinal cord of the cat. Brain Res. *166:* 273–382 (1979a).

Zieglgänsberger, W.; Tulloch, I.F.: The effects of methionine- and leucine-enkephalin on spinal neurones of the cat. Brain Res. *167:* 53–64 (1979b).

Zimmermann, M.: Neurophysiology of nociception. Int. Rev. Physiol. Neurophysiol. *10:* 179–221 (1976).

Zimmermann, M.: Ethical guidelines for investigations of experimental pain in conscious animals. Pain *16:* 109–110 (1983).

Subject Index

Aβ fibers
  of dorsal horn 101, 130
  in peripheral nerve stimulation-evoked inhibition 182, 210
  spinothalamic tract cell response to 159
Aδ fibers
  capsaicin effect on 50
  cold receptors of 33
  of dorsal horn 93, 99
    lamina I 93, 94
    marginal zone 94, 128
  innervating heart 66, 267
  of mechanical nociceptors 25–28
    C polymodal nociceptors vs. 28, 32
    dendritic ramification of, in marginal zone 118
    fatigue from repeated stimulation of 27
    heat stimulation of 33
    in hyperalgesia 45–47
    inactivation of 27, 28
    pain thresholds for 38
    receptive fields for 25, 26
    stimulus velocity and position detection by 25
  in nociceptor innervation 264
  pain from 264
  in peripheral nerve stimulation-evoked inhibition 182, 183, 210
  pricking sensation from 7, 35, 37
  of respiratory system 67
  in spinal cord, course of 100
  spinothalamic tract cell response to 159, 166
  termination of 140, 268
  VPL nucleus neurons responding to 232
Abdominal pain 68
Acetylcholine
  in mechanoreceptor activation 50
  muscle nerves affected by 58
  in nociceptor activation 31, 50
  spinothalamic tract cells in response to 183, 210
Achilles tendon 234
Aching
  cold stimuli for 42
  cutaneous sensation of 34
  muscle pain 60
Acid phosphatase
  capsaicin effect on 51, 52, 109
  of dorsal root ganglion cells 109
  with peripheral nerve sectioning 280
  in terminals for primary afferent fibers 101, 140
Acids in C polymodal nociceptor activation 31
Acupuncture 183, 210
Acute pain 3, 5, 279
Allodynia 137
Anal canal 68
Analgesia
  acupuncture and 183, 210
  capsaicin-induced 51, 52
  from cordotomy 146, 152, 153
  from dorsal column stimulation 206
  via neurotransmitter manipulation 270
  spinomesencephalic tract cells in 196
  transcutaneous nerve stimulation and 183, 210
Anesthesia
  cortical effects of 246, 277, 278
  spinothalamic tract cells in response to 175
  thalamic effects of 277, 278
  thalamic posterior complex with 240
Angina pectoris 66
Angiotensin II 106
Animal models
  for antidromic activation technique 148
  of ascending nociceptive tracts 146
  for capsaicin administration 51, 52
  of dorsal horn orientation 94
  ethical issues relating to 280
  of hyperalgesia 46, 47
  of joint nociceptors 61
  of lateral cervical nucleus 197
  of Lissauer's tract 90, 91
  mechanical nociceptors of, Aδ 25
  of medial part of posterior complex 239
  of nociceptive responses of neurons in VPL nucleus 225–238
    cat 232–235
    monkey 225–232
    rat 235–238
  of nociceptor correlation to pain 38, 40–44
  of nucleus submedius 244
  for pain research 21

Animal models (continued)
of postsynaptic dorsal column pathway 202
of receptive fields of cold afferent fibers 23
of segregation of nociceptive afferent fibers 268
of somatosensory cerebral cortex 246–257
cat SII cortex 256, 257
monkey SI cortex 247–251
monkey SII cortex 255, 256
rat SI cortex 251–254
for spinoreticular tract neuron mapping 185
of spinothalamic tract cell projection to thalamus 172
of substance P distribution 108
of substantia gelatinosa cell types 123
of ventral posterior lateral nucleus 216
Antenna-type neurons 132
Antibodies to neurofilament protein 105
Antidromic activation technique 148
Arboreal cells of substantia gelatinosa 120
Archipelagos 217
Arousal response 1, 5, 275
Arthritic joints 280
Arthritis 61, 62
Articular nerve 83
Ascending nociceptive tracts 14–16
dorsally situated 196, 197
historical overview of 145–147
postsynaptic dorsal column 202–206
cells of origin of 202
destination of terminals of 202–204
identification of neurons of 205
organization of 202–204
in pain transmission 205, 206
response properties of cells of 205
spinocervical 197–202
cell response properties of 200–202
cells of origin of 197–199
destination of terminals of 199, 200
identification of neurons of 200
lateral cervical nucleus of 197
organization of 199, 200
in pain transmission 205, 206
spinomesencephalic 194–196
cells of origin of 194
electrophysiological response properties of cells of 195, 196
identification of neurons of 195
nuclei of termination of, in midbrain 195
organization of, in spinal cord and brain stem 195
in pain transmission 195, 196
spinoreticular 185–193
cells of origin of 185–191
electrophysiological response properties of neurons of 193
identification of neurons of 193
nuclei of termination of, in reticular formation 191, 192
organization of, in spinal cord and brain stem 191
in pain transmission 193
spinothalamic 147–185
A and C fiber volleys affecting cells of 159–166
anesthesia effect on cells of 175
capsaicin effect on cells of 175
cells of origin of 147–152
identification of cells of 157–159
inhibition of, after peripheral nerve stimulation 181–183
natural forms of stimulation affecting cells of 166–172
organization of, in spinal cord and brain stem 152–154
in pain transmission 183–185
pharmacology of cells of 183
projection of, to medial thalamus 172–175
receptive field organization for 177–181
thalamic nuclei of termination of 154–157
Aspartic acid 105
ATP
of dorsal root ganglion cells 109
spinothalamic tract cells in response to 183
Attentional mechanisms 259

Barbiturates
cortical nociceptive neurons affected by 250, 251
recordings of deeper layers of dorsal horn with 129
spinothalamic tract cells in response to 175, 209
substantia gelatinosa cell stimulation with 129
Baroreceptors 66
Basal ganglia 241, 242
Biliary system 68
Bladder nociceptors
characteristics of 73, 77
spinothalamic tract cell response to 172
Blood supply
C fiber polymodal nociceptors and 32
to dorsal horn 138
to frontal cortex 215
Bradykinin
in angina 66
chemoreceptors affected by 66
deactivator of 53
in dorsal horn lamina IV cell stimulation 133
in dorsal horn lamina V cell stimulation 134
in mechanoreceptor activation 50, 66
muscle afferents affected by 56
in muscle receptor activation 55

in nociceptor activation 31, 50
in nociceptor sensitization 52
spinothalamic tract cell response to 167
thalamic posterior complex neurons responding to 240
visceral mechanoreceptors affected by 69, 70
VPL nucleus neurons in response to 234, 261
Brain metabolism 214
Brain stem
in pain transmission 5, 20
spinocervical tract organization in 199, 200
spinomesencephalic tract organization in 195
spinoreticular tract organization in 191
spinothalamic tract organization in 152–154, 172
Bronchioles 67
Burning pain
from C polymodal nociceptors 35, 37
from capsaicin 50
cold stimuli for 44
cutaneous sensation of 34
from electrical stimulation 37, 38
from ventrobasal complex stimulation 215
Burns 45

C fibers
anesthesia and 175
capsaicin effect on 50
cold receptors of 33
of cutaneous nerves 38
flare reaction from 266
of heart 66, 267
of lamina I 93, 99
of lamina II 99, 129
of mechanical nociceptors 33
of muscle receptors 56
neurons of VPL nucleus and 228
in nociceptor innervation 264
pain from 264
in peripheral nerve stimulation-evoked inhibition 182, 183, 210
in plasma extravasation 266
of polymodal nociceptors 28–34
in anal canal innervation 68
blood supply and 32
burning pain from 35, 37
capsaicin effect on 50, 51
chemical stimuli of 31
cold stimuli of 32
of cutaneous nerves 38
heat stimuli for 30, 31, 265
in hyperalgesia 45–47, 53
itch-provoking stimuli for 40
muscle nerves vs. 58
noxious mechanical stimuli of 28–30
pain thresholds for 38
spinothalamic tract cells and 175
termination zones for, mapping of 269
of testis 73
unmyelinated 28
in spinocervical tract cell stimulation 200
spinothalamic tract cell response to 159, 166
of substantia gelatinosa 94, 128
termination of, in dorsal horn 140, 268
Capsaicin
analgesia from 51, 52
behavioral effects of 52
J receptors affected by 67
laminae I and II of dorsal horn affected by 99
neurotransmitter depletion with 109
in nociceptor activation 50
spinothalamic tract cell response to 175, 209
substance P depletion with 108
Carboxypeptidase B 53
Cardiac pain 66
Cardiac sympathetic nerve 168
Cardiovascular system nociceptors 66
Carrageenan 61
Central cells
of lamina II of dorsal horn 119, 120, 142
of laminae III and IV of dorsal horn 132
Central lateral nucleus
cerebellar connection with 241, 258
in pain 258
spinothalamic tract ending in 241
Central nervous system
in angina 66
in hyperalgesia 54, 265
respiratory nociceptors and 67
Central terminals of afferent endings 102, 103, 140, 269
in laminae I and II 104
in laminae III and IV 101
Centre median of thalamus 241
Cerebellum
central lateral nucleus input from 241, 258
in nociceptive transmission modulation 273
in pain sensation 273
VPL nucleus projections to 222
Cerebral cortex
anesthetic effects on 277, 278
central lateral nucleus projections to 241, 258
damage to 213
electrical stimulation of 213
lesions of 17
medial part of posterior complex projecting to 239
nociceptive columns in 246
nociceptive transmission to 246–263
in cat SII region 256, 257
in monkey SI region 247–251

Cerebral cortex
  nociceptive transmission to (continued)
    in monkey SII region 255, 256
    in rat SI region 251–254
    responses of neurons in 246–257
  nucleus submedius projecting to 244
  pain columns in 257
  in pain transmission 5, 21, 213, 214, 257–259
  peripheral nerve sectioning affecting 280
  reticular nucleus of thalamus and 245
  SI region of
    columnar organization of 222
    in monkey 247–251
    neurons of VPL nucleus projecting to 246
    organization of neurons of 257
    in pain 257–259
    in rat 251–254
    ventral posterior lateral nucleus projection to 221
  SII region of
    area proper of 259
    in cat 256, 257
    caudal vs. rostral 255
    in monkey 255, 256
    neurons of VPL nucleus projecting to 246
    ventral posterior lateral nucleus projection to 221
  SIV region of 257
  in thalamic syndrome 259
Cerebrum 214
Cervical nucleus, lateral 197
Cervicothalamic tract 216
Chemical stimuli
  for activation and sensitization of nociceptors 49–53
  of angina 66
  for C fibers 264
  for C polymodal nociceptors 31, 264, 265
  for muscle afferents and receptors 55, 56
  neurons of thalamic posterior complex responding to 240
  neurons of VPL nucleus responding to 234, 261
  for nociceptive endings 265, 266
  pain sensation from 40
  spinothalamic tract cell response to 167
  for testicular polymodal nociceptors 73
Chemoreceptors
  bradykinin effect on 66
  of gastrointestinal tract 68
Cholecystokinin
  capsaicin effect on 51, 52, 109
  in dorsal root ganglion neurons 106, 109
  in nociceptive cell excitation 138
  in nociceptive cell inhibition 138
Chromatolysis 148
Chronic pain 6
  models of 279, 280
  treatment of 279
Cingulotomies 215
Clarke's column 151
Cold
  C polymodal nociceptors affected by 32
  end-bulbs in, Krause's 22
  high- vs. low-threshold receptors of 33, 34
  pain from 34, 42, 44
  receptive fields for receptors of 22, 23, 33
  sensation of, from skin stimulation 22, 23
  spinothalamic tract cell response to 146, 167, 208
  ventrobasal complex stimulation and 215
Congenital insensitivity to pain 13, 20, 280
Convergence-projection theory 64
Cordotomy
  analgesia from 146, 152, 153
  ascending nociceptive tracts with 145
  distribution of loss of pain following 15
  marginal cells of dorsal horn with 116
  Melzack-Wall theory on 14
  nociceptive ascending tracts from 14
  pain following 147, 211
  percutaneous 16
  pharmacological substitutes for 275
  spinothalamic tract in 145
  temperature sensation with 15
Coronary artery occlusion 66, 267
Cough receptors 67
Cutaneous nerves 38
Cutaneous nociceptors
  Aδ mechanical 25–28
  animal pain sensation correlated to responses of 40–44
  C mechanical 33
  C polymodal 28–32
  in electrical stimulation of peripheral nerve fibers 37, 38
  evidence for 264
  in first and second pain 36, 37
  human pain sensation correlated to responses of 38–40
  in hyperalgesia 44–48
  interneuron stimulation from 137
  mechanoreceptors vs. 24
  neurons of VPL nucleus and 228
  in pain referral 64
  in pain sensation 34–49
  types of 33, 34
Cutaneous pain 34–49

Deafferentation pain 185, 215
Denervation hypersensitivity 3
Diffuse noxious inhibitory control system 180, 209

Dopamine
in nociceptive cell inhibition 138
spinothalamic tract cells in response to 183, 210
Dorsal column nuclei 203, 204
peripheral nerve sectioning affecting 280
postsynaptic dorsal column pathways ending in 246
VPL neurons affected by cells of 224
Dorsal column pathway, postsynaptic
cells of origin of 202
destination of terminals of 202–204
identification of neurons of 205
organization of 202–204
in pain transmission 205, 206
response properties of cells of 205
thalamic nuclei ending 245
Dorsal horn
axoaxonic synapses of 104, 105
axonal plexus orientation in 94
blood supply to 138
cell types in 112–135
dendroaxonic synapses of 105
dendrodendritic synapses of 104
glomeruli of 101
intermediate region of 134, 135, 149
interneurons of
diffuse noxious inhibitory controls for 180
noxious stimuli for 135–137
of substantia gelatinosa 124
lamina I of 138
Aδ fibers of 93, 94, 268
activation of cells of 129, 130
axon orientation in 140
C fibers of 93, 268
capsaicin effect on 99
cells of
projecting to medial part of posterior complex 239
projecting to nucleus submedius 244
types of 112–120
central terminals in 104
enkephalin-containing cells of 117
granular vesicles and endings in 101, 102
lamina II distinguished from 117
with Lissauer's tract lesions 139
postsynaptic dorsal column pathway cells of 202
primary afferent fibers of 94, 96–98
receptive fields of neurons of 129
spinomesencephalic tract cells of 194, 211
spinoreticular tract cells of 185
spinothalamic tract cells of 174, 175, 207, 217
lamina II of 120–132, 138
Aδ fibers of 93, 268
activation of cells of 129, 130
avian pancreatic polypeptide of 124
C fibers of 268
capsaicin effect on 99
cell types of 120–132
central terminals in 104
granular vesicles and endings in 101, 102
lamina I distinguished from 117
with Lissauer's tract lesions 139
low-threshold cells of 129
neurotransmitters of neurons of 124, 142
nociceptive-specific neurons of 129
postsynaptic dorsal column pathway cells of 202
primary afferent fibers of 94, 96, 98
receptive fields of neurons of 129
somatostatin of 124
substance P of 105, 124
synaptic endings of cells of 271
variations of 97
wide dynamic range cells of 129
lamina III of 132, 133, 138
cell types of 132, 133
central terminal endings of fibers in 101
central terminals in 104
hair follicle afferents in 269
lamina IV vs. 132
neuron projections from 143
postsynaptic dorsal column pathway cells of 202
primary afferent fibers of 96–98
receptive fields of neurons of 129
spinocervical tract cells of 199
spinothalamic tract cells of 207
variations of 97
lamina IV of 132, 133, 138
cells of
projecting to medial part of posterior complex 239
types of 132, 133
central terminal endings of fibers in 101
lamina III vs. 132
neuron projections from 143
spinocervical tract cells of 199
spinomesencephalic tract cells of 211
variations of 97
lamina V of 134, 138
Aδ fiber termination in 268
cells of
projecting to medial part of posterior complex 239
types of 134
neuron classification in 143
reticulated zone of 134
spinocervical tract cells of 185, 199
spinothalamic tract cells of 217
lamina VI of 134, 138
laminae VII–X of 134, 135
laminations of 80, 138
Lissauer's tract afferents in 79, 80

Dorsal horn (continued)
  marginal zone of 79, 80
    Aδ fibers of 94
    axon orientation in 140
    cells of 112
    noxious stimuli for neurons of 270
    orientation of fibers in 94
    in sacral and coccygeal segments 97
    spinothalamic tract cells of 149
  naloxone binding in 111
  nociceptive afferent fibers of
    arborization of synaptic endings of 94
    central terminal ending of 101
    degeneration of 94
    distribution of 98
    as flame-shaped arbors 98, 99
    granular vesicle terminals of 101
    input system 78–144, 267–272
    neurotransmitters associated with 105–111
    opiate receptors on 111
    round vesicle terminals of 101, 102
    segregation of 86–88, 139, 268
    spinal cord entry of 267
    substance P of 108
    synaptic endings of 100–105
    target zones for 93
  nucleus proprius of 80
    primary afferent fibers of 98
    spinothalamic tract cells of 149
    substance P of 108
  pharmacologic responses of afferent fibers of 137, 138
  presynaptic dendrites of 105
  substantia gelatinosa of 80, 82
    acid phosphatase of 109
    activation of cells of 128, 129
    afferent fiber approach to 139
    afferent fiber endings in 94
    axoaxonic synapses of 104
    axon orientation in 140
    axonal projections of neurons of 124
    cells of 119, 120
    central cells of 132
    central terminals of 101
    functional types of cells of 271
    interneurons of 124
    neuronal perikarya of 124
    neurotransmitters of neurons of 124, 142, 272
    orientation of neuropil in 94
    primary afferent fiber termination in 98
    primary afferent terminals entering 93
    recording of extracellular activity of 127, 128
    rhizotomy effect on 94, 95
    thermoreceptor dendritic arborization in 118
Dorsal root
  antidromic vasodilation from stimulation of 266
  ganglia of 83, 103
  ganglion cells of 82, 83
    amino acid transmitters of 141
    neuronal markers of 106
    in nociceptive transmission system 105
    nucleotide transmitters of 109
    peptides of 106, 109
    receptors of 82
    substance P of 105–109
    tritiated amino acid injection into 96, 97
  historical perspectives on 78
  horseradish peroxidase labeling of 98, 99
  lesions of 78, 138, 268
  nociceptive afferent fiber entry into 82, 83, 267
  nociceptive afferent fibers from 92
  pain from stimulation of 78
  sectioning of 95
  segregation of large and small afferent fibers in 86–88
  sensory function of 138
Dynorphin 106
Dysesthesias, central 196

Edinger-Westphal nucleus 195
Electrical stimulation
  burning pain from 35
  of cerebral cortex 213, 258
  of neurons of VPL nucleus 233
  of peripheral nerve fibers 37, 38
  pricking pain from 35
  of spinomesencephalic tract cells 196
  of substantia gelatinosa cells 129
Electroencephalographic changes with pain 214
Enkephalin
  in dorsal root ganglion neurons 106
  of marginal layer of dorsal horn 117
  in nociceptive cell inhibition 138
  spinothalamic tract cells in response to 183, 210
  of stalked and islet cells 124
  of substantia gelatinosa neurons 142
Epilepsy
  pain as aura in 214
  seizures with 213
Ergoceptors 56, 68

Facial pain 214
Fascia in muscle pain 60
Fatigue
  of Aδ mechanical nociceptors 27
  of C polymodal nociceptors 28, 31
Frontal cortex 214, 215

GABA
in nociceptive cell inhibition 138
spinothalamic tract cells in response to 183, 210
of substantia gelatinosa neurons 142
Ganglia
basal 241, 242
of dorsal root 83, 103
Gastrin-releasing peptide 106
Gastrocnemius 83
Gastrointestinal tract nociceptors 67–70
Gate control system 2, 3
Geniculate nucleus, medial 239
Genitourinary tract nociceptors 71–73
Glomeruli in dorsal horn 101, 125, 126
Glutamic acid 105
in lamina II cell activation 130
spinothalamic tract cells in response to 183, 210
Glutamic acid decarboxylase 269
Glycine
in nociceptive cell inhibition 138
spinothalamic tract cells in response to 183, 210
Golgi tendon organs 61
Granular vesicle terminals of afferent fibers 101, 102, 104, 140

Hair follicle afferents 22, 98, 99, 269
Heat 9
in Aδ mechanical nociceptor stimulation 27, 28, 42
C polymodal nociceptors and 30, 39, 40, 42, 175
hyperalgesia and 184
intralaminar complex neurons responding to 244
in magnitude estimates of pain 48
SI cortex neurons responding to 249
spinothalamic tract cell response to 167
spread of sensitization from 54
testicular polymodal nociceptors in response to 73
thalamic posterior complex neurons responding to 241
time course of responses to injury from 46
ventrobasal complex stimulation and 215
VPL nucleus neurons responding to 230
Histamine
antagonist of 53
in mechanoreceptor activation 50
muscle afferents affected by 56, 58
in nociceptor activation 31, 50
Historical perspectives
on ascending nociceptive tracts 145
on nociceptive afferent input to dorsal horn 78–82
on nociceptive transmission to thalamus and cerebral cortex 213–215
on nociceptors 7, 52
Horseradish peroxidase 98, 148, 149
Horsley-Clarke coronal plane 255
Hyperalgesia
Aδ fibers in 45, 47, 268
anatomical distribution of 44, 53, 54
C polymodal nociceptors in 45–47, 265
from capsaicin 50
central nervous system in 265
with heat damage 46–48
nocifensor nerves and 53
primary vs. secondary 44, 45
spinothalamic tract cells in 183, 184, 210
from thermal damage 184
Hypersensitivity, denervation 3
Hypoalgesia 47

Immunocytochemistry
enkephalin identification in dorsal horn by 117
for neurotransmitter identification 124, 272
of substance P distribution 108
Indomethacin 53
Inflammation
of genitourinary system 73
joint 61, 62
nociceptors in 266
Interneurons
diffuse noxious inhibitory controls and 180
muscle afferent stimulation of 135
noxious inputs affecting 135–137
of reticular nucleus of thalamus 245, 258
stalked cells as 129
of substantia gelatinosa 124
visceral afferent stimulation of 136
Intralaminar nuclei 241–244
Islet cells
of lamina II of dorsal horn 120, 124–126, 142, 143
of laminae III and IV of dorsal horn 132
Itching 12
chemical and 40
stimuli provoking 265

J receptors 67
Joint inflammation 61, 62
Joint nociceptors
dorsal horn neuron response to 144
evidence for 264
in joint pain 61, 62
neurons of VPL nucleus and 228
in proprioception 60
stimulation of 61
Joint pain 61, 62

Kaolin 61
Knee joint 60, 61
Krause's end-bulbs 22

Larynx 67
Lateral cervical nucleus 197
Learning mechanisms 259
Limiting cells of substantia gelatinosa 119, 120, 142
Lissauer's tract 13
  anatomy of 88–90
  composition of 78, 79, 138, 268
  interruption of 97, 268
  lesions of 92, 139
  marginal cells of 116
  in pain 79
  primary afferent fibers of 78, 98, 138, 139
  in spinothalamic tract cell activation 153
  substance P of 105, 108
  in substantia gelatinosa activation 129
Lobotomies 215
Lucifer yellow 130
Lung nociceptors 66, 67

Marginal cells 112, 115–119
Mechanical nociceptors 25–28
  C polymodal nociceptors vs. 28, 32
  dendritic ramification of, in marginal zone 118
  fatigue from repeated stimulation of 27
  heat stimulation of 33
  in hyperalgesia 45–47
  inactivation of 27, 28
  pain thresholds for 38
  receptive fields for 25, 26
  stimulus velocity and position detection by 25
Mechanoreceptors
  bradykinin effect on 66, 69, 70
  cardiovascular system 66
  chemical activation of 50
  in cutaneous nerves of cat 38
  cutaneous nociceptors vs. 24
  of gastrointestinal tract 68
  of knee joint 60
  in mechanical stimulation of skin 24, 35
  of primates 92
  thermal stimulation of 35
  urethral 73
Mechanosensitive neurons 21
Mechanothermal nociceptors 35
Medial dorsal nucleus 241
Medial lemniscus 239
Medial part of posterior complex 239–241
  in alerting and orienting behavior 263
  anesthesia and 240
  cortical projection of 239
  medial lemniscus input into 246
  in pain 258
  spinothalamic tract input into 239
  stimuli for neurons of 240–241
Meissner corpuscle
  in tactile function 22
  tapping sensation from 13
Mepyramine 53
Merkle cell 13
Mesenteric receptors 68
Methysergide 53
Microneurography 264
Midbrain
  pain from stimulation of 215
  spinocervical tract projections to 200
  spinomesencephalic tract termination in 195
  spinoreticular tract cell projection to 190
Morphine 138
Motivational-affective information 1–3, 5
Multireceptive cells 142
Muscle nociceptors
  dorsal horn interneuron excitation from 135
  evidence for 264
  group III 54–56
  group IV 56–59
  mechanoreceptors, *see* Mechanoreceptors
  role of, in pain 60
  in spinocervical tract cell stimulation 200, 201
  spinothalamic tract cell response to 166
Muscle pain 59
Muscle stretch 56, 58, 59, 228
Mycobacterium butyricus 280

Naloxone
  dorsal horn binding of 111
  in spinothalamic tract cell pharmacology 183
Narcotics 279
Nausea 68
Neurofilament protein 105
Neuromas 279
Neurotensin
  dorsal horn identification of 124
  in nociceptive cell excitation 138
  spinothalamic tract cells in response to 183
  of substantia gelatinosa neurons 142
Neurotransmitters
  assignment of, to functional categories of neurons 270
  of dorsal root ganglion cells 105–111
  immunohistochemistry in identification of 124, 272
  of lamina II neurons 124
  of spinothalamic tract cells 275
Nociceptive afferent fibers
  of dorsal horn 78–144, 120–132
    arborization of synaptic endings of 94
    central terminal ending of 101
    degeneration of 94
    distribution of 98
    as flame-shaped arbors 98, 99

granular vesicle terminals of 101
input system 78–144, 267–272
neurotransmitters associated with 105–111
opiate receptors on 111
round vesicle terminals of 101, 102
segregation of 86–88, 139, 268
spinal cord entry of 267
substance P of 108
synaptic endings of 100–105
target zones for 93
of dorsal root 82–84
of Lissauer's tract 88–92
recurrent sensibility from 268
of ventral horn 134, 135
of ventral root 85, 86
Nociceptive pain 3
Nociceptive tracts 273–276
ascending 14–16
dorsally situated 196, 197
historical overview of 145–147
postsynaptic dorsal column 202–206
spinocervical 197–202
spinomesencephalic 194–196
spinoreticular 185–193
spinothalamic 147–185
Nociceptive transmission 7–21
cerebellum in 273
to cerebral cortex 246–263
cat SII cortex 256, 257
monkey SI cortex 247–251
monkey SII cortex 255, 256
rat SI cortex 251–254
criteria for identification of neurons in 18, 19
experimental approaches to investigation of 17, 18
higher centers of 17
input system for 267–272
modulation of 93
nociceptive tracts for 14–16, 273–276
nociceptors in 7–13
plasticity of 279, 280
thalamocortical mechanisms for 276–279
to thalamus 213–246
in cat 232–235
historical overview of 213–215
intralaminar nuclei in 241–244
medial part of posterior complex in 239–241
in monkey 225–232
nucleus submedius in 244
in rat 235–238
reticular nucleus in 245
ventral posterior lateral nucleus in 216–225
Nociceptors
activity evoked in 1
in cardiovascular system 66
chemical effects on 49–53
cutaneous, *see* Cutaneous nociceptors
definition of 23
evidence for 264
gastrointestinal tract 67–70
genitourinary tract 71–73
historical perspectives on 7, 22
in inflammation 266
innervation of 264
joint
dorsal horn neuron response to 144
evidence for 264
in joint pain 61, 62
neurons of VPL nucleus and 228
in proprioception 60
stimulation of 61
motivational-affective information from 1, 2
muscle
dorsal horn interneuron excitation from 135
evidence for 264
group III 54–56
group IV 56–59
role of, in pain 60
in spinocervical tract cell stimulation 200, 201
spinothalamic tract cell response to 166
neurotransmitters of 109
postsynaptic dorsal column cells 205
respiratory system 66, 67
in response to damaging stimuli 23, 24
in response to stimuli threatening damage 23
sensitization of 24, 25
chemicals in 49–53
hyperalgesia and 53, 54
sensory discriminatory processing of 1, 2
in spinocervical tract cell stimulation 200, 201
spinomesencephalic tract cells 195
spinoreticular neurons 193
transmission system of 7–13
ventral root afferent fibers as 86
visceral 63–65
Nocifensor nerves 53
Noise 270
Norepinephrine
in nociceptive cell inhibition 138
spinothalamic tract cells in response to 183, 210
Noxious stimuli
of Aδ mechanical nociceptors 27, 28, 264
for ascending nociceptive tracts 145
attention to 259
for burning pain 35
of C polymodal nociceptors 28
for centre median-parafascicularis complex 241

Noxious stimuli (continued)
for cortical nociceptive neurons 215, 250
definition of 23
frontal cortex blood flow with 215
for interneurons of dorsal horn 135–137
for intralaminar complex neurons 244
for joint nociceptors 61
for lamina I cells of dorsal horn 118
for marginal layer neurons of dorsal horn 142, 270
of muscle, fascia and tendon 60
for pricking pain 34, 35
for reticular nucleus of thalamus neurons 245
for SI cortex neurons 247, 249
for SII cortex neurons 255
for spinomesencephalic tract 147
for spinoreticular tract 147, 193, 211
for spinothalamic tract cell 146, 166–172, 208, 209
in substantia gelatinosa cell activation 128
in substantia gelatinosa neuron classification 143
for testicular polymodal nociceptors 73
for thalamic posterior complex neurons 239
for VPL nucleus neurons 228
for wide dynamic range neurons 270
Nucleus cuneiformis 195
Nucleus of Darkschewitz 195
Nucleus gigantocellularis 192, 210
Nucleus gracilis 204
Nucleus intercollicularis 195
Nucleus interfascicularis hypoglossi 192, 210
Nucleus medullae oblongatae centralis 191, 210
Nucleus parafascicularis 151
Nucleus paragigantocellularis lateralis and dorsalis 192, 210
Nucleus pontis centralis caudalis 192, 210
Nucleus pontis centralis oralis 192
Nucleus proprius 80, 138
Aδ fiber projection to 140
primary afferent fibers of 98
spinothalamic tract cells of 149
substance P of 108
Nucleus of Roller 192, 210
Nucleus subcoeruleus ventralis 192
Nucleus submedius 156, 157, 244
lesion of 278
in pain 258
Nucleus supraspinalis 191, 210

Opiate receptors 111
Orbitofrontal cortex 244
Orientation behavior 278

Pacinian corpuscle
joint afferents terminating in 61
vibration sensation from 13

Pain
from Aδ fibers 264
abdominal 68
aching 34
acute 3, 5, 279
allodynia 137
of angina pectoris 66
arousal from 214
attention and learning related to 259, 263
as an aura 214
burning
from C polymodal nociceptors 35, 37
from capsaicin 50
cold stimuli for 44
cutaneous sensation of 34
from electrical stimulation 37, 38
from ventrobasal complex stimulation 215
from C fibers 264
C polymodal nociceptors in 265
from capsaicin activation of nociceptors 50
cardiac 66
central 183, 184, 210
central lateral nucleus in 258
centre median-parafascicularis complex in 242
chemicals in transduction of 49, 50
chronic 6
models of 279, 280
treatment of 279
from cold 34, 42, 44
congenital insensitivity to 13, 20, 280
after cordotomy 147
cortex in transmission of 213, 257–259
cutaneous nociceptors in sensation of 34–49
animal correlations 40–44
electrical stimulation of peripheral nerve fibers 37, 38
for first and second pain 36, 37
human correlations 38–40
hyperalgesia 44–48
deafferentation 185, 215
dorsal root lesions preventing 138
from dorsal root stimulation 78
electroencephalographic changes with 214
facial 214
first vs. second 36, 37
free nerve endings in 22
gastrointestinal nociceptors in 70, 71
genitourinary nociceptors in 73
joint 61, 62
Lissauer's tract in 79
lobotomy for 215
magnitude estimates of 48
motivational-affective aspects of 17
muscle 59, 60
from nervous system injury 6
neuron types and 20
nociceptive 3

nucleus submedius in 258
phantom limb 214
postcentral gyrus lesions and 213
posterior complex of thalamus in 245, 258
postsynaptic dorsal column pathway in transmission of 205, 206
pressure nerve endings and 54, 55
pricking
from Aδ nociceptors 35, 37
cold stimuli for 44
cutaneous sensation of 34
from electrical stimulation 37, 38
stimuli for 34, 35
reaction to 1, 2
receptive fields of skin in 265
hypotheses of, in visceral disease 266, 267
spinothalamic tract cells in 172, 184
referred 63–65
research on 21
respiratory nociceptors in 67
reticular nucleus of thalamus in 245, 258
sensation of 5
sensory-discriminative processing of 17, 257
skin stimulation causing 22
spinal cord in transmission of 20
spinocervical tract in transmission of 205, 206
spinomesencephalic tract cells in transmission of 195, 196
spinoreticular neurons in 193
spinothalamic tract in transmission of 183–185, 210
splanchnic 64
testicular 73
thalamic 214
thalamus in transmission of 257–259
from ventral root stimulation 78, 86
from ventrobasal complex stimulation 215
visceral 63
Pancreatic polypeptide, avian 124, 142
Parafascicular nuclei 241
Periaqueductal gray 273, 276
Peripheral nerve
electrical stimulation of 37, 38
inhibition of spinothalamic tract from stimulation of 181–183, 210
microstimulation of 264
sectioning of 279, 280
Peroneal nerve 181, 182
Phantom limb pain 214
Plantar nerve 228
Pleides 217
Postcentral gyrus 213
Posterior complex of thalamus 239–241
in alerting and orienting behavior 263
anesthesia and 240
cortical projection of 239
medial lemniscus input into 246
in pain 258
spinothalamic tract input into 239
stimuli for neurons of 240, 241
Potassium
in angina 66
in C polymodal nociceptor activation 31
muscle afferents affected by 56
muscle receptor activation from 55
VPL nucleus neurons in response to 234
Pressure
pain nerve endings with 54, 55
receptors for 55
SI cortex neurons responding to 247, 248
VPL nucleus neurons responding to 229, 230
Pricking pain
from Aδ nociceptors 35, 37
cold stimuli for 44
cutaneous sensation of 34
from electrical stimulation 37, 38
stimuli for 34, 35
Projection cells 116, 141
Proprioception 60
Propriospinal cells 141
Prostaglandins
from heart 66
muscle afferents affected by 56
in nociceptor sensitization 52
synthesis inhibitor of 53

Receptive fields
for Aδ mechanical nociceptors 25, 26
ameboid 129
of biliary tract 69
for C polymodal nociceptors 28
for cold fibers 22, 23, 33
of cortical neurons 222
for joint nociceptors 61
of laminae I–III neurons 129
for muscle nerves 56
for neurons of intralaminar complex 244
for neurons of thalamic posterior complex 239, 240
for neurons of SI cortex 247, 249
for neurons of SII cortex 255
for neurons of VPL nucleus
in cat 234
in monkey 227, 229, 230
in rat 238
pain referral area 38
with peripheral nerve sectioning 280
for postsynaptic dorsal column cells 204
for spinocervical tract cells 200
for spinoreticular neurons 193, 211
for spinothalamic tract cells 169, 172, 177–181, 209
for substantia gelatinosa cells 129
for ventral root afferent fibers 85

Receptors
of biliary tract 68, 69
in blood vessel innervation 265
capsaicin effect on 50, 51
cardiac 66
cough 67
of dorsal root ganglion cells 82
J 67
lung 66, 67
mesenteric 68
muscle 54–60
group III 54–56
group IV 56–59
role of, in pain 60
nociceptors, *see* Nociceptors
opiate 111
pharmacological manipulation of 270
pressure 55
skin 22
specificity of 22
subcutaneous, activation of 34, 44
urethral 73
Referred pain 63–65
hypotheses of, in visceral disease 266, 267
spinothalamic tract cells in 172, 184
Referred tenderness 267
Research
future 264–280
on nociceptive transmission system 17, 18
on pain mechanism 6
Respiratory system nociceptors 66, 67
Reticular-activating system 274
Reticular formation
central lateral nucleus input from 241
mesencephalic 245
rhombencephalic 245
spinoreticular tract cell projection to 185, 190–192
spinothalamic tract projections to 184, 207
thalamic projection of 245
Reticular nucleus of thalamus 245, 258
Retroambiguous nucleus 191, 210
Retroinsular cortex
in alerting and orienting behavior 259, 263
medial part of posterior complex projecting to 239, 261
Rexed's laminae 80
Rhizotomy, dorsal
acid phosphatase with 109
axon labeling with 98
central neuron activity change with 279
degeneration of primary afferent fibers with 94
distribution of axons following 96, 97
in pain relief from ventral root stimulation 138
substance P depletion with 108
synaptic endings with 102
Round vesicle terminals of primary afferent fibers 101, 102, 104
Ruffini endings 22, 61

Saphenous nerve 232
Sciatic nerve 109
Seizures 213
Sensory-discriminative information 1–3, 5
Serotonin
antagonist of 53
in mechanoreceptor activation 50
muscle afferents affected by 56
in muscle receptor activation 55
in nociceptive cell inhibition 138
in nociceptor activation 50
spinothalamic tract cell response to 167, 183, 210
Single fiber recording technique 22
Skin
Aδ mechanical nociceptors of 26
C polymodal nociceptors of 28
electrical stimulation of 37, 38
injury to, with hyperalgesia 44
receptive fields of cutaneous nociceptive fibers on 265
sensations from stimulation of discrete spots on 22
specialized sensory receptor organs in 22
Soleus muscles 234
Somatosensory cortex
cat SII 256, 257
monkey SI 247–251
monkey SII 255, 256
nociceptive responses of neurons in 246–257
rat SI 251–254
plasticity of 280
Somatostatin
capsaicin effect on 50–52, 109
in dorsal root ganglion neurons 106, 109
in nociceptive cell inhibition 138
receptor types containing 272
spinothalamic tract cells in response to 183
of substantia gelatinosa neurons 142
Spinal cord
dorsal horn of, *see* Dorsal horn
dorsal roots of, *see* Spinal roots, dorsal
in pain transmission 20
projection from, to thalamus 145
spinomesencephalic tract organization in 195
spinoreticular tract organization in 191
spinothalamic tract organization in 152–154
ventral horn of
cell types of 134, 135
spinothalamic tract cells in 149, 207
ventral roots of, *see* Spinal roots, ventral
Spinal nerves 83

- Spinal roots
  - dorsal
    - antidromic vasodilation from stimulation of 266
    - ganglia of 83, 103
    - ganglion cells of 82, 83
      - amino acid transmitters of 141
      - neuronal markers of 106
      - in nociceptive transmission system 105
      - nucleotide transmitters of 109
      - peptides of 106, 109
      - receptors of 82
      - substance P of 105–109
      - tritiated amino acid injection into 96, 97
    - historical perspectives on 78
    - horseradish peroxidase labeling of 98, 99
    - lesions of 78, 138, 268
    - nociceptive afferent fiber entry into 82, 83, 267
    - nociceptive afferent fibers from 92
    - pain from stimulation of 78
    - sectioning of 95
    - segregation of large and small afferent fibers in 86–88
    - sensory function of 138
  - ventral
    - composition of 85, 86
    - historical perspectives on 78
    - motor function of 138
    - nociceptive afferent fiber entry into spinal cord via 267
    - pain from stimulation of 78, 86
    - primary afferent fibers of 139
    - substance P of 105, 108
- Spino-olivary tracts 273
- Spinocerebellar tract 273
- Spinocervical tract
  - cells of 132
    - identification of 200
    - muscle afferent stimulation of 135
    - origin of 197–199
    - response properties of 200–202
  - destination of terminals of 199, 200
  - at lateral cervical nucleus 197
  - organization of 199, 200
  - thalamic nuclei ending 245
- Spinocervicothalamic system 197, 245, 246
- Spinomesencephalic tract
  - cells of origin of 194, 211
  - cordotomy effect on 145
  - electrophysiological response properties of cells of 195, 196
  - identification of neurons of 195
  - noxious stimuli for 147
  - nuclei of termination of 195
  - organization of, in spinal cord and brain stem 195
  - in pain transmission 195, 196
  - thalamic nuclei ending 245
- Spinoreticular tract
  - cells of
    - origin of 185–191
    - as spinothalamic neurons 210
  - cordotomy effect on 145
  - electrophysiological response properties of neurons of 193
  - identification of neurons of 193
  - noxious stimuli for 147
  - nuclei of termination of 191, 192
  - organization of, in spinal cord and brain stem 191
  - in pain transmission 193
  - problems in research on 275
  - thalamic nuclei ending 245
- Spinothalamic columns 260
- Spinothalamic tract
  - cells of 64
    - anesthesia effect on 175
    - in awake behaving animals 275
    - capsaicin effect on 175
    - characteristics and classification of 273, 274
    - vs. cortical nociceptive neurons 250
    - cutaneous receptive field for 169
    - decussation of 207
    - deep 167
    - distribution of 148–150
    - high threshold 166, 167
    - identification of, in physiological experiments 157–159
    - inhibition of, after peripheral nerve stimulation 181–183
    - of lamina I 129
    - of lamina III 132
    - low threshold 166
    - multireceptive 166, 167
    - muscle afferent stimulation of 135
    - neurotransmitters of 275
    - nociceptive-specific 166, 167
    - noxious stimuli for 146, 208, 209
    - of origin 147–152
    - in pain referral 172
    - pharmacology of 183
    - projecting to lateral thalamus 150, 151
    - projecting to medial thalamus 150, 151, 172–175
    - receptive field organization for 172, 177–181
    - responses of, to A and C fiber volleys 159–166
    - responses of, to natural forms of stimulation 166–172
    - thermoreceptor 146, 147
    - wide dynamic range 166, 167
  - cordotomy effect on 145
  - course of 208

Spinothalamic tract (continued)
historical perspectives on 145–147
intralaminar complex ending 239, 241–244
medial part of posterior complex ending 239–241
nucleus submedius ending 239, 244
organization of, in spinal cord and brain stem 152–154
in pain transmission 183–185
reticular formation projections of 184
reticular nuclei of thalamus ending 239, 245
thalamic columns and termination of 257
thalamic nuclei of termination of 154–157
VPL nucleus in termination of 216–238
Splanchnic nerve 135, 136
Splanchnic pain 64
Stalked cells of substantia gelatinosa 120, 124–126, 142, 143
Stereotaxic surgery 4
Stilling's nucleus 151
Stimuli
chemical, *see* Chemical stimuli
itch-provoking 265
noxious, *see* Noxious stimuli
tactile
for SI cortex neurons 250
for SII cortex neurons 255
for spinothalamic tract cells 146, 166, 167, 184
for thalamic posterior complex neurons 240
for VPL nucleus neurons 227
thermal
of Aδ mechanical nociceptors 27, 28, 35
for burning pain 35
of C polymodal nociceptors 30, 31, 35, 39, 40, 264
for dorsal horn marginal layer neurons 118, 142
of mechanoreceptors 35
muscle afferents affected by 57
of thermoreceptors 35
Striatum 241
Subcoeruleus complex 192, 210
Subnuclei multiformis and densocellularis 241
Substance P
in antidromic vasodilation 266
capsaicin effect on 50–52, 109
distribution of, in dorsal horn 101
of dorsal root ganglion cells 105, 106
granular vesicle endings with 101
of Lissauer's tract 105, 108
in nociceptive cell stimulation and inhibition 137, 138
in nociceptor sensitization 52, 53
with peripheral nerve sectioning 280
spinothalamic tract cells in response to 183, 210
of substantia gelatinosa neurons 142
Substantia gelatinosa of dorsal horn, *see* Dorsal horn, substantia gelatinosa
Subthalamus 239
Sural nerve
Aδ fibers of 38
antidromic activation of 54
C fibers of 38
myelinated vs. unmyelinated axons of 83
spinothalamic tract cells in response to stimulation of 182
Surgery, stereotaxic 4
Synaptic endings of primary afferent fibers 100–105, 141, 268

Tactile stimuli
for SI cortex neurons 250
for SII cortex neurons 255
for spinothalamic tract cells 146, 166, 167, 184
for thalamic posterior complex neurons 240
for VPL nucleus neurons 227
Temperature
ascending nociceptive tract transmission of 146
cerebral cortex damage affecting sensibility of 213
cordotomy affecting sensation of 15
muscle afferents affected by 56, 57
postcentral gyrus lesions and sense of 213
skin stimulation with sensation of 22, 23
spinothalamic tract cells in response to 184
VPL nucleus neurons in response to 236
Tenderness
cutaneous 63
referred 267
Tendons in muscle pain 60
Testis nociceptors 73, 267
Thalamic columns 224
Thalamic pain 214
Thalamic syndrome 214, 259
Thalamus
anesthetic effects on 277, 278
dorsal nuclei of 203, 204
lesions of 215
peripheral nerve sectioning and 280
postsynaptic dorsal column pathways ending in 246
VPL neurons affecting 224
intralaminar complex of
lesions of 214, 215
spinothalamic tract ending in 241–244
lesions of 17
nociceptive transmission to 213–263
in cat 232–235
historical overview of 213–215
intralaminar nuclei in 241–244
medial part of posterior complex in 239–241
in monkey 225–232

nucleus submedius in 244
in rat 235–238
thalamic nuclei and spinothalamic tract in 216–225
thalamic reticular nucleus in 245
VPL nuclei in 216–238
in pain transmission 5, 21, 213, 257–259
posterior complex of 239–241
in alerting and orienting behavior 263
anesthesia and 240
cortical projection of 239
medial lemniscus input into 246
in pain 258
spinothalamic tract input into 239
stimuli for neurons of 240, 241
spinal cord projection to 145
spinoreticular tract cell projection to 185, 190
spinothalamic tract termination in 150, 151, 154–157, 172–175, 208
in temperature sensibility 213
touch sense via 213
ventral posterior lateral nucleus of, *see* Ventral posterior lateral nucleus
ventral posterior medial nucleus of
lesion of 214
neuron organization around 235
in pain 257, 258
ventromedial lesion of 215
Thermal nociceptors 35
Thermal stimuli
of Aδ mechanical nociceptors 27, 28, 35
for burning pain 35
of C polymodal nociceptors 30, 31, 35, 39, 40, 264
for dorsal horn marginal layer neurons 118, 142
of mechanoreceptors 35
muscle afferents affected by 57
of thermoreceptors 35
Thermoreceptors
dendritic arborization of, in substantia gelatinosa 118
spinothalamic tract 146, 147
stimuli for 35, 43
Thermosensitive neurons 21
Touch 146
cerebral cortex damage affecting 213
hair follicle afferents in 22
Meissner's corpuscles in 22
neurons of VPL nucleus responding to 229, 230
sensation of, from skin stimulation 22
spinocervical tract cells in 200
spinothalamic tract cell mediation of 167
Trachea 67
Tractotomies, subcaudate 215
Transcutaneous nerve stimulation 183, 210
Transmission cells 3
Trigeminal nucleus 93
Trigeminal system 275

Ureter 73
Urethra 73

Vagus nerve 67
Vasoactive intestinal polypeptide
capsaicin effect on 52
in dorsal root ganglion neurons 106
Ventral horn
cell types of 134, 135
spinothalamic tract cells in 149, 207
Ventral posterior lateral nucleus
axon labeling of 222
in cat 216, 232–235
cells of, vs. cortical nociceptive neurons 250, 251
cerebellar projections of 222
lesions of, in pain relief 214
medial lemniscus input into 246
in monkey 217
nociceptive responses of neurons in 225–238
in cat 232–235
in monkey 225–232
in rat 235–238
nociceptive-specific cells of 260, 261
periaqueductal gray pathway to 276
projections of, to SI and SII regions of cortex 221
in rat 216, 235–238
reticular nucleus of thalamus effect on 259
small vs. large neurons of 260
spinal cord projection to 216, 217
spinocervicothalamic pathway ending in 245, 246
spinothalamic tract projection to 216–225
wide dynamic range cells of 260, 261
Ventral posterior medial nucleus
lesion of 214
neuron organization around 235
in pain 257, 258
Ventral root
composition of 85, 86
historical perspectives on 78
motor function of 138
nociceptive afferent fiber entry into spinal cord via 267
pain from stimulation of 78, 86
primary afferent fibers of 139
substance P of 105, 108
Ventralis intermedius 221
Ventrobasal complex 150
lesions of 214
reticular formation projections to 245
SII region of cortex and 259
stimulation of 215

Visceral nociceptors
  characteristics of 63–65
  dorsal horn interneuron excitation from 135, 136
  evidence for 264
  spinothalamic tract cell response to 168
  in spinocervical tract cell stimulation 200, 201
Visceral pain 63
Visual system 278, 279

Waldeyer cells 112, 141
Warmth 9
  Ruffini endings in 22
  sensation of, from skin stimulation 22, 23
Wide dynamic range cells 118, 142
  of posterior complex of thalamus, medial part of 240
  response properties of 270, 271
  of somatosensory cortex 248, 251, 254
  of spinocervical tract 276
  of ventral posterior lateral nucleus 230, 235, 238

Zona incerta of subthalamus 239